Future of Energy

Technologies, Economics and Politics

Series Editors

Manfred Hafner, Johns Hopkins University SAIS Europe, Bologna, Italy

Giacomo Luciani, Sciences Po, Paris School of International Affairs, Paris, Switzerland

The *Future of Energy* books series reports on future trends in energy and offers new insights into transition pathways and their implications for more efficient and decarbonizing energy systems. It offers an outlet for teaching and research as well as for professional experience on energy transitions both at local and regional levels, covering technological, economic, and political aspects and their interplay. It welcomes theoretically sound and empirically robust monographs and edited volumes from various disciplines and approaches, as well as textbooks and books written by professionals on topics such as energy technologies and innovations, energy economics, energy policy, geoeconomics and geopolitics of energy, energy governance, political economy of energy, international energy relations, energy security and diplomacy, global energy systems, and social aspects of the energy transition. All titles in the series are peer-reviewed.

André Estermann · Marius Schrade ·
Louise Anderson
Editors

European Electricity Market Coupling

A Practitioner's Guide

Springer

Editors
André Estermann
50Hertz Transmission GmbH
Berlin, Germany

Marius Schrade
50Hertz Transmission GmbH
Berlin, Germany

Louise Anderson
Zurich, Switzerland

Future of Energy
Technologies, Economics and Politics
ISBN 978-3-031-86314-1 ISBN 978-3-031-86315-8 (eBook)
https://doi.org/10.1007/978-3-031-86315-8

This Springer imprint is published by the registered company Springer Nature Switzerland AG
The registered company address is: Gewerbestrasse 11, 6330 Cham, Switzerland

Preface

The “European Electricity Market Coupling—A Practitioner’s Guide” provides the reader with a manual on how to navigate Europe’s Internal Electricity Market (IEM). It is therefore timely that we release this important guide given that electricity markets and infrastructure are centre stage for policy makers and industry leaders, as concerns around energy security and affordability increase globally.

Over the past thirty years, Europe’s electricity markets have undergone significant changes and have shown resilience and adaptability in the face of various challenges.

The story began back in the 90s when the European Union’s Member States agreed to open their national electricity markets to competition. The first (1996) and second (2003) energy packages set this liberalization into law and mandated the introduction of competition into the retail markets.

The third energy package (2009) became the cornerstone of Europe’s Internal Electricity Market and introduced a number of reforms including the unbundling of energy supply and generation from transmission networks. The areas for network code development, including the Capacity Allocation and Congestion Management Guideline, were set out in the third energy package. These EU network codes and guidelines significantly accelerated the harmonization and integration of national electricity markets.

In the past five years, there have been two further energy packages. The fourth energy package in 2019 (also known as “Clean Energy for all Europeans”) aimed to accelerate the clean energy transition, while also spurring growth and job creation. Meanwhile, the fifth energy package in 2021 (also known as “Fit for 55”) was designed to align the EU’s energy targets with the new European climate ambitions for 2030 and 2050.

The five packages and their legislative requirements have been implemented during a time of geopolitical and economic upheaval for the EU. In the past five years alone we have experienced a series of systemic shocks leading us into a full scale climate and energy crisis.

COVID-19, the global pandemic, had a significant impact on electricity consumption across the bloc due to confinement measures, particularly affecting the industrial and commercial sectors. Electricity wholesale prices fell during 2019 and 2020, as consumption reductions coincided with an increase in renewable generation and a decrease in commodity prices.

As the EU emerged from the pandemic in the second half of 2021, wholesale gas prices increased significantly, and as a consequence electricity prices, due to an increase in economic activity, driving increased demand globally. Other factors included lower gas supplies, and unfavourable weather conditions for renewable power generation. Throughout all of this, Europe's electricity markets continued to deliver, with benefits of approximately EUR 34 billion from electricity cross-border trade in 2021.

With energy prices already on the rise since mid-2021, Russia's invasion of Ukraine further compounded Europe's energy security and affordability challenges. By Q3 2022, Europe's electricity prices were over 200% more expensive than twelve months earlier. The high gas prices were impacting electricity prices due to the marginal pricing system in operation in the electricity market, which set the price for electricity on the basis of the last fuel to meet the electricity demand. This fuel was usually gas in 2022.

The high prices kicked off a contentious debate in Europe questioning whether Europe's internal energy market was functioning properly. In September 2022, during her annual State of the Union speech, Ursula von der Leyen, President of the European Commission, announced that the EU power markets were no longer fit for purpose and promised a comprehensive reform of the bloc's electricity market. The Commission proposed such a reform in March 2023 which focused on areas such as increasing the liquidity of long-term markets, an extension of regulated retail prices to households and SMEs during crises and new rules on access to renewables.

In the interim, in addition to energy efficiency and consumption reduction measures, some member states began implementing financial support mechanisms for consumers and utilities, as well as windfall profit taxation approaches. Other more structural options being considered was the capping of the electricity market price and the division of the market per technology.

Some of the tensions on European energy markets receded in 2023, resulting in a decrease in wholesale energy prices from the record highs observed in 2022. Nonetheless, retail energy prices continued to rise in 2023 following the lifting of some of the financial support mechanisms implemented in 2022.

In the midst of all this, 2023 saw the 10th anniversary of the Single-Day-Ahead Coupling (with 98.6% of EU consumption now coupled) and the 5th anniversary of the cross-border intraday initiative (with over 242 million trades executed in the first five years of the initiative).

In April 2024, the European Parliament officially adopted the new reforms for the EU electricity market. Given the myriad of new regulations and directives introduced in the last five years, there is now an expectation that energy sector stakeholders will focus on implementation, particularly given the 2024 EU elections and the introduction of a new Commission.

This practitioner's handbook is therefore important because it provides a unique collection of chapters covering all aspects of European market coupling, while also addressing third countries to the EU and their plans to join market coupling. Such a collection will be invaluable onboarding material for the next generation of energy professionals, accelerating the familiarization process for graduates trying to understand the complexities of the world's largest interconnected system. This guide should also serve as a source of practical information for regions outside of the EU that want to embark on a similar mission as Europe's Internal Electricity Market.

The layout of the document is described underneath and enables the reader to efficiently access the different sections depending on their interest.

Part I introduces the framework and main principles of market coupling, including the relevant EU legislation, geographical organization, the different roles of the actors, and illustrative examples.

Part II describes the beating heart of market coupling, namely the capacity calculation by Transmission System Operators (TSOs), the procedures to ensure a safe and secure operations of day-ahead and intraday markets and the role of TSOs and Regional Coordination Centres (RCCs) therein, as well as the algorithms and systems used in both market timeframes.

Part III highlights the link between the results of the trading algorithms and their physical and financial implications by explaining the concepts of shipping, nominations, clearing, and congestion income distribution.

Part IV reports on developments in different regions of Europe.

Part V sheds light on the contractual framework between parties involved in market coupling, as well as the regulatory roles of national regulatory authorities and the Agency for the Cooperation of Energy Regulators (ACER).

Part VI emphasizes the importance of transparency from the perspective of the data platform provider and one of its users, a market party.

The beauty of this subject of electricity market coupling is that it is an ever-evolving machine which will render a future extension and update of this book necessary and worthwhile.

Inside Europe, as mentioned, the implementation of the Electricity Market Design reform as well as the highly anticipated update of the Capacity Allocation and Congestion Management network code, and their implications will be felt over the next years. At the same time another type of grid is taking shaping in Europe, the offshore grid that comes with its own unique set of challenges with regards to market design and system operations.

Following Brexit, the UK had to leave the European Internal Electricity Market (IEM) as of January 2021, going back to the less efficient process of explicit auctions. Nonetheless, policymakers from both sides of the Channel are in talks about the future set-up of electricity energy exchanges.

In a similar vein, in March 2024, the European Council authorized the European Commission to negotiate an agreement on the integration of Switzerland into the internal energy market as part of a broader package. Similarly to the UK, Switzerland is currently only connected to the internal energy market via explicit auctions, although it is geographically in the centre of Europe, and its power grid is highly interconnected to its neighbours.

Lastly, Japan and India, the third- and fifth-largest economies on this planet respectively, are taking steps to liberalize their energy markets and to set-up comparable market designs to the European one.

This book should be regarded as an important contribution to collect and summarize the different aspects of electricity market coupling for both professionals and policymakers to support a safe and robust market design for an energy transition that can increase competitiveness, enhance security of supply and combat climate change.

We would like to thank Gerard Doorman and Knut Eggenberger for their very valuable comments to the first drafts of this book.

Further, we would like to express our gratitude towards Accenture, E-Bridge Consulting, Magnus Energy, N-Side SA, 50 Hertz Transmission GbmH, Bird & Bird, and enspired for their support in making this publication open access, thus guaranteeing that this book reaches a wider audience, contributing to the advancement of integrating electricity markets not only in Europe but also globally.

Berlin and Zurich
December 2024

André Estermann
Marius Schrade
Louise Anderson

Contents

Part I Introduction to Market Coupling

1 **Framework of Market Coupling** 3
Ksenia Tolstrup

2 **Principles of Market Coupling** 31
Peter Willis and Louise Anderson

Part II Operations and Algorithms of Market Coupling

3 **Capacity Calculation** 53
Reinhard Kaisinger and Iacopo Bertelli

4 **Role of TSOs in the European Market Coupling Operations** 79
Marius Schrade

5 **Regional Coordination Centres and Their Role in European Market Coupling** 99
Jakub Władysław Glegoła

6 **Day-Ahead and Intraday Auctions Market Coupling Algorithm** 119
Mehdi Madani and Martin Starnberger

7 **Continuous Intraday Market Coupling Algorithm** 139
Suzanna Vogeler

Part III Physics Meets Finance

8 **Cross-Zonal Shipping and Nominations** 155
Jean-Michel Reghem

9 **Clearing and Settlement** 177
Robert Wand

10 **Congestion Income Distribution** 201
Karel Sebesta and Martin Palkovsky

Part IV News from the Regions

11 Day-Ahead Flow-Based Capacity Calculation and Market Coupling in Core CCR 219
Ferenc Nagy, Melinda Nagy, Luca Toth, Ágnes Takacsne Esze, Ákos Arnold, Dániel Diveny, and Gábor Szathmari

12 Day-Ahead Flow-Based Capacity Calculation in CCR Nordic 259
Pieter Schavemaker

13 The Energy Community 279
Lisa-Marie Mohr

Part V Legal Framework of Market Coupling

14 The Contractual Framework from a TSO Point of View 305
Vicky Blana

15 The Role of National Regulatory Authorities in Market Coupling 325
Nico Schoutteet

Part VI Importance of Data

16 Transparency in European Electricity Markets and ENTSO-E Transparency Platform 339
Nikolaj Nåbo Andersen and Jayaram Anandha

17 A Market Player View 353
Thomas König and Sarah Pischler

Glossary 359

References 367

About the Editors

Dr. André Estermann has more than 20 years of experience in energy sector and currently leads at the German TSO 50Hertz Transmission GmbH a team in system operations responsible for cross-border activities. Furthermore, he is the TSO Chairman of the Market Coupling (MCSC), in the Single Day-Ahead Coupling (SDAC) and Single Intraday Coupling (SIDC) Steering Committees. André holds a Ph.D. in economics at the European University Viadrina Frankfurt, Oder and received an MBA at the University of Southampton, UK. Moreover, he holds a degree in international politics from the University of Potsdam.

Marius Schrade is an Expert in European Market Coupling projects at 50Hertz Transmission GmbH, one of the four transmission system operators in Germany, within the System Operations Concepts and Analytics department since 2019. At 50Hertz, he started his career as project lead for the integration of the German–Czech and German–Polish border into Single Day-Ahead Coupling, which was completed in 2021. Since then, he focuses on cooperation between TSOs and NEMOs in the Market Coupling Steering Committee where he also lead a task force on day-ahead simulation software. Marius holds a Double Degree in European Economics from the Universities of Maastricht in The Netherlands and Louvain-La-Neuve in Belgium.

Louise Anderson currently works with Hitachi Energy as an Assistant Vice President in the Government and Institutional Relations department. Previously, Louise spent 5 years managing the World Economic Forum's Electricity Industry Community, while also launching the WEF's global initiative on Accelerating Industrial Clusters. Louise started her career working in a power plant before moving to work for EirGrid, Ireland's Transmission System Operator, managing high voltage asset outage planning until 2014. Louise also spent some time in Belgium working for ENTSO-E on different aspects of the implementation of the CACM Guideline. Louise has a bachelor's degree in Mechatronics and an MBA specializing in International Business.

Part I

Introduction to Market Coupling

Framework of Market Coupling 1

Ksenia Tolstrup

Abstract

This chapter introduces the key building blocks of the electricity market in Europe. From the First Energy Package in 1996 to the Fifth Energy Package in 2021, Europe's single, integrated electricity market has been evolving over the last three decades. The Guideline on Capacity Allocation and Congestion Management (CACM Guideline) is also introduced, which ensured that market coupling became the mandatory means of allocating cross-zonal capacity in the day-ahead and intraday timeframes. Furthermore, this chapter describes the key aspects of the integrated market design from bidding zones and market coupling to price formation and congestion income. Additionally, the enablers of market coupling including Transmission System Operators, Regional Coordination Centres, and Nominated Electricity Market Operators are explained. The chapter closes by discussing current and upcoming developments in market coupling.

1.1 Key EU Energy Packages Shaping European Market Integration

Aiming to create a single, integrated electricity market with fair market access and a high level of consumer protection, as well as adequate levels of interconnection and generation capacity, the EU has taken several steps over the last three decades.

As illustrated in Fig. 1.1, it started with mandating the liberalization of the power sector with the aim to break up the monopolies that historically ran national

K. Tolstrup (✉)
Magnus Energy, Naarden, The Netherlands
e-mail: k.tolstrup@magnus.nl

A. Estermann et al. (eds.), *European Electricity Market Coupling*, Future of Energy,
https://doi.org/10.1007/978-3-031-86315-8_1

energy markets. A stepwise process of unbundling business activities into transmission, distribution, generation, and retail was put in motion. With the First Energy Package (entry into force 1996), unbundling of accounting was mandated.

In 2003, with the Second Energy Package, legal unbundling was imposed. National Regulatory Authorities (NRAs) were established to oversee the activities

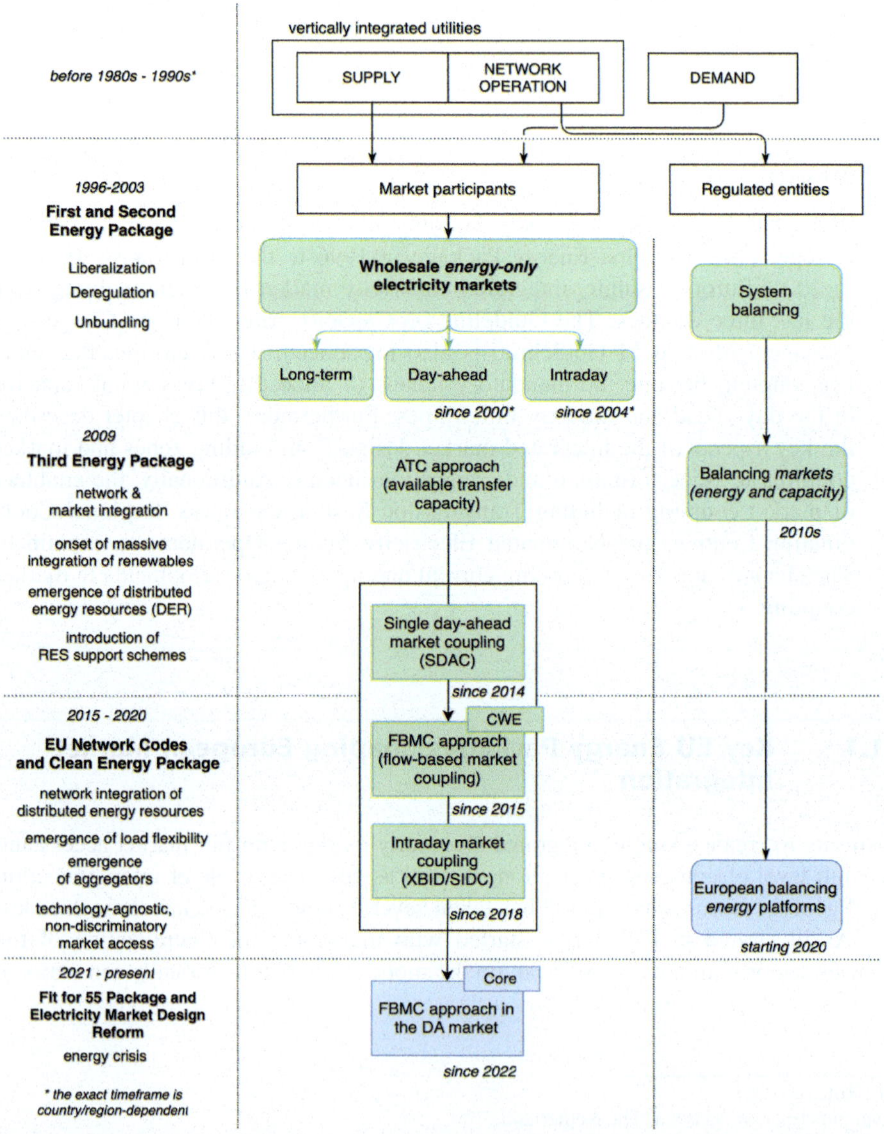

Fig. 1.1 Evolution of electricity markets and the relevant EU regulatory packages. *Source* Adapted from Poplavskaya (2021)

of the newly formed Transmission System Operators (TSOs). As member states implemented the requirements of the First and Second Energy Packages, the power sector became composed of both regulated entities, such as transmission and distribution system operators (DSOs), whose networks represent natural monopolies, and deregulated market participants in wholesale and retail markets.

The Third Energy Package (2009), and the guidelines that derived from it, made market coupling obligatory and created a framework for TSOs working together to compute cross-zonal capacities in a coordinated manner.

In 2019, the Clean Energy for all Europeans (CEP) arrived, further paving the road for EU energy market integration. At the heart of this package are the proper functioning of the internal market and its competitiveness by mandating market-based principles for operating the electricity market. In this sense, it also lays the groundwork for the access of all technologies on the supply and demand side in the electricity markets and enables a non-discriminatory and technology-neutral approach for market participation across borders.

To date, the latest energy package, the "Fit for 55" (fifth) package, entered into force in 2021. This package is set to align EU energy targets with its climate ambitions.

Finally, the recent energy crisis of 2022 had a profound impact on the integrated electricity market and put it through a true acid test. This crisis, among other drivers, spurred the review of the market design by the Agency for Cooperation of Energy Regulators (ACER) and later the Electricity Market Design Reform that was adopted in June 2024, as will be discussed in the last section of this chapter.

The main focus of this chapter is the Third Energy Package and the Clean Energy Package since they have shaped market coupling into what it is today and formed the foundation for the EU's electricity market design and its entities.

1.1.1 The Third Energy Package (2009)

The main objective of the Third Energy Package was to pave the way for the completion of the EU's Internal Energy Market (IEM). To achieve that, it included provisions on the unbundling of suppliers from network operators and strengthening cross-border TSO cooperation in Europe.

Each Energy Package is made up of directives and regulations. The main legal documents found in the Third Energy Package pertaining to the electricity sector include:

- Directive 2009/72/EC of the European Parliament and of the Council of 13 July 2009 concerning common rules for the internal market in electricity and repealing Directive 2003/54/EC
- Regulation (EC) No 714/2009 of the European Parliament and of the Council of 13 July 2009 on conditions for access to the network for cross-border exchanges in electricity.

Thus, the Third Energy Package pushed unbundling further and required TSOs to be fully independent from their parent companies (often incumbent vertically integrated utilities) through ownership unbundling (Articles 9 and 14, Directive 2009/72/EC). This implied the effective separation of appointed system operators from any generation or supply activities to prevent potential conflict of interest and ensure non-discriminatory access to electricity networks. To verify that the unbundling process has indeed been completed, an additional step required that TSOs be certified (Article 3, Regulation 714/2009).

In addition, the package mandated the independence of NRAs from governments and industries. It established ACER, as laid down in Regulation (EC) No 713/2009 of the European Parliament and of the Council of 13 July 2009 establishing an Agency for the Cooperation of Energy Regulators. The Regulation tasked NRAs with developing framework guidelines, which are non-binding and serve as a basis for the process on network code and guideline development.

The Third Energy Package further established the European Network of Transmission System Operators for Electricity (ENTSO-E) as the coordinating body among European TSOs[1] (Fig. 1.2). Following the adoption of the framework guidelines drafted by ACER, ENTSO-E is tasked with drafting network codes and guidelines. The draft network codes and guidelines then each go through a *comitology* process before becoming legally binding in the entire EU. In line with this process, ACER provides its "reasoned opinion and recommendations" on each network code to ENTSO-E and submits it to the European Commission (Article 6, Regulation 713/2009). ACER is moreover tasked with monitoring network code implementation and analysing the progress in terms of facilitation of market integration, competition, and non-discrimination (Article 6(6), Regulation 713/2009).

Figure 1.3 outlines the Network Codes and Guidelines which have been developed following the framework of the Third Energy Package. Subsequent NCs and GLs are now being developed and amended based on the Clean Energy Package, in particular Regulation (EU) 2019/943 of the European Parliament and of the Council of 5 June 2019 on the internal market for electricity (so-called Electricity Regulation). It is worthwhile to mention that an important change brought by the IEM Regulation is that the European Commission may now either request ENTSO-E or the EU DSO Entity in cooperation with ENTSO-E, to submit a proposal for a Network Code.

In the context of (physical) power market integration, the most relevant piece of legislation is the Capacity Allocation and Congestion Management (CACM) Guideline (Regulation 2015/1222), which will be addressed in more detail in the next Section.

Requirements on transparency were also strengthened and later further specified in Commission Regulation 543/2013 of 14 June 2013 on submission and publication of data in electricity markets (Regulation 543/2013). Also called Transparency

[1] Note that ENTSO-E also includes non-EU members.

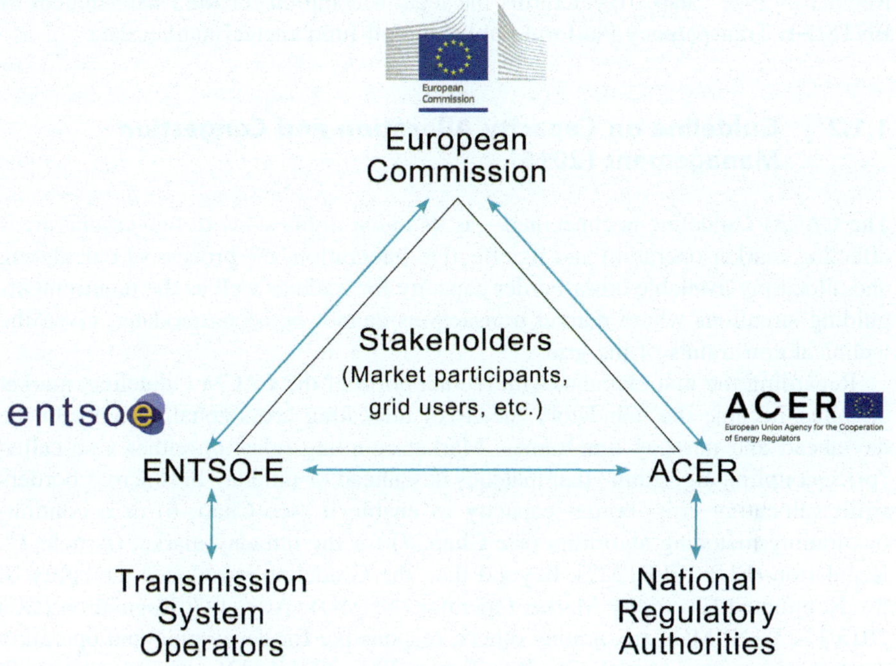

Fig. 1.2 Entities involved in the development of Network Codes. *Source* Author's own illustration

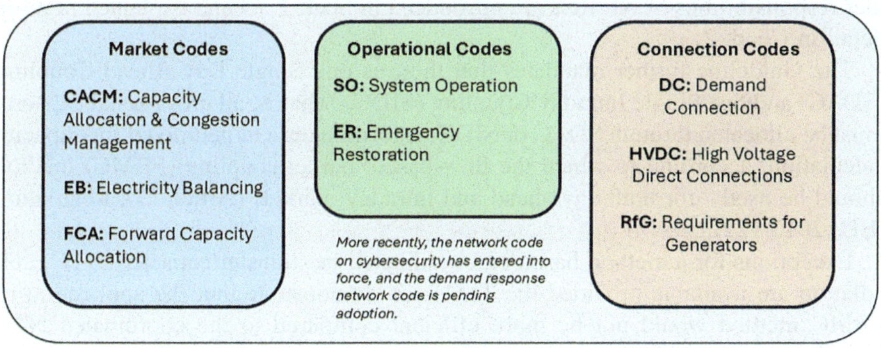

Fig. 1.3 Overview of EU's electricity network codes and guidelines (status October 2023). *Source* Author's own illustration

Regulation (see Chap. 16), it forms the legal foundation for the establishment of ENTSO-E Transparency Platform publishing all fundamental market data.

1.1.2 Guideline on Capacity Allocation and Congestion Management (2015)

The CACM Guideline is concerned—as its name implies—with two crucial areas affecting market operation and its effective integration, the process of calculating and allocating available cross-border capacity for trade as well as the requirements guiding situations where market transactions cannot be accommodated given the technical constraints of the grid.

Regarding the first element, with the adoption of the CACM Guideline, market coupling became the mandatory means of allocating cross-zonal capacity in the day-ahead and intraday timeframes. Market coupling relies on either a so-called "price coupling algorithm" that matches day-ahead or intraday bids across borders while allocating cross-border capacity to enable it (see Chap. 6) or a continuous trading matching algorithm (see Chap. 7) for the intraday market (Article 37, Regulation (EU) 2015/1222). Beyond that, the Guideline introduced the entity of the Nominated Electricity Market Operator (NEMO) (Arts. 4–7, Regulation (EU) 2015/1222). NEMOs are, among others, responsible for developing and operating the two algorithms (Article 36, Regulation (EU) 2015/1222) and—together with TSOs—carrying out market coupling. For this, NEMOs were tasked with a joint implementation of Market Coupling Operator (MCO) functions. The role and further responsibilities of NEMOs are introduced in Sect. 1.3.3 and explained in more detail in Chap. 2.

The Guideline further mandates that there is one Single Day-Ahead Coupling (SDAC) and one Single Intraday Coupling (SIDC). That is, all cross-zonal capacity must be allocated through SDAC or SIDC. It formulates a target model for capacity calculation, according to which the flow-based market coupling (FBMC) method should be used—for both day-ahead and intraday markets (Article 20, Regulation (EU) 2015/1222).

Exceptions for a method based on coordinated net transfer capacity (NTC) calculation are available provided the TSOs can demonstrate that the application of FBMC method would not be more efficient compared to the coordinated NTC method (Article 20(7), Regulation (EU) 2015/1222).

The stepwise implementation approach since 2015 has led to first the introduction of FBMC in Central Western Europe (CWE), including Austria, Belgium, France, Germany, and the Netherlands. On 8 June 2022, FBMC went live in the Core Capacity Calculation region, spanning the largest area of the EU. For an analysis of the effects see Chap. 11.

Concerning the second element, the CACM Guideline holds provisions on three means of congestion management:

- redispatch and countertrade (short-term, within a day)
- cross-zonal capacity calculation (short-term, two day-ahead to day-ahead)
- bidding zone review (mid- to long-term, lead time of a few years).

The CACM Guideline lays the groundwork for the review process of bidding zone configurations (Article 32, Regulation (EU) 2015/1222) that would ideally enable a way to tackle structural congestion that exacerbates a misalignment between the (nodal) physics of the grid and the zonal markets based on a copperplate assumption. See Sect. 1.2.1 for a description of bidding zones.

In the context of this handbook, we focus on cross-zonal capacity calculation and allocation (further detailed in Chap. 3).

The EU regulation creates a framework for the so-called "target model" of the internal electricity market, which relies on the zonal model, SDAC, SIDC, and further implementation of FBMC—all meant to facilitate a common market with efficient price formation. In the depicted target model for market coupling, TSOs provide cross-zonal capacities and NEMOs operate the market:

- The starting point on the capacity *calculation* side is the determination of the Capacity Calculation Regions (CCRs, Fig. 1.4) (Article 15, Regulation (EU) 2015/1222). The CCRs then serve as the defined geographical areas within which TSOs first develop a methodology, implement it, and then compute cross-zonal capacities in a coordinated fashion across a set of bidding zone borders following the previously developed methodology.
- The starting point on the capacity *allocation* side is the designation of NEMOs. Following the establishment of NEMOs, the MCO plan was put in place. This plan serves as the basis for the cooperation of NEMOs in carrying out SDAC and SIDC and ensuring that the results of the calculation are shared among all power exchanges in a non-discriminatory fashion (Recital 5, Regulation (EU) 2015/1222). With the governance in place, NEMOs were required to draft a set of methodologies fundamental to market coupling (e.g. on the algorithms for SDAC and SIDC, minimum and maximum prices, gate opening and closure time of cross-zonal intraday markets, etc.).

A Capacity Calculation Region (CCR) is a geographical region in which TSOs cooperate to compute cross-zonal transmission capacities. A CCR is defined as a set of bidding zone borders (note: not bidding zones or TSO control areas). Hence, one bidding zone or TSO can be a member of several CCRs, while each bidding zone border is associated with only one CCR. See Chap. 3 for more details.

Fig. 1.4 Capacity Calculation Regions in Europe. *Source* Author's own illustration

1.1.3 The Clean Energy for All Europeans Package (2019)

The Clean Energy Package (CEP) consists of several legislative documents, of which two are most relevant in terms of giving another layer of definition to the European internal market for electricity:

Regulation (EU) 2019/943 of the European Parliament and of the Council of 5 June 2019 on the internal market for electricity (so-called "Electricity Regulation").

Directive (EU) 2019/944 of the European Parliament and of the Council of 5 June 2019 on common rules for the internal market for electricity and amending Directive 2012/27/EU (so-called "Electricity Directive").

The revised Electricity Regulation (2019/943) contains crucial provisions pertaining to the integration of the European electricity market. This regulation introduced the "70% minRAM rule", which requires TSOs to provide at least 70% of the remaining available margin (RAM) capacity for cross-zonal trade (Article 16, Regulation 2019/943). The Regulation cites "uncoordinated curtailment of interconnector capacities" that can significantly limit cross-border trade as well as insufficient transparency towards market participants as the reason for the introduction of the 70% criterion (Recital (27) Regulation 2019/943). It thus discourages TSOs from using cross-border capacities to solve internal congestion thus prioritizing internal exchanges over cross-border ones (Article 16(8), Regulation 2019/943).

The adoption of this rule has created a lot of debate where one of the arguments against it were centred about the arbitrary nature of the value of 70%. Additionally, according to ENTSO-E Market Report 2023 (ENTSO-E, 2023), ENTSO-E continues to question whether a general minimum cross-zonal trading margin is appropriate for enhancing European market integration. While ENTSO-E supports the general optimization of the use of trading capacities, the organization highlights that further analysis on the economic and technical impact of the 70% rule is needed, with a focus on system security, economic efficiency, and decarbonization targets.

The CEP provides for two ways of (temporarily) derogating from the 70% rule:

- Until 2026, TSOs who cannot cope with providing 70% due to internal, structural congestion, may develop an Action Plan. This Action Plan must list measures on how to overcome these internal congestions and must contain a linear trajectory reaching 70% by 2026 (defining a ramp from the time of adoption of the Action Plan to 2026).
- Next to Action Plans, temporary derogations from the 70% rule may be granted by NRAs for individual periods not exceeding one year in situations where the TSO can demonstrate that the application of the rule would endanger operational security. Targeting short-term obstacles that prevent TSOs from providing 70%, derogations must be consulted with other NRAs before they can be approved and oblige the TSO to develop a methodology and projects to achieve a long-term solution (Article 16(9) Regulation 2019/943).

NRAs are responsible for assessing the compliance of their TSO(s) with the 70% rule. Not being able to provide 70% (or any relevant value based on the linear trajectory after adopting an Action Plan) can trigger a discussion on bidding zone reconfiguration.

With the aim to specify further how the 70% rule is to be interpreted and applied, ACER issued a recommendation detailing the application of the rule in August 2019. In this recommendation, the term "minimum margin available for cross-zonal trade" (minMACZT) was introduced and serves as the "yardstick" for assessing whether the rule has been complied with.

Even though it is not explicitly stated in the regulation or in ACER's recommendation, the (unofficial and) applicable timeframe for the 70% rule is (currently) the day-ahead timeframe.

In addition, the CEP introduced the following changes:

- It imposed additional provisions guiding the bidding zone review (BZR) process (Article 14, Regulation 2019/943), such as obliging ENTSO-E to report on structural congestion, its location and frequency every 3 years (Article 14(2)), obliging "all relevant TSOs" to provide a proposal for a methodology for a BZR by 5 October 2019 (Article 14(5)). Note that these are ultimately Member States that decide whether the proposed adjustment should be progressed, or the present configuration maintained.
- Regulation 2019/943 (Arts. 34–43) introduced the role of Regional Coordination Centres (RCCs) to facilitate regional cooperation among TSOs. It defined their main tasks, including coordinated capacity calculation and security analyses as well as creation of common grid models (Article 37). RCCs will be described in more detail in Sect. 1.3.2.
- Finally, it laid the foundation for the establishment of System Operation Regions (SORs) in Article 36 of the Electricity Regulation within which RCCs perform their tasks. Note that SORs were defined several years later in ACER's Decision 5/2022 to include TSOs "*that have been designated or assigned with responsibilities which are relevant for system operation, such as, but not limited to: calculation of capacity, assessment of needed remedial actions to ensure security of the whole system, coordination of all the outages*" (Article 3, Annex I, ACER, 2022a). The SORs defined in the ACER Decision are shown in Fig. 1.5.

1.2 Geographical Scope of the European Integrated Market Design

1.2.1 Bidding Zones

Geographically, the electricity market in the EU is based on the so-called "zonal model". The zonal model implies that there is one single price per bidding zone and market time unit (MTU). At the same time, the grid is considered a copperplate, i.e. no congestion is assumed to occur *within* the bidding zone but only between them.

A bidding zone is commonly defined as the geographic (grid) area within which trade of electricity is not constrained by any capacity limitations. This means that a zonal model is a rough approximation of the actual power grid, which requires multiple assumptions. At the time of writing, bidding zone borders predominantly coincide with the control area borders of TSOs and/or member states (with the notable exceptions of Norway, Sweden, and Italy, as is shown in Fig. 1.6).

Capacity does not need to be allocated within a bidding zone whereas bidding zones are coupled via market coupling. The coupling happens via allocation of

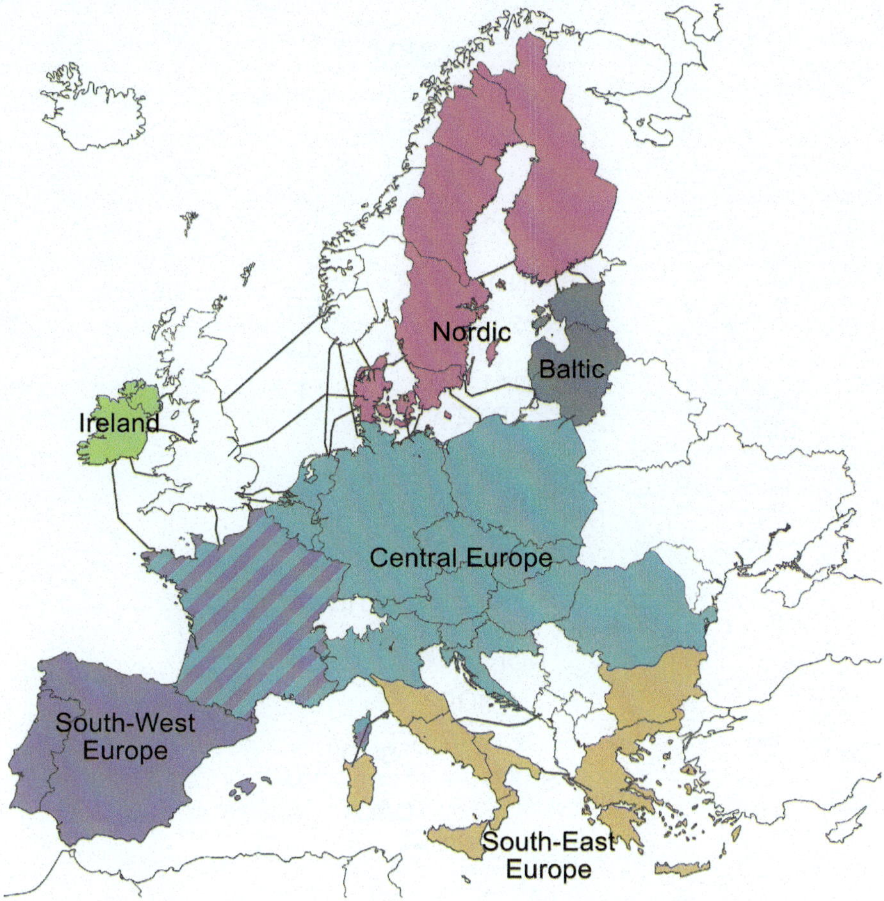

Fig. 1.5 System operations regions in Europe. *Source* Author's own illustration

cross-zonal capacity available for power to be traded across bidding zones. Cross-zonal capacities determine how much trade can happen between bidding zones. If cross-zonal capacity is abundant, price convergence is going to be achieved, that is, two or more bidding zone would have the same market price. In contrast, price divergence generally increases with limited levels of cross-zonal capacities.

Market coupling ensures that the prices between bidding zones are coupled—either through auctions or continuous trading. Thus, market coupling reflects possible congestion *between* bidding zones. Hence, the bidding zone delineation determines how efficiently congestion is addressed through market-based allocation.

Note that several other power systems (e.g. most of the US, Australia) use the nodal market model to manage congestion. A nodal approach relies on a detailed model of the physical grid whereas the market clearing is done at substation level.

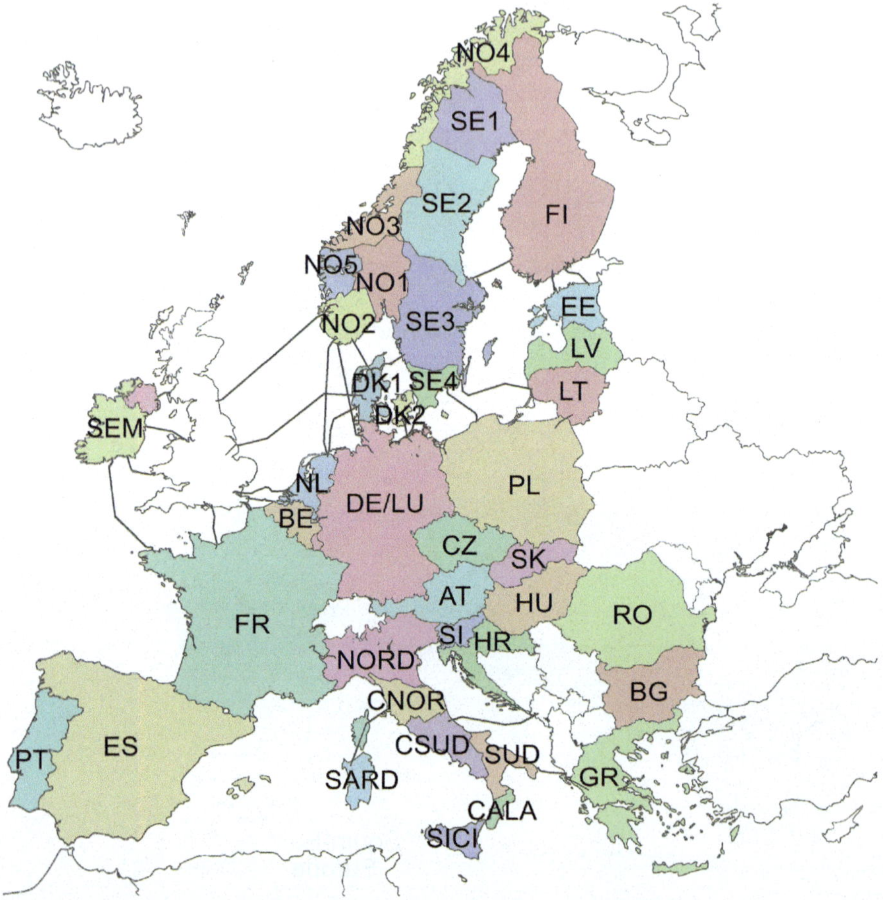

Fig. 1.6 Bidding zones in the EU and Norway. *Source* Author's own illustration

It thus represents physical properties of the individual components as well as laws of physics governing the flows in a meshed AC grid. Consequently, capacity allocation is done on all lines going in and out of each node. The result of this market model is different prices at each node affected by congestion.

The zonal vs. nodal approach for Europe has been subject of years of debate with proponents and opponents on both sides. It became particularly acute due to increasing instances of internal congestion within bidding zones, e.g. due to a simultaneous infeed of large generation volumes from variable renewables in a single grid area. As a result, the market outcome and market efficiency may be weakened as at least some of the trades may not be feasible and would require the TSO to intervene by redispatching (i.e. changing the dispatch) of some assets to relieve congestion.

From the bidding zone external perspective, a telling sign of internal congestion is the presence of so-called loop or transit flows. In the first case, generation in country A that is also consumed in country A flows through country B. In the second case, generation in country A is consumed in country B but to get there it flows through country C. Such a situation is problematic as it may lead to congestion in a country that is not involved in the trade in the first place. For instance, up until October 2018, Germany and Austria formed part of a single bidding zone. Frequent transit flows through their Eastern European neighbours exacerbating the local grid situation were cited as the main reason for the bidding zone split of the two countries, which introduced limits to their cross-border exchanges (e.g. Hurta et al., 2022).

The main argument for the nodal approach appears to be better alignment with the reality of the grid and thus its ability to price congestion in the market prices. Some of the arguments often mentioned against it in the European context is a high computational complexity, unresolved governance issues, liquidity problems of nodal markets, as well as low political acceptance of having different prices within the same country.

Although there remains majority support for keeping the zonal model, the EU regulation provides for mechanisms to ensure that overall market efficiency is preserved while structural congestion is addressed. For this purpose, the CACM Guideline and the CEP defined a bidding zone review (BZR) process (Arts. 32–34 Regulation (EU) 2015/1222), meaning that the current bidding zone delineation is not set in stone. Based on provisions in the CACM Guideline, the participating TSOs are required to formulate a methodology and assumptions for the review process. In the second step and after conducting the comparison of the status quo with alternative configurations, they are to publish a consultation and submit a joint proposal to the participating Member States (Article 32, Regulation (EU) 2015/1222). The CACM Guideline further includes the minimum set of assessment criteria such as change in operational security, overall market efficiency as well as bidding zone stability and robustness over time (Article 33, Regulation (EU) 2015/1222). Within this process, the bidding zone configuration review is completed every couple of years and can be launched by ACER, NRAs, TSOs themselves, or a Member State (Article 32(1)). Finally, the current configuration is to be assessed by ACER every three years based on the technical and market reports drafted by ENTSO-E.

In its latest related Decision 11/2022 ACER proposed alternative bidding zone configurations for Germany, France, Italy, the Netherlands, and Sweden (Annex I, ACER, 2022b). The publication of the BZR is expected in summer 2024.

1.2.2 Market Timeframes, Market Coupling, and Price Formation

The CACM Guideline covers two timeframes: day-ahead and intraday. The day-ahead market is the most liquid market in Europe and, as a result, provides the most

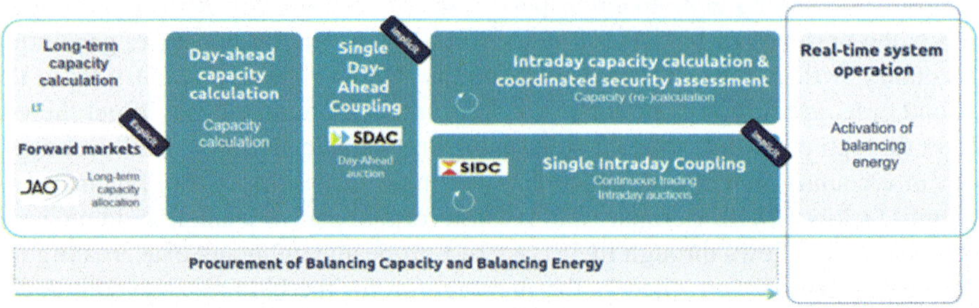

Fig. 1.7 Sequence of market timeframes. The focus of this chapter, day-ahead and intraday markets and the respective capacity calculation processes are marked in teal. *Source* Author's own illustration

robust price signal to buyers and sellers. The traded volumes in the intraday market have been increasing over the last decade in parallel with the growing penetration of variable renewable generation and the need to balance portfolio positions closer to real time. As Fig. 1.7 shows, the electricity market in fact consists of a number of interlinked markets in different timeframes, from long-term to real time. Note that forward, as well as balancing, markets are out of scope of this book.

In the **day-ahead** timeframe:

- The allocation of cross-zonal capacities is performed in one pan-European auction. This auction is held at 12:00 CET D-1 (hence the name "day-ahead") daily and determines the day-ahead clearing prices for all participating bidding zones for all 24 h of the upcoming day from 0:00 to 24:00. That is, the market time unit (MTU) in the day-ahead market is equal to one hour defining the shortest time unit for which a market price is determined. Note that there is only one single clearing price per bidding zone and MTU—there are no side payments or other differentiation between the price that sellers receive, and the price buyers pay (Article 42, Regulation (EU) 2015/1222).
- Single-day-ahead coupling (SDAC) has been introduced as a mechanism to achieve market price setting and simultaneous capacity allocation for all coupled bidding zones thus maximizing social welfare across the EU (and Norway) (Article 2(26), Regulation (EU) 2015/1222). For this, it uses a so-called price coupling algorithm (EUPHEMIA) whose objective function is to maximize economic surplus (Article 38, Regulation (EU) 2015/1222). All the buyer and seller bids submitted by the gate closure time at 12:00 CET D-1 form part of a common merit order—as long as sufficient cross-border capacity is available. In this regard, social welfare can be broken down in three components: (1) producer surplus, (2) consumer surplus, and (3) congestion income. The latter is calculated from the price difference between two zones generated by the congestion between them and its volume, which is meant to facilitate TSOs' investment in improved interconnection. When determining the clearing prices in SDAC,

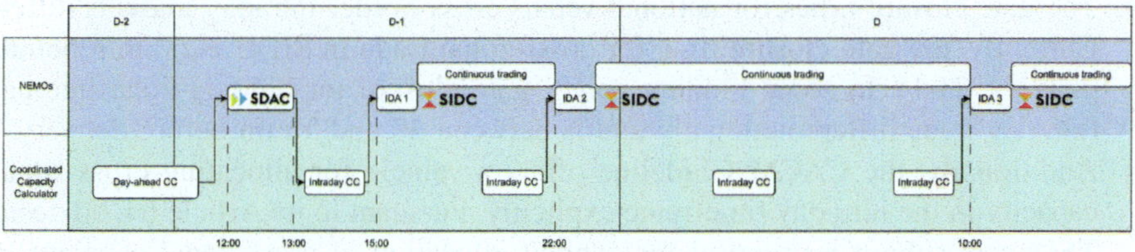

Fig. 1.8 Sequence of market coupling timeframes. *Source* Author's own illustration

marginal pricing rule is applied, i.e. the marginal production/seller bid sets the price (i.e. the bid that is just needed to cover demand at a given price).

In the **intraday** timeframe:

- Single Intraday Coupling (SIDC) was similarly introduced by the CACM Guideline (Article 2(27)) and refers to the process during which bids are matched continuously and capacity is allocated simultaneously while observing cross-border constraints. While SDAC is based on one pan-EU auction with marginal pricing, SIDC is (for now solely) based on continuous trading with direct matching (known as Cross-Border Intraday, XBID). In contrast to auctions, continuous trading matches bids and offers constantly and instantly, provided that a matching set of bids and offers are available and, if the bids and offers originate from two different bidding zones, sufficient cross-zonal capacity is available. This also means that—in contrast to SDAC—no optimization takes place in the continuous market and, as a consequence, the pricing rule is pay-as-bid and there is not one single clearing price per bidding zone. Note that, in continuous trading, allocated cross-zonal capacity does not receive any congestion income. Instead, a power trader sees only allowed bids that can be accommodated based on available cross-zonal capacities. For instance, in a situation where all capacity was allocated to the day-ahead market for export, only capacity in the import direction would be available in the intraday market.
- To address the missing congestion income in SIDC, the CACM Guideline requires a methodology on intraday cross-zonal capacity pricing (Article 55). This methodology was adopted by ACER in 2019 and foresees the introduction of three intraday (implicit) auctions (IDAs) with an MTU of 15 min.[2] These three auctions are to take place at 15:00 CET D-1, 22:00 CET D-1, and 10:00 CET D (for the second half of day D, see Fig. 1.8). To conduct these auctions, continuous trading is temporarily on hold to avoid double allocation of cross-zonal capacity. At the time of writing, the go-live of the three IDAs is planned for Q2 2024.

[2] Note that national intraday auctions have been used in Austria, Belgium, France, Germany, Italy, The Netherlands, and Spain.

- The gate closure times for national versus cross-border intraday markets differ. Currently, the gate closure time for cross-zonal trade in SIDC is 60 min before delivery (H-1). In some bidding zones, gate closure for bidding zone internal intraday trade differs nationally and is between 30 and 5 min before delivery.
- Additionally, the CACM Guideline offers a vehicle for allocating cross-zonal capacity in the intraday timeframe explicitly. Pursuant to its Article 64, adjacent TSOs can submit a proposal for explicit allocation of cross-zonal capacity in the intraday timeframe to the relevant NRAs for approval. Such an arrangement refers to a specific border, should be temporary and gradually phased out.

The CACM Guideline allows the use of two methods for capacity calculation that are aimed at maximizing transfer capacity available for trade while observing operational security constraints:

Coordinated net transfer capacity approach (CNTC) (primarily for regions with "less interdependent" cross-zonal capacity, Recital 7, Regulation (EU) 2015/1222)

Flow-based approach (FB), which is based on a more detailed representation of the grid by considering power transfer distribution factors and available margins on critical network elements (CNEs) thus "expanding the envelope" of the available cross-border capacity. FBMC is part of the EU Target Model and thus should ultimately substitute CNTC.

Since 2022, all borders within Core CCR use FBMC (see also Chap. 11) whereas capacity is calculated using CNTC in the rest of SDAC. Currently, SDAC allows for a combined NTC and flow-based allocation ("hybrid coupling"). In contrast, SIDC so far only allows allocation of NTCs. However, the CACM Guideline foresees flow-based allocation also in SIDC. An interested reader is invited to see more details on the two approaches in Chap. 3.

1.3 Key Enablers of European Electricity Market Integration

1.3.1 Transmission System Operators

TSOs, as regulated entities in charge of power system operation, were established through the process of unbundling that started in the 1990s (see also Fig. 1.9). They are regulated due to the fact that they hold the monopoly of access to the transmission grid. Hence, they are obliged to ensure non-discriminatory access to the transmission infrastructure to grid users. Activities of European TSOs are coordinated within ENTSO-E, which at the time of writing includes 40 TSOs from 36 countries, including outside the EU.

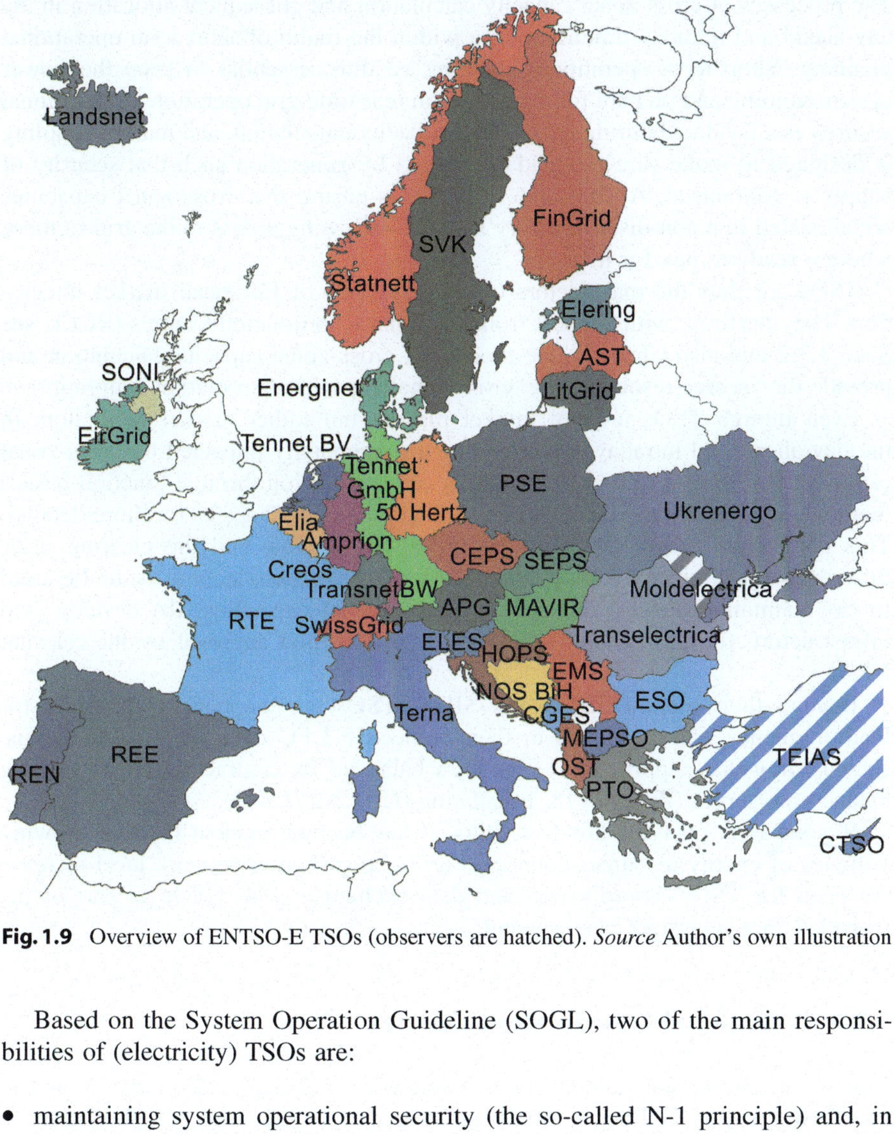

Fig. 1.9 Overview of ENTSO-E TSOs (observers are hatched). *Source* Author's own illustration

Based on the System Operation Guideline (SOGL), two of the main responsibilities of (electricity) TSOs are:

- maintaining system operational security (the so-called N-1 principle) and, in the event of a contingency, returning to the normal state as soon as possible using remedial actions[3] and
- ensuring security of supply, including security of (long-term) production and demand, of grid availability and of continuous grid balancing.

[3] Pursuant to Article 2(13), Regulation (EU) 2015/1222, remedial actions refer to "any measure applied by a TSO or several TSOs, manually or automatically, in order to maintain operational security", such as redispatching, countertrading, or topological actions.

The processes of cross-zonal capacity calculation and consequent allocation in the day-ahead and intraday timeframes are within the realm of short-term operational planning. Short-term operational planning is, thus, essential to keep the power system running and prepare for challenges in real-time grid operations. Operational security is considered during cross-zonal capacity calculation, and market coupling is designed to make sure the load is covered by generation such that security of supply is guaranteed. At the same time, TSOs ensure that cross-zonal capacities are allocated in a non-discriminatory fashion—allowing access to the grid to those who are ready to pay for it.

TSOs are thus the main actors on the grid side of European market integration. They perform, with support from Regional Coordination Centres (RCCs, see Sect. 1.3.2 and also Chap. 5), the process of cross-zonal capacity calculation and provide the "scarce resource" (i.e. cross-zonal capacities) to market coupling.

Even though TSOs are not market players but rather market facilitators in the day-ahead and intraday markets, they are financially impacted by cross-zonal capacity allocation. From cross-zonal capacity allocation through (auction-based) market coupling, TSOs receive congestion income (see Chap. 10 for more details). This congestion income depends on the allocated flow and the clearing price difference on a given bidding zone border. Congestion income is to be used to (a) maintain firmness[4] of allocated cross-zonal capacities, (b) develop grid infrastructure further, and (c) reduce grid tariffs (upon approval by the relevant NRA).

Beyond their role in SDAC and SIDC, TSOs are the main responsible for load–frequency control (LFC) in their respective LFC areas or blocks, that is, safeguarding that supply and demand are balanced by contracting and activating balancing reserves (Article 118, Regulation (EU) 2017/1485).

Beyond that, over the last five years, it has become evident that the growing volumes of energy resources connected on the distribution network level increase the need for TSOs' coordination and data exchanges with DSOs as part of the so-called "active system management".

1.3.2 Regional Coordination Centres

Regional Coordination Centres (RCCs) emerged from Region Security Coordinators (RSCs) introduced by the SOGL as an actor with functions complimentary to those of individual TSOs, namely for the matters of regional relevance. They are fully owned by TSOs; typically, by the TSOs with the intention of being serviced by the respective RCC(s). There are, however, a few TSOs who are shareholders of more than one RCC. Participation in at least one RCCs is mandatory for TSOs.

[4] According to Article 2(44) Regulation (EU) 2015/1222, firmness refers to "a guarantee that cross-zonal capacity rights will remain unchanged and that a compensation is paid if they are nevertheless changed".

In the context of cross-zonal capacity calculation, the RCCs assume the role of the Coordinated Capacity Calculator (CCC) (Article 2(11), Regulation (EU) 2015/1222). This role implies that RCCs support the TSOs within a given CCR with the process of computing cross-zonal capacities while accounting for the TSOs' operational security limits (Recital 20, Regulation 2019/943).

Beyond cross-zonal capacity calculation, RCCs are also responsible for coordinated security analysis, regional week-ahead, and day-ahead adequacy forecasts as well as regional outage planning, among others (Article 37, Regulation 2019/943). Note that RCCs do not play a role in the process of allocating cross-zonal capacities.

Figure 1.10 shows the current RCCs and their respective geographic coverage (based on their shareholder TSOs). Note that 50 Hertz is a shareholder of both TSCnet and Coreso; Terna of both Coreso and SEleNe CC; and Transelectrica of both TSCnet and SEleNe CC. In some CCRs, two RCCs perform the task of capacity calculation (e.g. Core and Italy North), either by splitting tasks or by following a rotational scheme of the main and a back-up CCC.

1.3.3 Nominated Electricity Market Operators

The entity of the Nominated Electricity Market Operator (NEMO) was created by the CACM Guideline (Articles 4–7). The main role of the NEMOs is to operate, maintain, and further develop SDAC and SIDC. However, they do not engage in the long-term timeframes (year-ahead and month-ahead) and are not obliged to operate both SDAC and SIDC simultaneously. That is, there are NEMOs-only active in the day-ahead timeframe but not in the intraday timeframe. Typically, NEMOs were power exchanges before seeking to become a NEMO.

After concluding a designation process with competent NRAs or ministries, a (power) exchange can become a NEMO (Arts. 4–5 Regulation (EU) 2015/1222). Designations are timeframe-specific (day-ahead and/or intraday) and may be limited in time before re-designation is needed. Once an entity has been designated as a NEMO in (at least) one EU Member State, this NEMO can use the so-called "passport" to become active in any other EU member state. When "passporting" into another EU member state, the NEMO does not have to be (re-)designated locally and merely needs to notify its intention of wanting to be active to the relevant NRA or ministry. In this way, EU regulation ensures a level-playing field among NEMOs regardless of the country of operation.

The CACM Guideline introduces competition between NEMOs or, more specifically, between the power exchange function of NEMOs. The rationale for this was to have NEMOs compete via the fees that they charge the market participants trading on their platforms. In member states with several competing NEMOs, so-called "Multi-NEMO Arrangements" (MNAs) have been agreed upon between the relevant TSOs and NEMOs and implemented. At the time of writing, MNAs are in place in Central Western Europe (Austria, Belgium, France, Germany, and the Netherlands), Poland, the Baltic countries, and the Nordic countries, including

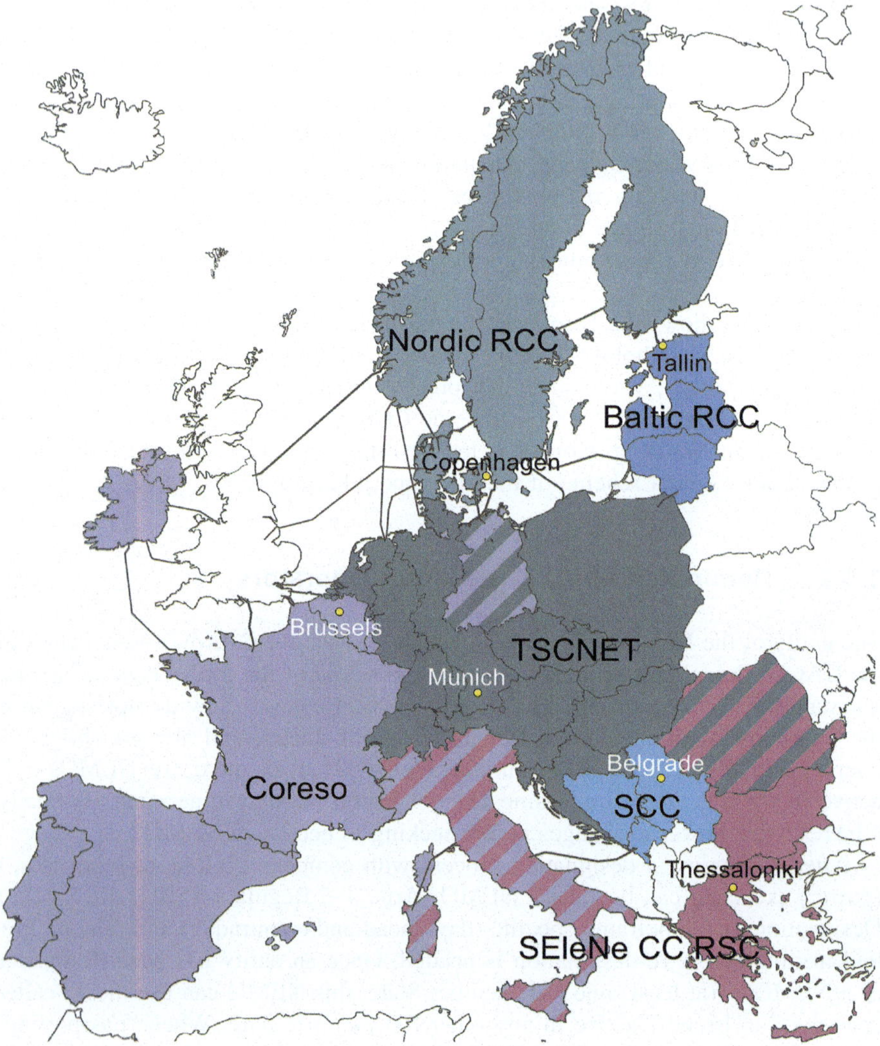

Fig. 1.10 Regional Coordination Centres in Europe. *Source* Author's own illustration

Norway. Figure 1.11 shows whether a Member State allows for competition among NEMOs or has a single monopoly NEMO. Note that some member states allow for competing NEMOs, but there is only one NEMO active and, at this stage, no MNA implemented. This is, for example, the case in Croatia and Slovenia. It is important to keep in mind that the NEMO status is required if cross-zonal capacity is to be allocated. Operating markets within a bidding zone (i.e. without allocating cross-zonal capacity), does not require the NEMO status (e.g. auctions within a bidding zone).

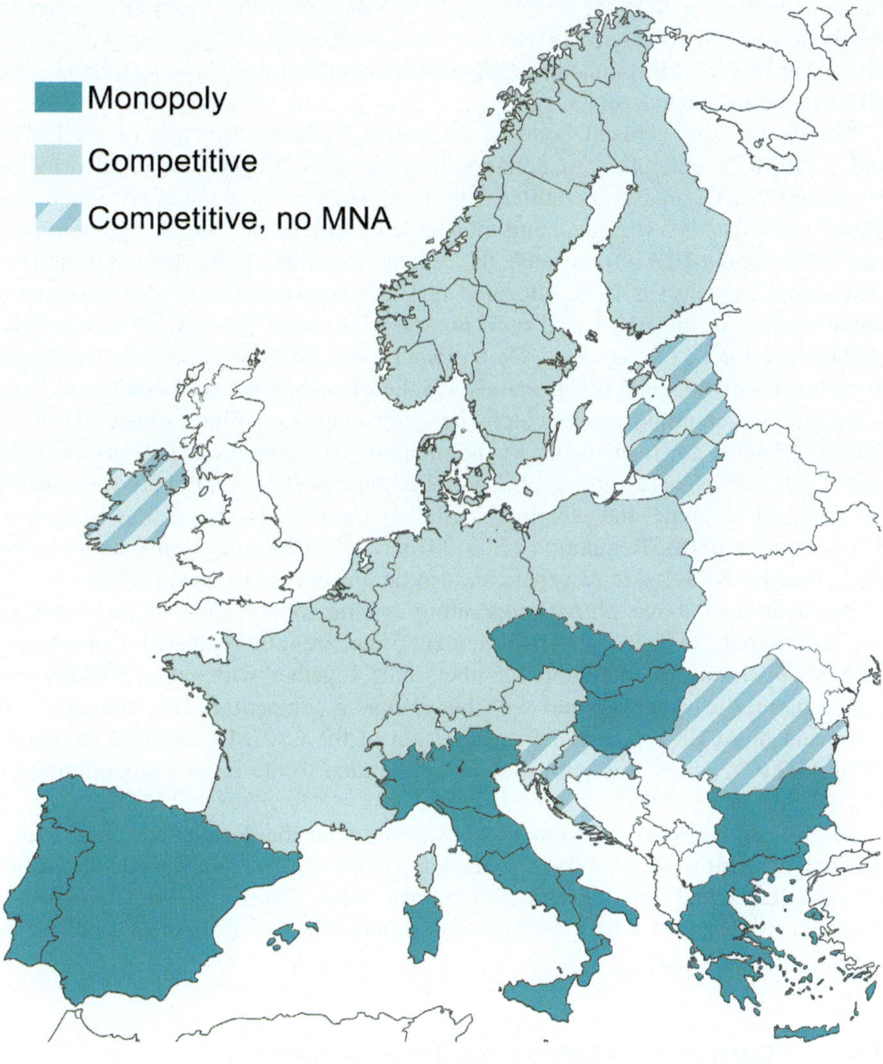

Fig. 1.11 NEMO market type in the EU and Norway. *Source* Author's own illustration

At the same time, in order to enable SDAC and SIDC, NEMOs take over the function of Market Coupling Operators (MCOs)[5] that are in charge of using the MC algorithm for matching and allocating orders, of cooperating and sharing relevant information with TSOs in joint procedures (Article 7, Regulation (EU) 2015/

[5] The CACM Guideline does in fact allow for the Commission, in cooperation with ACER, to "create or appoint" a single regulated MCO (Recital 15, Regulation (EU) 2015/1222), a provision which did produce a lot of debate between ACER and NEMOs.

1222) as well as with other NEMOs in a non-discriminatory fashion in order to jointly perform MCO functions (Recital 5 and Article 6(4), Regulation (EU) 2015/1222). For this, NEMOs must keep separate accounts for their power exchange and MCO functions (Article 6(1)(c)).

The competitive element between the power exchange function of a NEMO, and a NEMO's obligation to collaborate with other NEMOs to jointly realize SDAC and SIDC, creates a conflict within one and the same company. Operating SDAC and SIDC is a regulated and monopolistic task (hence, the designation process to become a NEMO), because the "scarce resource" in market coupling (i.e. cross-zonal capacity) is to be allocated in a fair, transparent way that maximizes social welfare in the EU, based on a non-discriminatory process. To ensure this, NEMOs are forced to cooperate. On the other hand, NEMOs are power exchanges in competition with each other, mainly via their fees and areas of operation.

In the post-coupling process step (i.e. after-market coupling, either SDAC or SIDC), NEMOs are responsible for nominating cross-zonal capacity to the relevant TSOs (i.e. physical delivery). Central counter parties (CCPs) are responsible for financial shipping and clearing, including transferring congestion income to TSOs (Article 68(8), Regulation (EU) 2015/1222). This role is in practice often performed by NEMOs, however it can also be outsourced to third parties.

As mentioned above, NEMOs operating as a monopoly, and—so far—without any activity outside their respective Member State, are fully regulated. Competitive NEMOs operating in the same Member State together with other NEMOs are both a partner to work together with but also as a competitor. This situation was addressed in ACER's proposal for a revision of the CACM Guideline (so called "CACM 2.0"), which was published and submitted to the European Commission in 2022.

ACER maintains an overview on its website with the designated, passporting, and active NEMOs per EU Member State and Norway. This overview distinguishes between day-ahead and intraday timeframes. Also, the All NEMO Committee keeps an overview of which NEMO is designated in which member state on its website (see Fig. 1.12).

1.4 Current and Upcoming Developments and the Future of Market Coupling

During the energy crisis of 2022 provoked by the war in Ukraine and the shortages of natural gas, the coupled European electricity markets were put through a major test. Low availability of nuclear power generation in France and negative weather effects during an exceptionally hot summer contributed to exacerbating the already complicated situation. These events raised serious questions about the fundamental value of European electricity market integration and whether the market design as it is known today was still fit for purpose.

Among other challenges, the crisis led to the reduction of cross-border capacities in some Member States to reduce the impact of extreme prices, several

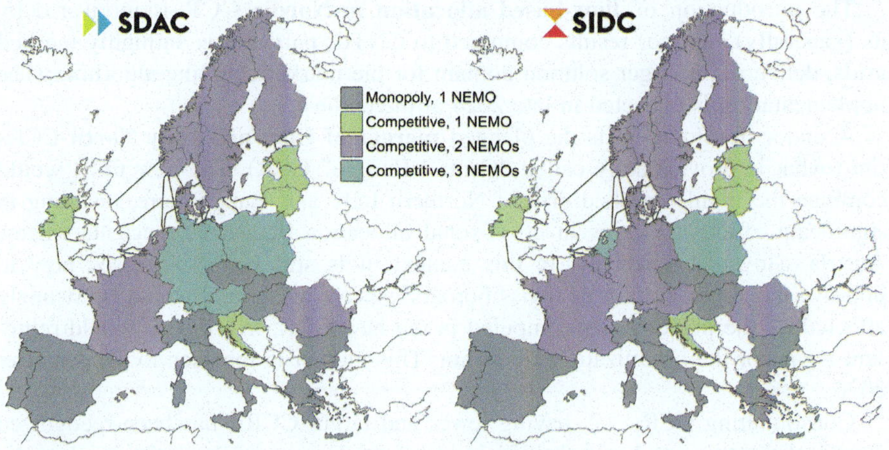

Fig. 1.12 Number of NEMOs operating in each country; per timeframe (Day-ahead and Intraday). *Source* Author's own illustration

instances of hitting the price ceiling and thus increase of maximum market prices and the introduction of several emergency measures to tackle price hikes. Despite multiple critical voices and drastic proposals, the value of cross-border cooperation and coupled electricity markets as well as their crucial role in averting the worst effects of the crisis was recognized by stakeholders across the board.[6] This is reflected in the current proposal for Electricity Market Design Reform,[7] which preserves current market design and the main pillars of the European electricity market functioning. The adoption of the final reform is expected at the beginning of 2024.

Beyond that, market coupling expected is evolved further in 2024–2027 in a number of important ways.

In **SDAC**, the most important developments concern the introduction of the 15-min MTU, flow-based allocation in Nordic CCR, Advanced Hybrid Coupling as well as the merger of Core and Italy North CCRs.

Currently, the cross-zonal markets work on an hourly granularity, while several national markets already employ 15-min granularity in XBID. The introduction of the **15-min MTU** in cross-zonal markets results is meant to accommodate more dynamic (and volatile) generation and consumption patterns and allow for more flexible trading. At the same time, higher granularity is expected to put a strain of EUPHEMIA while some implementation challenges still exist with regard to accommodating multiple-MTU bids, such as exclusive or linked block bids. The go-live of the 15-min MTU is expected for Q1 2025.

[6] A detailed analysis of the measures adopted during the crisis can be found in ACER (2023b).

[7] Reform of electricity market design: Council and Parliament reach deal—Consilium (europa.eu).

The introduction of **flow-based allocation in Nordic CCR** is motivated by its (generally) superior results compared to CNTC, particularly in highly meshed grids, delivering a larger solution domain for the market coupling algorithm. The implementation is expected in late 2024 at the earliest.

A major development is the planned **merger of Core and Italy North** CCRs (into what is provisionally called "Central Europe" CCR). The new CCR would combine the highly meshed grid of Northern Italy and that of Core, resulting in significant efficiency gains in cross-zonal allocation between France, Germany, Austria, Slovenia, and Italy. In this context, it is still being discussed how to improve the consideration of Swiss borders, since Switzerland would be strongly affected by the power flows computed in the new CCR, in which it would represent a gap for the optimization algorithm. This merger is foreseen to happen after 2026.

Consolidating CCRs, i.e. having fewer and larger CCRs, has been recognized as one of the ways to boost the efficiency of the European electricity market. An important complementary measure is to improve the interaction between CCRs. **Advanced hybrid coupling** (AHC) involves consideration of the external borders of a CCR more similarly to internal ones, such that cross-CCR flows can be optimized to a certain extent. AHC is particularly relevant for the borders with HVDC interconnectors between Core and Nordic CCRs. For reference, the current solution is called Simple Hybrid Coupling, in which a *forecasted* flow across an HVDC interconnector is considered as a load (or generator, for imports) at the border. As a result, rather than being optimized, it is a fixed input. Currently, the expected go-live is set to mid-2025, but is highly dependent on the timely go-live of previous implementations. Further details can be found in Chap. 3.

In **SIDC**, the most relevant developments include the introduction of Intraday Auctions and flow-based allocation (currently only available in day-ahead), resulting in the intraday market becoming more similar to the current day-ahead market.

Three **Intraday Auctions** (IDAs) will take place, at 15:00 D-1, 22:00 D-1, and 10:00 D, lasting 30 min, during which continuous trading will be suspended. Important reasons for the introduction of IDAs are the pricing cross-zonal capacity and allowing TSOs to collect congestion income. Both elements are absent in the current XBID solution. IDAs will go-live on a 15-min MTU granularity, expected in Q2 2024.

Similarly, to the introduction of flow-based in Nordic CCR for the DA market mentioned above, **FB allocation in intraday** is expected to result in higher social welfare in Europe. Currently, the go-live is expected after 2025.

The development of offshore hybrid systems also introduce a new layer of complexity in terms of how they fit into a bidding zone model and how the 70% rule might apply to so-called offshore bidding zones.

Another project under development, with lower impact on the markets yet very important for reasons of operational security is the Regional Operational Security Coordination (**ROSC**). This encompasses a series of operational security assessments and sessions to coordinate the dispatch of remedial actions needed to relieve grid congestion. ROSC is aimed at increasing the operational security of the European grid and at distributing the costs of the remedial actions applied more fairly. At the moment, ROSC is planned for implementation at the end of 2025.

Finally, market integration and coupling does not stop with short-term wholesale markets. Increasing amounts of variable renewable generation are being progressively integrated in the energy system, while the supply of flexibility is relatively limited. In this context, **system balancing** is becoming ever more challenging, and cross-border cooperation ever more essential for European power system stability. These developments already created the impetus for TSOs to cooperate for imbalance netting and joint procurement of Frequency Containment Reserve (FCR), as well as to create European balancing energy platforms (PICASSO, MARI, TERRE). The latter has been mandated by the European Balancing Guideline and the roadmaps for their go-live and TSO accessions were formulated. The next stepping stone for the integration of the balancing market is the regional procurement of balancing capacity, based on the methodologies approved by ACER in July 2023 (ACER, 2023a). These methodologies, including harmonized market-based allocation of cross-zonal capacity, are expected to be jointly implemented by ENTSO-E, TSOs, and RCCs within two years.

Bibliography

Association for the cooperation of the European transmission system operators (ENTSO-E). (2023). *Market Report 2023.* https://eepublicdownloads.blob.core.windows.net/public-cdn-container/clean-documents/nc-tasks/ENTSO-E_Market_Report_2023.pdf. Accessed June 2024.

European Commission. (2013). *Commission Regulation (EU) No 543/2013 of 14 June 2013 on submission and publication of data in electricity markets and amending Annex I to Regulation (EC) No 714/2009 of the European Parliament and of the Council. OJ L 163, 15.6.2013*, pp. 1–12. https://eur-lex.europa.eu/legal-content/EN/TXT/?uri=CELEX%3A32013R0543

European Commission. (2015). *Commission Regulation (EU) 2015/1222 of 24 July 2015 establishing a guideline on capacity allocation and congestion management. OJ L 197, 25.7.2015*, pp. 24–72. https://eur-lex.europa.eu/legal-content/EN/TXT/?uri=CELEX%3A32015R1222

European Commission. (2017). *Commission Regulation (EU) 2017/1485 of 2 August 2017 establishing a guideline on electricity transmission system operation. OJ L 220, 25.8.2017*, pp. 1–120. https://eur-lex.europa.eu/legal-content/EN/TXT/?uri=CELEX%3A32017R1485

European Parliament and European Council. (2009a). *Directive (EU) 2009/72/EC of the European Parliament and of the Council of 13 July 2009 concerning common rules for the internal market for electricity and repealing Directive 2003/54/EC. OJ L 211/55, 14.8.2009.* https://eur-lex.europa.eu/legal-content/EN/TXT/PDF/?uri=CELEX:32009L0072&from=BG

European Parliament and European Council. (2009b). *Regulation (EC) No 714/2009 of the European Parliament and of the Council of 13 July 2009 on conditions for access to the network for cross-border exchanges in electricity. OJ L 211/55, 14.8.2009.* https://eur-lex.europa.eu/LexUriServ/LexUriServ.do?uri=OJ:L:2009:211:0015:0035:EN:PDF

European Parliament and European Council. (2019a). *Regulation (EU) 2019/943 of the European Parliament and of the Council of 5 June 2019 on the internal market for electricity. OJ L 158, 14.6.2019*, pp. 54–124. https://eur-lex.europa.eu/legal-content/EN/TXT/?uri=CELEX%3A32019R0943&qid=1715683977934

European Parliament and European Council. (2019b). *Directive (EU) 2019/944 of the European Parliament and of the Council of 5 June 2019 on common rules for the internal market for electricity and amending Directive 2012/27/EU (recast). OJ L 158, 14.6.2019*, pp. 125–199. https://eur-lex.europa.eu/legal-content/EN/TXT/?uri=celex%3A32019L0944

European Union Agency for the Cooperation of Energy Regulators (ACER). (2022a). *Definition of System Operation Regions in accordance with Article 36 of the Regulation (EU) 2019/943 of the European Parliament and of the Council of 5 June 2019 on the internal market for electricity, Decision.* https://www.acer.europa.eu/sites/default/files/documents/Individual%20Decisions_annex/ACER%20Decision%2005-2022%20-%20Annex%20I%20-%20Definition%20of%20SORs_0.pdf

European Union Agency for the Cooperation of Energy Regulators (ACER). (2022b). *Decision on the alternative bidding zone configurations to be considered in the bidding zone review process. Annex 1. List of alternative bidding zone configurations to be considered in the bidding zone review, Decision.* https://www.acer.europa.eu/sites/default/files/documents/Individual%20Decisions_annex/ACER%20Decision%2011-2022%20on%20alternative%20BZ%20configurations%20-%20Annex%20I%20-%20rectified.pdf

European Union Agency for the Cooperation of Energy Regulators (ACER). (2023a). *Decision No 11/2023 on the TSOs' proposal for the harmonised cross-zonal capacity allocation methodology, Decision.* https://acer.europa.eu/sites/default/files/documents/Individual%20Decisions/ACER_Decision_11-2023_on_Harmonised_Cross-Zonal_Capacity_Allocation_Methodology.pdf .

European Union Agency for the Cooperation of Energy Regulators (ACER). (2023b). *Security of EU electricity supply.* https://acer.europa.eu/sites/default/files/documents/Publications/Security_of_EU_electricity_supply_2023.pdf

Hurta, A., Žilka, M., & Freiberg, F. (2022). Impact of the splitting of the German-Austrian electricity bidding zone on investment in a grid-scale battery energy storage system deployed for price arbitrage with gray and green power in Austrian and German day-ahead power markets. *Energy Reports, 8*, 12045–12062. https://doi.org/10.1016/j.egyr.2022.09.045

Poplavskaya, K. (2021). *Balancing and redispatch: The next stepping stones in European electricity market integration: Improving the market design and the efficiency of the procurement of balancing and redispatch services* (PhD thesis). https://doi.org/10.4233/uuid:fb8c99cc-24d6-4718-8986-95833ffc1f49

Dr. Ksenia Tolstrup is a Principal at Magnus Energy. She co-leads Magnus Energy's Technical Advisory practice and is heavily involved in several projects related to electricity market and regulatory analysis, balancing, flexibility, and hydrogen transition. Ksenia has almost a decade of experience in the energy sector—in consulting and in research. As a Senior Research Engineer at AIT Austrian Institute of Technology she focused on the analysis and simulation of electricity markets in all timeframes, from forward to balancing and congestion management, led multiple national and international projects and authored over 30 top-journal articles, conference papers and project reports. In 2021 she obtained her PhD cum laude in energy economics from Delft University of Technology.

Principles of Market Coupling

2

Peter Willis and Louise Anderson

Abstract

This chapter describes market coupling in detail and guides the reader through the evolution of market coupling in Europe. The chapter lays out the underlying market coupling principles and highlights the benefits of EU market coupling and cross-border interconnectors. Single Day-Ahead Coupling and Single Intraday Coupling processes are introduced at a high level, as well as the roles and responsibilities of major market participants. The Terms, Conditions, and Methodologies, which are developed to ensure a harmonized approach to market coupling, as well as the market coupling timeframes, are briefly introduced. Finally, a number of high-level examples are provided to illustrate how market coupling works.

2.1 Introduction to Market Coupling

The reform of electricity markets in the EU has been a long and complex process, starting with the first energy package in the 1990s, and culminating with the fifth energy package or "Fit for 55" in 2021 which aims to align the EU's energy and climate targets for 2030 and 2050. As of December 2023, the European Parliament and Council have provisionally agreed on the reform of the EU's electricity market design to help build a renewables-based energy system, lower energy bills, and

P. Willis (✉)
Bird & Bird, London, UK
e-mail: peter.willis@twobirds.com

L. Anderson
Zurich, Switzerland

A. Estermann et al. (eds.), *European Electricity Market Coupling*, Future of Energy,
https://doi.org/10.1007/978-3-031-86315-8_2

better protect consumers from price spikes. This could be a precursor for the sixth energy package.

These energy packages could be seen as milestones. They have delivered on market reforms, with successes ranging from: unbundling of monopoly network businesses from competitive production and supply businesses; opening up of retail markets; formation of independent regulatory agencies; establishment of independent verification relating to transmission and distribution pricing; and creation of a competitive EU-wide electricity market through national and regional coupling.

2.1.1 What Is Electricity Market Coupling?

Electricity market coupling refers to the integration of different electricity markets into a single market, allowing for the trading of electricity across borders. More concisely, market coupling is the merging of individual, national markets to make possible the trade of electricity across a large geographical area (Schönheit et al., 2021). Interconnections between market zones and internal transmission networks facilitate these exchanges. The coupling of electricity markets is a key EU objective, to achieve a more efficient and integrated regional electricity market. This has multiple benefits including ensuring access to the lowest prices, and enhanced security of supply.

To couple electricity markets across a region, there are political, technical, financial, and collaboration perspectives to consider. Political consensus on a target model was a key starting point. The EU's target model which was agreed between 2006 and 2009 includes a forward market, a day-ahead and intraday market and a balancing market. Day-ahead and intraday markets are planned to be closely coupled. [There are also elements of market coupling in forward and balancing markets.]

From a technical perspective, cross-border infrastructure must be built out in order to make cross-border capacities available to trade.

From a financial perspective, money must be available to build large-scale cross-border electricity infrastructure, but also a liquid market place must be in place to ensure that buyers and sellers can depend on the realization of their bids and orders. Another important aspect is clear rules around settlement.

From a collaboration perspective, countries must work together to manage changes in socio-economic benefits from market coupling, e.g. as can be seen in the shift from the Net Transfer Capacity market coupling approach towards the Flow-Based approach in the Nordic countries, discussed in Chap. 12.

2.1.2 Principles of Market Coupling

Electricity Market Coupling has emerged as a pivotal concept, revolutionizing the way electricity markets operate. Market coupling in the European electricity market aims to create a more integrated and efficient market, with increased competition and price convergence. Thus, it creates benefits for both consumers and producers, while also incentivizing investment in new infrastructures and renewable energy sources.

Electricity market coupling must adhere to some fundamental principles:

- Price Convergence: Any approach to market coupling must promote price convergence, meaning that prices for electricity will become more similar across different countries (or regions), and ideally price differences between neighbouring markets will be eventually eliminated. This should create a level-playing field for electricity producers and consumers, promoting competition, and lowering costs for end users.
- Increased Competition: Any form of market coupling must increase competition in the electricity market, allowing producers to compete across a wider geographical area and incentivizing them to increase efficiency and put in place measures to reduce costs.
- Market Efficiency: Any effective market coupling approach should reduce the market inefficiencies and distortions that arise from the fragmentation of electricity markets. In addition, through the elimination of market segmentation, market integration should aim to create a larger, more liquid market which reduces price discrepancies. As a concrete example, by pooling resources and coordinating electricity trading across different countries, market coupling can improve resource utilization and reduce the need for expensive reserve capacity.
- Investment Incentives: Electricity market coupling should create incentives for investment in new electricity infrastructure. Expanding and maintaining a robust cross-border transmission infrastructure is vital for the success of market coupling. Cross-border infrastructure, supported by a resilient transmission backbone, should optimize cross-border electricity flows, thereby reducing price differences between markets and increasing the potential market for electricity producers. This will help ensure that there is adequate generation capacity to meet future demand and can also incentivize investment in renewable energy sources by helping reduce curtailment.
- Regulatory Framework: Achieving regulatory harmonization across a region can be a complex and time-consuming process. Nonetheless, the success of any market coupling project depends on the development of a harmonized regulatory framework that facilitates cross-border trade and ensures fair competition. This framework should comprise principles and rules, including for cross-border transmission, market design, and settlement mechanisms. Effective regulatory oversight (including market monitoring) is also essential to reduce the risk of market manipulation and the exercise of market power by large market participants.
- Cross-Border Trading: Any market coupling approach should be based on cross-border electricity trading. Cross-border trading fosters cooperation among neighbouring countries, allowing them to optimize resource utilization, and enhance energy security. Implicit auctions are an important part of cross-border electricity trading. Implicit auctions enable market participants to submit bids

for cross-border capacity, through a process which simultaneously optimizes both the energy exchange and cross-border capacity allocation. This ensures efficient use of interconnectors and minimizes congestion.

- Enhanced Energy Security and Grid Stability: Market coupling should create interconnected markets which can rely on each other during supply shortages or unexpected disruptions, enhancing energy security by reducing dependency on a single energy source or supplier. Market coupling should also enhance grid stability by enabling cross-border flows and supporting the efficient balancing of supply and demand.

2.2 European Electricity Market Coupling Overview

Electricity market coupling can be quite nuanced. It is generally not a choice between having a fully coupled electricity market with no influence from national borders or fully individual disconnected power systems. Europe's internal electricity market is almost 30 years old; however, cross-border lines and trading arrangements for electricity have existed for much longer. EU countries also trade electricity with third countries, e.g. Switzerland, the UK, and West Balkan countries. The Federal Council of Switzerland is pursuing an electricity agreement with the EU at time of writing to allow full access for Switzerland to the single market for electricity.

2.2.1 Background

Market coupling in Europe kicked off in 1993 when a common spot market was introduced between Norway and Sweden removing border tariffs for trade of electricity. Nord Pool was formally established in 1996 as the first international power exchange between Norway and Sweden. In 1998, Finland joined the Nordic spot market (Nord Pool) and in 1999 West Denmark joined the Nordic spot market, followed by East Denmark in 2000. In 2006, the French, Belgian, and Dutch day-ahead markets coupled. Extensions followed including Central West Europe.

The go-live of price coupling in North-Western Europe (NWE) in 2014 was a major milestone for European market integration. This coupling project was an initiative by the Transmission System Operators (TSOs) and Power Exchanges in North-Western Europe including 50Hertz, Amprion, Elia, Energinet, RTE, Statnett, and others. This was also the first use of the pan-European Price Coupling of Regions (PCR) solution for the simultaneous calculation of market prices and flows on interconnectors, based on the Euphemia algorithm (as described in Chap. 6).

2.2.2 Formalization Through CACM

Price Coupling of Regions (PCR) was a project of European Power Exchanges and other market participants to develop a single price coupling solution to calculate electricity prices across Europe both respecting the limits of cross-border assets and on a day-ahead basis. PCR is now the key mechanism for pan-European electricity market coupling. PCR is based on a single uniform algorithm for calculating electricity prices, decentralized management of data, and individual responsibility for market areas (i.e. power exchanges would be responsible for their own market areas and PCR Matcher and Broker Service calculates the different market and reference prices). PCR simultaneously determines volumes and prices in all relevant bidding zones based on the marginal pricing principle, aiming to ensure efficient allocation of cross-border transmission capacity and thereby maximizing social welfare.

The launch of PCR was a crucial step to achieve a harmonized European electricity market, and was eventually incorporated into Single Day-Ahead Coupling (SDAC). The interested reader can find more information on SDAC in Chap. 6. According to ENTSO-E (2024b), 98.6% of EU consumption is now coupled under SDAC and the daily average value of matched trades is approximately EUR200 mil.

Market coupling, based on SDAC and Single Intraday Coupling (SIDC), was formalized and harmonized across the EU through the Capacity Allocation and Congestion Management (CACM) Regulation 2015/1222. However, the process started back in 2011 when ACER's Framework Guidelines on Capacity Allocation and Congestion Management for electricity (ACER, 2021) were adopted. The Framework Guidelines included provisions for the forthcoming development of the CACM Network Code, centred around cross-zonal capacity calculation and allocation in the day-ahead and intraday timeframes. Based on these Framework Guidelines, ENTSO-E was tasked to develop the Network Code on Capacity Allocation and Congestion Management. Once developed, the draft network code was submitted to ACER for an opinion which resulted in some amendments prior to re-submission. ACER submitted a recommendation to the European Commission to adopt the Network Code subject to specific conditions. The European Commission further revised the CACM Network Code, which was then finally adopted and entered into force as a Commission guideline in August 2015.

2.2.3 Major Market Participants and Their Roles

The term "market participant" refers to any entity actively involved in the European electricity market including producers, consumers, and aggregators. It is assumed that market participants will act in good faith and will bid into the market based on their "true" preferences. In addition, market coupling requires the involvement and collaboration of a number of important participants, with specific roles and responsibilities:

- **NEMO**—NEMOs play a crucial role in delivering on an internal European electricity market. A NEMO is a nominated electricity market operator, designated by the competent authority to perform specific tasks such as day-ahead and/or intraday market coupling. NEMOs are designated nationally in accordance with Articles 4–6 of the CACM Guideline, and their roles are defined in Article 7. NEMOs facilitate efficient electricity exchange by coordinating activities between different market participants, including TSOs, power exchanges, and other stakeholders. NEMOs are discussed further in Chaps. 1, 4, 8, and 9.
- **MCO**—The function of the Market Coupling Operator (MCO), carried out by the NEMOs, is to operate the two price coupling algorithms (day-ahead and intraday) to match orders across bidding zones, and to allocate cross-zonal capacity based on network capabilities and clearing prices. To carry out this role, the MCO must process inputs from NEMOs and TSOs and use these inputs to operate a central algorithm, matching bids, constraints, and capacities. The MCO must then validate and share outputs such as the matched trades, clearing prices, and scheduled exchanges, with NEMOs and TSOs. The MCO role of the NEMOs is defined in Article 7(2) of the CACM Guideline. The operation of the algorithms is explained further in Chaps. 6 and 7.
- **CCP**—The Central Counter Party (CCP) plays an important role in electricity market coupling, ensuring smooth and secure transactions through its clearing and settlement role. The CCP acts as an intermediary in electricity markets, essentially standing between buyers and sellers and guaranteeing the fulfilment of trades, even if one party defaults. By assuming counterparty credit risk, the CCP enhances market stability and reduces the impact of defaults. The CCP is discussed further in Chaps. 8 and 9.
- **Shipping Agent (SA)**—A shipping agent is defined, according to the CACM Guideline as the entity or entities responsible for transferring net positions between different central counter parties. When trades occur in the electricity market, they result in net positions, either surplus or deficit. These net positions must be transferred between different CCPs involved in the market. The shipping agent ensures that these net positions are accurately moved from one CCP to another, thus ensuring the efficient functioning of the market. It is useful to note that two CCPs may appoint a shipping agent to handle energy exchanges between them across national borders. The Shipping Agent is discussed further in Chaps. 8 and 9.
- **TSO**—A Transmission System Operator (TSO) is responsible for operating the electricity transmission system. Under the CACM Guideline, TSOs also provide inputs (figures for available capacity and constraints) to the coordinated capacity calculators. The TSOs' role in the market coupling process is set out in Article 68 of the CACM Guideline. The TSO role is explained further in both Chap. 1 and Chap. 4.
- **Coordinated Capacity Calculator (CCC)**—Based on the inputs received from TSOs, a CCC is responsible for calculating and coordinating cross-zonal capacity in specific regions. It submits the cross-zonal capacities to each TSO within its designated capacity calculation region for validation. The CCC's calculations

are essential for ensuring efficient electricity trade across borders and managing congestion. The CCC role in the CACM Guideline is carried out by the Regional Coordination Centres (RCCs), described further in Chap. 5.

2.2.4 Introduction to Market Coupling

A number of studies carried out in recent years highlight the significant benefits of market coupling in Europe:

- The EU's Agency for the Cooperation of Energy Regulators (ACER) released its final assessment of the EU wholesale electricity market design in April 2022 (ACER, 2022), as requested by the European Commission. ACER concluded that in 2021, cross-border trade delivered an estimated EUR34 billion of benefits while helping to smoothen price volatility. In addition, higher market integration and cross-zonal capacities resulted in more cross-border competition and reduced opportunities to exert market power. All of this lowers the energy bill over the long-term. Nonetheless, ACER pushed for further improvements to the electricity market including improving the liquidity of forward markets and continued integration of markets, emphasizing that more European independence and energy security requires more interdependence.
- Meanwhile, in 2018, the Regulatory Assistance Project estimated (Baker et al., 2018), based on studies by Booz & Company in 2013 and the European Commission in 2016, the potential increase in social welfare due to full market coupling in Europe could be between €16 and €43 billion annually by 2030, depending on a number of factors including how Europe optimizes its generation portfolio, the development of adequate interconnector capacity, and the deployment of demand response. It is worthwhile to re-emphasize that beyond generation and demand response, the development of interconnection capacity will play a huge role in maximizing welfare benefits from market coupling.
- Finally, in 2013, Booz & Company notes a number of studies in literature which attempt to calculate the potential benefits from the flow-based capacity allocation method (described in more detail in Chaps. 11 and 12). Even back in 2007, the results showed an approximate EUR200 million increase in social welfare due to implementing the flow-based method in the Netherlands, Belgium, France, and Germany, increasing by almost a factor of three if that was extended across the whole of the European Union (Booz & Co, 2013).

In 2023, Bruegel reported on the benefits of market coupling by separating these benefits into techno-economic and managerial-governance. According to Bruegel, and as illustrated in Fig. 2.1, from a techno-economic perspective optimizing the design and operation of multiple electricity systems together rather than individually and increasing renewable penetration will result in less CO_2 emissions, less volatile short-term prices, reduced need for reserve capacity and flexibility, as well

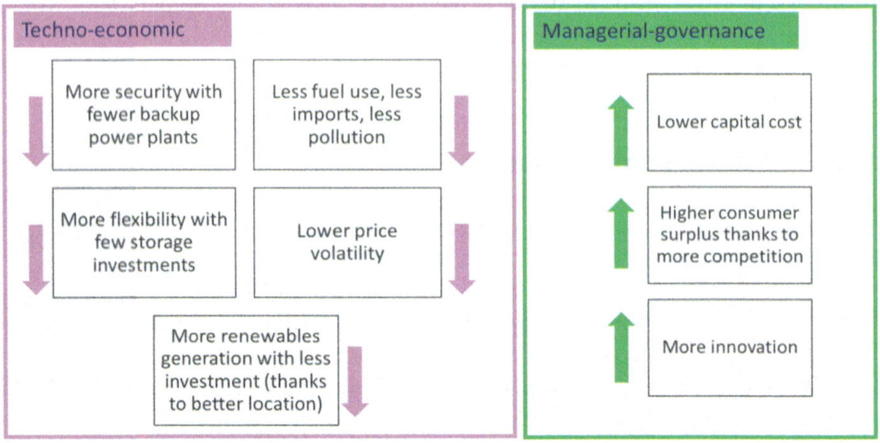

Fig. 2.1 Benefits of market coupling. *Source* Zachmann et al. (2024)

as cost savings, enhanced resilience, and less need for grid investments (Zachmann et al., 2024).

A salient example of the energy security/resilience benefits provided by market integration in Europe can be seen in France. During 2021/2022, a combination of corrosion issues and long periods of hot weather resulted in many nuclear reactors across France going offline for extended periods, causing a shortfall of 81 TWh of nuclear generation representing about 20% of French demand. France went from being a net exporter of electricity to being a net importer of electricity, facilitated by electricity market coupling across the region.

Meanwhile, from a managerial-governance perspective (Zachmann et al., 2024) shows how market coupling can deliver on a higher consumer surplus, more innovation thanks to increased competition and a level-playing field, as well as lower capital costs, thanks to increased regulatory harmonization delivering on more certainty, thus amplifying investor confidence and hence lowering the cost of capital for clean technologies.

2.2.5 Benefits of Cross-Border Interconnectors

Cross-border interconnectors play a crucial role in enhancing energy systems, facilitating market coupling, and fostering cooperation between nations. They offer multiple advantages including:

- Energy Security: By connecting grids across borders, countries can share energy resources. This resilience ensures stable electricity supply even during emergencies.

- Access to Clean and Affordable Electricity: Interconnectors facilitate wider access to clean and affordable electricity. Larger power systems can integrate greater volumes of renewable energy, which can vary based on weather conditions.
- Market Expansion and Investment Attraction: Integrating markets creates an expanded customer base and can reduce the curtailment of renewables, attracting investors.
- Efficiency through Economies of Scale: Expanding power systems across borders allows developers and market participants to take advantage of economies of scale. Access to more varied electricity generation resources can lower total operating costs.
- Price Convergence: Cross-border interconnectors contribute to ever-closer alignment of electricity prices in European countries. By optimizing transmission capacities, electricity exchanges automatically calculate the optimum use of resources, benefiting consumers.

To achieve these benefits, it is essential to not only build connected physical infrastructure but also to establish agreements between jurisdictions and coordinate stakeholders. Political will, technical requirements, and institutional frameworks are key to successful cross-border power systems. The two examples below shine a light on the negative impact of Brexit[1] on 2 recent interconnectors between the UK and Denmark and the UK and Belgium, highlighting the crucial role of political collaboration in optimizing markets.

Case Study—Viking Link

The Viking Link interconnector is approximately 765 km long and allows electricity to be exchanged between Denmark and Great Britain and went into operation in December 2023. It is interesting to note that, according to Energinet, the Danish TSO, initially the link will operate at a maximum of 800 MW, which is lower than the expected 1400 MW. This is based on the capacity of the Danish electricity grid which will become overloaded above 800 MW.

Since Brexit, the British market is no longer part of the Single Day-Ahead Coupling (SDAC). Therefore, trading of capacity on the Viking Link is based on explicit allocation, instead of implicit allocation. In the European electricity market, explicit allocation refers to a specific method of handling transmission capacity on cross-border interconnections. In explicit auctions, the transmission capacity on an interconnector is auctioned separately and independently from the marketplaces where electrical energy is auctioned. This means that the capacity is allocated through dedicated auctions, distinct from the auctions for electricity itself. These auctions occur at different time horizons, including annual, monthly, and daily auctions.

[1] Brexit refers to the withdrawal of the UK from the European Union.

While explicit allocation is a stable approach, it is not very efficient from a societal perspective. It tends to result in unused capacity and can also result in the redistribution of value from the interconnector owners (which in Denmark's case are the consumers) to market participants. This makes it less attractive for consumers to contribute to such large infrastructure investments. In fact, according to one particular author (Pindstrup, 2024), with explicit allocation on the Viking Link, approximately 25–30% of the potential interconnector revenue is not realized to the benefit of consumers but is left with the market participants.

Case Study—Nemo Link

A similar explicit allocation mechanism is in place on the Nemo Link Limited interconnector. The Nemo Link is a 1000 MW HVDC submarine power cable between Belgium and the UK. It is a joint venture between Elia, the Belgian TSO, and the British company National Grid. The link has been fully operational since 2019.

Initially, when the Great Britain bidding zone was still a part of the European Single Day-Ahead Coupling mechanism, implicit allocation would have been used. However, since the Great Britain bidding zone has been dissolved, the Nemo Link is now subject to explicit allocation.

This provides an opportunity to compare the impact of implicit (2019 and 2020) and explicit allocation (2021–2023) on the Nemo Link based on 5 years of data. Using this data (Schoutteet, 2024), was able to highlight that the complexities of the explicit allocation process (which requires that capacity and energy are traded separately), resulted in inefficiencies. Whereas under the implicit mechanism in 2019 and 2020, in operation before the Great Britain bidding zone was dissolved, reverse flows (or flows which go against the market spread i.e. from the higher priced to the lower-priced zone) only occurred 0.003% of the time. Post 2020, reverse flows were observed 17.2% of the time. This ultimately negatively affects consumers, not only from Great Britain but also from Belgium.

2.3 How Does It All Work Together?

As mentioned earlier, market coupling systems exist both in day-ahead trading and intraday markets. This section provides a very brief high-level description of day-ahead and intraday coupling.

2.3.1 Single Day-Ahead Coupling

The process for Single Day-Ahead Coupling consists of three phases: Pre-coupling, coupling, and post-coupling.

Pre-coupling: From several months in advance until gate closure:

The NEMOs and TSOs use the processes and procedures defined in Sect. 2.4 Terms, Conditions, and Methodologies (TCMs) below to generate the inputs (mainly cross-border capacity and constraints, and orders from market participants to the NEMOs to purchase and sell power) which feed into the coupling equations.

The TSOs provide inputs for the capacity calculation process to the RCCs (Article 29(1) of the CACM Guideline).

The RCCs then calculate cross-zonal capacity (Article 29(2)-(10) of the CACM Guideline) and send it to the TSOs for validation (Article 29(11) of the CACM Guideline).

The TSOs validate the capacity calculation and send it to the other regional TSOs and the RCCs (Article 30(1) and (2) of the CACM Guideline), which in turn send the validated capacities and constraints to the NEMOs (Articles 30(3) and 46 of the CACM Guideline).

The NEMOs receive the constraints and validated cross-zonal capacity (Articles 39(1)(a) and (b) of the CACM Guideline) and publish them. They also receive the orders from market participants (Article 39 and 40 of the CACM Guideline).

Coupling: Shortly after gate closure:

The NEMOs, carrying out the role of MCO, take the inputs from the pre-coupling phase (Article 47 of the CACM Guideline) and operate the EUPHEMIA market coupling algorithm (Article 48(1) of the CACM Guideline), determining cross-border transfers and prices and scheduled exchanges, and maximizing price convergence.

Post-coupling: Once the prices and net positions have been determined:

The MCO delivers the results to the NEMOs (i.e. information to allow them to determine the order status) (Article 48(1)(b) of the CACM Guideline) and deliver the clearing price and net position per bidding zone to the TSOs, RCCs, and NEMOs (Article 48(1)(a) of the CACM Guideline).

The TSOs verify that the results correspond to the constraints and validated capacity (Articles 39(4) and 48(2) of the CACM Guideline).

The NEMOs verify the results and deliver them to market participants (Article 48(3) and (4) of the CACM Guideline).

The CCPs clear and settle matched orders for their relevant markets (Article 68(1) of the CACM Guideline). They also settle the net financial position between bidding zones and nominate scheduled exchanges to interconnectors (Article 68(3) of the CACM Guideline).

The shipping agent may play a role, if appointed, to act as a counterparty CCP between CCPs for energy exchanges. The shipping agent also collects and delivers congestion income to TSOs (Article 68(6)-(8) of the CACM Guideline).

The SDAC process is illustrated through examples in Sect. 2.6 below.

2.3.2 Single Intraday Coupling

SIDC follows similar principles and is summarized rather more briefly than SDAC below.

The “XBID Program” started as a joint initiative by Power Exchanges (EPEX SPOT (including former APX and Belpex), GME, Nord Pool, and OMIE) and

TSOs from 11 countries, to create a coupled integrated intraday cross-border market. It was launched in 2018 with 10 local implementation projects which delivered continuous trading of electricity across Austria, Belgium, Denmark, Estonia, Finland, France, Germany, Latvia, Lithuania, Norway, the Netherlands, Portugal, Spain, and Sweden.

The XBID Platform has now been renamed as the Single Intraday Coupling (SIDC), designed to enable continuous cross-border trading across Europe.

SIDC has many benefits, including that it:

- enables market participants to balance their positions up to one hour prior to delivery.
- reduces the need for reserves and associated costs while allowing enough time for carrying out system operation processes for ensuring system security.
- promotes competition.
- increases liquidity.
- makes it easier to share energy generation resources; and
- facilitates unexpected changes in consumption and outages.

SIDC is based on a common IT system with a Single Order Book (SoB), a Capacity Management Module (CMM) and a Shipping Module (SM). This system facilitates the continuous matching of orders from market participants from multiple Bidding Zones, provided that Cross-Zonal Capacity is available.

Orders entered by market participants for continuous matching in one country can be matched by orders similarly submitted by market participants in any other country within the project's reach as long as cross-zonal capacity is available. Both explicit (capacity only) and implicit (capacity and energy together) continuous trading is possible, and the solution is in line with the EU Target model for an integrated intraday market.

When market participants of each NEMO submit orders, they are put together in one Shared Order Book (SOB). In a similar way, TSOs make available all the intraday cross-border capacities in the Capacity Management Module (CMM).

This allows NEMOs to operate trading systems to show orders to market participants from other market participants in three groups:

- within the same NEMO;
- from other NEMOs in the same market area; and
- from other market areas if there is enough capacity available.

The data from the SOB and the CMM is enhanced with data from:

- the relevant TSO;
- the Central Counter Party (CCP); and
- shipping agent data from the Shipping Module (SM).

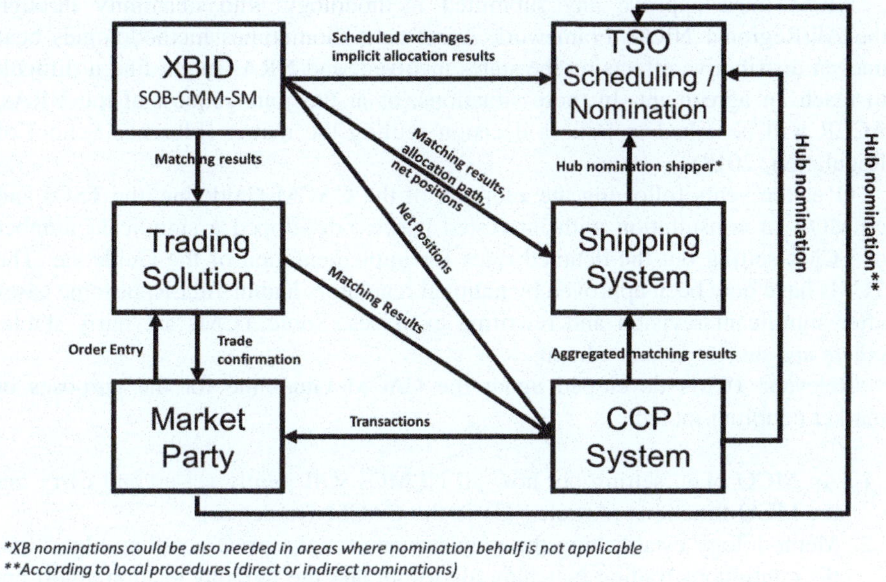

Fig. 2.2 Single Intraday Coupling process. *Source* ENTSO-E (2024a)

This enhanced data is then sent to relevant parties such as the NEMOs and TSOs.

The SIDC process according to ENTSO-E (2024a) is illustrated in Fig. 2.2.

2.4 Terms, Conditions, and Methodologies (TCM)

Since the TCMs for the CACM Guideline are built around market topics, the TSOs, as the main technical experts of the electricity markets, were appointed as the main drafters. In the case of the CACM Guideline, NEMOs are also under obligation to draft some of the TCMs. Once drafted, the TCMs are approved by the National Regulatory Authorities (NRAs) or ACER and eventually become binding on the TSOs and other addressed parties.

Therefore, while the drafting and approval of TCMs is performed at a regional or EU-wide level, the implementation and monitoring of the implementation is done, firstly, at national level, creating a more decentralized market regulation approach.

Depending on the particular provision of the CACM Guideline, either all EU TSOs or TSOs from a concerned region are required to cooperate and draft a methodology on a specific topic, by a certain deadline. Upon completion, the draft methodology must be submitted to either all NRAs or only to the NRAs from the concerned region, depending on the impact area of the topic dealt with in the methodology.

NRAs must approve any submitted methodology with unanimity through the All/Regional NRAs Framework Agreement. Sometimes methodologies bear uneven distributive effects on the states involved, and NRAs might find it difficult to reach an agreement. In these situations, or at the joint request of the NRAs, ACER will be asked to issue a decision settling the matter (Florence School of Regulation, 2018).

Over the years following the adoption of the CACM Guideline, the TSOs and NEMOs, in consultation with interested parties, developed a significant number of TCMs setting out the detailed rules for implementation of the Guideline. The TCMs have now been approved by national regulators and/or ACER, in some cases after significant revision and rewriting exercises. Some TCMs are fairly stable, others are subject to amendment.

Relevant TCMs developed under the CACM Guideline for the purposes of market coupling include:

1. An MCO plan, setting out how all NEMOs will jointly set-up and carry out the MCO functions (Article 7(3) of the CACM Guideline).
2. Methodology establishing the requirements for the price coupling algorithm, the continuous trading matching algorithm and the intraday auction algorithm (Article 37 of the CACM Guideline).
3. Terms and conditions for mandatory and optional products for SDAC and SIDC. Important to note that for each product type, the same attributes should be applied in all bidding zones. There should be no differentiation to ensure a fair market (Articles 40 and 53 of the CACM Guideline).
4. Minimum and maximum clearing prices for SDAC and SIDC, which also includes mechanisms for adjusting automatically the maximum and the minimum clearing prices (Article 41 and 54 of the CACM Guideline).
5. Methodology for a backup for MCO function, taking into account that there may be situations where the price coupling process is unable to produce results and therefore fallback solutions at a national and regional level will be required (Article 36 of the CACM Guideline).
6. Terms and conditions for intraday cross-zonal gate opening and closure time. It is useful to note that the intraday cross-zonal gate closure time shall be set in such a way that it: (a) maximizes market participants' opportunities for adjusting their balances by trading in the intraday market timeframe as close as possible to real time; and (b) provides TSOs and market participants with sufficient time for their scheduling and balancing processes to ensure network and operational security (Article 59 of the CACM Guideline).
7. Methodology for intraday cross-zonal capacity pricing, which establishes the framework for determining the price of intraday cross-zonal capacity in the European electricity market (Article 55 of the CACM Guideline).

8. Methodology for a day-ahead firmness deadline, which defines the deadline after which cross-zonal capacity for the day-ahead allocation becomes firm (Article 69 of the CACM Guideline).
9. Methodologies for complementary regional auctions which allow for the implementation of complementary regional intraday auctions within or between bidding zones in addition to the single intraday coupling solution if they do not have an adverse impact on the single intraday coupling (Article 63 of the CACM Guideline).
10. Methodologies for fallback procedures for each capacity calculation region. All TSOs from each CCR develop fallback procedures in the event that the coupling process is unable to produce results (Article 44 of the CACM Guideline).
11. Methodologies for the calculation of scheduled exchanges for SDAC and SIDC, which assist TSOs in the somewhat complex task of calculating schedule exchanges for each market coupling process (Articles 43 and 56 of the CACM Guideline).
12. Methodology for the distribution of congestion income, taking into account the general principles, goals, and other methodologies set out in the CACM Guideline (Article 73).
13. Definition of capacity calculation regions (CCRs). Interesting to note that Article 5(2) of Regulation (EU) 2019/942 introduced a new procedure for the approval of proposals for common TCMs where the CCRs proposal is now to be submitted directly to ACER (Article 15(1) of the CACM Guideline).
14. Methodology on generation and load data provision, developed so that specific data from generators and load units can be provided for each capacity calculation timeframe (Article 16 of the CACM Guideline).
15. Methodology on the development of the common grid model, taking into account the CACM Guideline and the System Operation Guideline (Article 17).
16. Methodologies on capacity calculation for the day-ahead and intraday market timeframes (Article 20 of the CACM Guideline).

2.5 Timeframes

Figure 2.3 provides an overview of different timeframes of the wholesale and balancing markets. In sequential order, energy can be traded year(s) before the delivery right up to the day after the actual delivery (for redispatching measures). Whereas in the day-ahead market energy is traded one day before real time, the intraday market enables market participants to correct their nominations on the delivery day itself (Amprion, 2024).

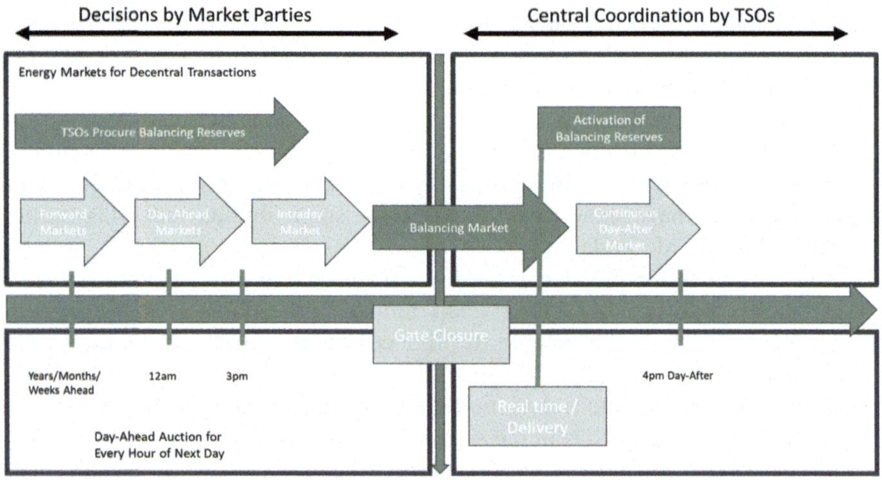

Fig. 2.3 Overview of different timeframes of the wholesale and balancing markets. *Source* ENTSO-E (2022)

2.6 Illustrative Examples

The following three examples illustrate how market coupling operates. The section compares at a basic level uncoupled markets, coupled markets, and partially coupled markets. Please note that Figs. 2.4, 2.5 and 2.6 are intended only to provide an illustration of the effect of market coupling and of changes in cross-border capacity on prices in connected markets.

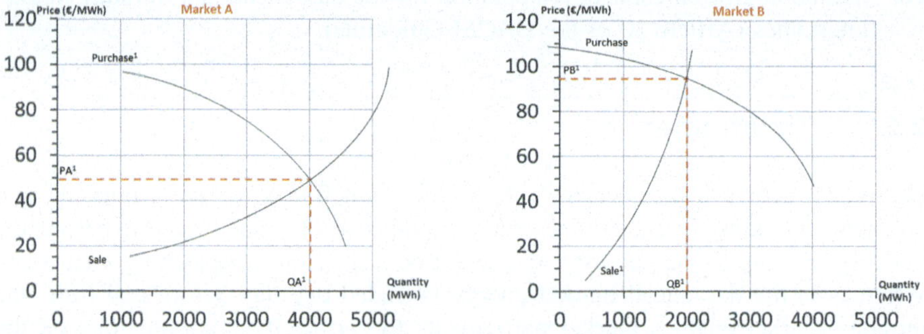

Fig. 2.4 Example of two uncoupled markets. *Source* Illustration by Peter Willis

2.6.1 Uncoupled Markets

Price curves PA^1QA^1 and PB^1QB^1 reflect purchase and sale orders within Market A and Market B, respectively, resulting in a price of €50/MWh in Market A and €95/MWh in Market B.

Purchasing market participants in Market A pay €50 × 4000 = €200,000 to CCP A. CCP A pays €200,000 to selling market participants in Market A.

Purchasing market participants in Market B pay €95 × 2000 = €190,000 to CCP B. CCP B pays €190,000 to selling market participants in Market B.

2.6.2 Coupled Markets

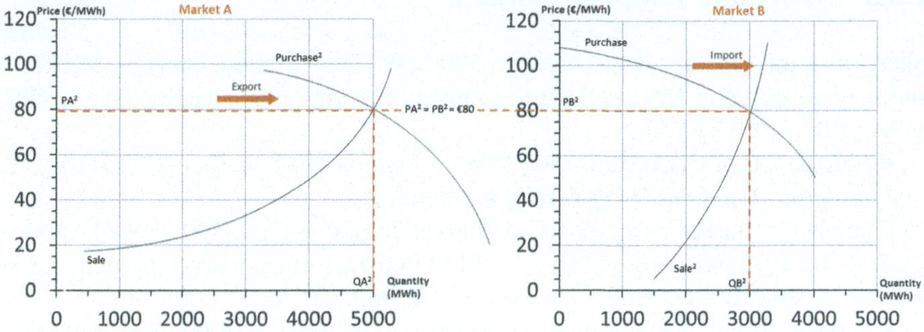

Fig. 2.5 Example of two coupled markets.
Source Illustration by Peter Willis

Cross-border interconnector capacity is 1000 MW. This is sufficient to allow the prices in Markets A and B to align. PA2 = PB2 = €80/MWh. The price in both markets is the same.

Physical: CCP A acts as a shipper by nominating 1000 MW of exports from Market A to Market B to the interconnector.

Financial (Article 68(3) of CACM Guideline):

Purchasing market participants in Market B pay €80 × 3000 = €240,000 to CCP B. CCP B pays €80 × 2000 = €160,000 to selling market participants in Market B (the domestic production). CCP B pays €80 × 1000 = €80,000 to CCP A (for the 1000 MWh of imported production).

Purchasing market participants in Market A pay €80 × 4000 = €320,000 to CCP A. CCP A pays €80,000 + €320,000 = €400,000 to selling market participants in Market A.

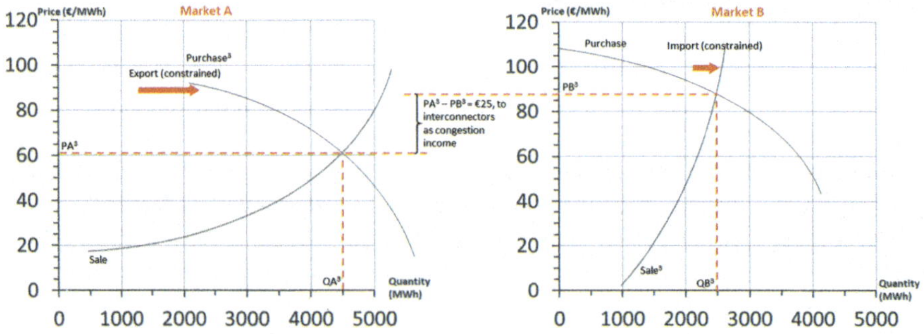

Fig. 2.6 Example of two partially coupled markets. *Source* Illustration by Peter Willis

2.6.3 Partially Coupled Markets

Interconnector capacity is restricted to 500 MW. The purchase curve in Market A and the sale curve in Market B move to reflect the available capacity. $PA^3 = €62/MWh$; $PB^3 = €87/MWh$.

Physical: CCP A nominates 500 MW of exports from Market A to Market B.

Financial (Article 68(3) of CACM Guideline):

Purchasing market participants in Market B pay $€87 \times 2500 = €217{,}500$ to CCP B. CCP B pays $€87 \times 2000 = €174{,}000$ to selling market participants in Market B. CCP B pays $€62 \times 500 = €31{,}000$ to CCP A.

Purchasing market participants in Market A pay $€62 \times 4000 = €248{,}000$ to CCP A. CCP A pays $€31{,}000 + €248{,}000 = €279{,}000$ to selling market participants in Market A (representing the total 4500 MWh produced, of which 4000 is used domestically and 500 exported).

Congestion Income: CCP B pays $(€87 - €62) \times 500 = €12{,}500$ to interconnector operator.

Bibliography

ACER. (2021). *Capacity allocation and congestion management.* Available at: https://documents.acer.europa.eu/en/Electricity/FG_and_network_codes/Pages/Capacity-allocation-and-congestion-management.aspx [Accessed 07 May 2024].

ACER. (2022). *ACER publishes its Final Assessment of the EU Wholesale Electricity Market Design.* Available at: https://www.acer.europa.eu/events-and-engagement/news/press-release-acer-publishes-its-final-assessment-eu-wholesale [Accessed 07 May 2024].

Baker, P., Hogan, M., & Kolokathis, C. (2018). *Realising the benefits of European market integration.* Regulatory Assistance Project.

Booz & Co. (2013). *Benefits of an integrated European energy market.*

ENTSO-E. (2022). *ENTSO-E Market Report 2022.* Available at https://ee-public-nc-downloads.azureedge.net/strapi-test-assets/strapi-assets/2022_ENTSO_E_Market_report_Web_836ec0a601.pdf [Accessed 07 May 2025]

ENTSO-E. (2024a). Available at: https://www.entsoe.eu/network_codes/cacm/implementation/sidc/ [Accessed 07 May 2024].

ENTSO-E. (2024b). *Single Day-ahead Coupling (SDAC)*. Available at: https://www.entsoe.eu/network_codes/cacm/implementation/sdac/ [Accessed 07 May 2024].
Florence School of Regulation. (2018). *Implementation challenges: Terms, conditions and methodologies*. Available at: https://fsr.eui.eu/implementation-challenges-i-terms-conditions-and-methodologies/ [Accessed 07 May 2024].
Pindstrup, M. (2024). *Opinion piece on Viking Link*. LinkedIn.
Schönheit, D., Kenis, M., Lorenz, L., Möst, D., Delarue, E., & Bruninx, K. (2021). Toward a fundamental understanding of flow-based market coupling for cross-border electricity trading. *Advances in Applied Energy, 2*, 100027 https://www.sciencedirect.com/science/article/pii/S2666792421000202
Schoutteet, N. (2024). *Opinion piece on NEMO Link*. LinkedIn.
Zachmann, G., et al. (2024). *Unity in power, power in unity: Why the EU needs more integrated electricity markets*. Bruegel.

Peter Willis is a partner in Bird & Bird's Competition & EU Law practice group based in London. Peter has over 30 years' experience of providing solutions for businesses in highly regulated and technically complex markets. He has an in-depth understanding of the interaction between competition rules and the complex technical, commercial, and regulatory arrangements governing the operation of energy networks and markets. He has experience of implementing Network Codes and Guidelines in the energy sector, and of challenging decisions of regulators. He has assisted clients to develop market rules and arrangements in a range of electricity markets, including capacity markets, forward markets, day-ahead, intraday, and balancing markets. He is well-known for his expertise on REMIT, and has advised on a number of investigations around Europe. His work also involves advising on disputes between businesses in which competition and regulatory issues arise, including competition follow-on damages claims. As well as authoring books and articles on EU and UK competition law, he is a regular contributor on competition and energy regulatory issues in the media and at conferences around Europe.

Louise Anderson currently works with Hitachi Energy as an Assistant Vice President in the Government and Institutional Relations department. Previously, Louise spent 5 years managing the World Economic Forum's Electricity Industry Community, while also launching the WEF's global initiative on Accelerating Industrial Clusters. Louise started her career working in a power plant before moving to work for EirGrid, Ireland's Transmission System Operator, managing high voltage asset outage planning until 2014. Louise also spent some time in Belgium working for ENTSO-E on different aspects of the implementation of the CACM Guideline. Louise has a bachelor's degree in Mechatronics and an MBA specializing in International Business.

Part II

Operations and Algorithms of Market Coupling

3 Capacity Calculation

Reinhard Kaisinger and Iacopo Bertelli

Abstract

Chapters 1 and 2 laid out the framework and guiding principles of the European electricity market. In this chapter, the basics of capacity calculation are discussed, i.e. how to determine how much (cross-zonal) transmission capacity is available for commercial exchanges. The geographical organization of the European Union into capacity calculation regions (CCRs) is discussed, and the succession of timeframes over which capacity is calculated. The capacity calculation process is described in detail: the inputs required, actors involved, and the different methodologies of capacity calculation used/applied in Europe. Then, the day-ahead capacity calculation methodology applied in Core CCR is explained through several important steps and concepts: non-costly remedial action optimization, individual validations, allocation constraints, evolved flow-based, advanced hybrid coupling. Finally, the future evolution of capacity calculation in Europe is explored.

3.1 Basic Principles of Capacity Calculation

As explained in Chap. 1, the amount of power that can be traded between bidding zones depends on the transmission capacity made available for cross-zonal exchanges. Cross-zonal capacity calculation is the process of determining cross-zonal capacities for market time units (MTUs) in the future. As market outcomes

R. Kaisinger (✉) · I. Bertelli
Magnus Energy, Naarden, The Netherlands
e-mail: r.kaisinger@magnus.nl

I. Bertelli
e-mail: i.bertelli@magnus.nl

A. Estermann et al. (eds.), *European Electricity Market Coupling*, Future of Energy,
https://doi.org/10.1007/978-3-031-86315-8_3

are not known, the process of computing cross-zonal capacities is based on forecasts of the market outcome, grid topology, generation, and load.

Capacity calculation is one of several available measures for congestion management. Other measures available to TSOs include:

- Short-term: Application of remedial actions (including redispatch and countertrade—both of which might be used in conjunction with the cross-zonal capacity calculation)
- Long-term: Adaptation of the bidding zone configuration (ideally, to reflect structural congestion)
- Long-term: Development of the network (reinforcing existing network elements and building new ones).

The capacity calculation process takes place in the environment provided by the grid, the bidding zone configuration, and the CCR determination (ACER, 2024).

In the context of this chapter, "capacities" are referred to as "cross-*zonal* capacities" rather than "cross-*border* capacities". This highlights that even in countries (or control areas) with multiple bidding zones, the same set of processes will apply for capacity calculation.

Furthermore, an "oriented bidding zone border" refers to a bidding zone border with a defined direction (e.g. A to B, but not B to A). In contrast, a "bidding zone border" is generally not oriented (i.e. A to B or B to A refer to the same bidding zone border).

"Net position" refers to a bidding zone's netted import and/or export (commonly assumed to be negative when a bidding zone imports and positive when a bidding zone exports).

This chapter focuses on the provision of cross-zonal capacities by TSOs and RCCs that serve as boundaries for the market coupling algorithm, as illustrated in Fig. 3.1.

3.1.1 Geographic Organization

As noted by European Commission (2015) in the CACM Guideline (see Chap. 1) requires that cross-zonal capacities are computed in a coordinated way based on a common set of assumptions. The groups within which TSOs cooperate to compute cross-zonal capacities in a coordinated manner are called "Capacity Calculation Regions" (CCRs). These CCRs are defined as sets of bidding zone borders (note: *not* as bidding zones or TSO control areas). Hence, one bidding zone or TSO can be a member of several CCRs, while each bidding zone border is associated with only one CCR.

As an example (see Fig. 3.2), Poland is a member of three CCRs—Core, Baltic, Hansa—while each bidding zone border is attributed to only one CCR (PL—DE/LU, PL-CZ, PL-SK to Core, PL-SE4 to Hansa, PL-LT to Baltic).

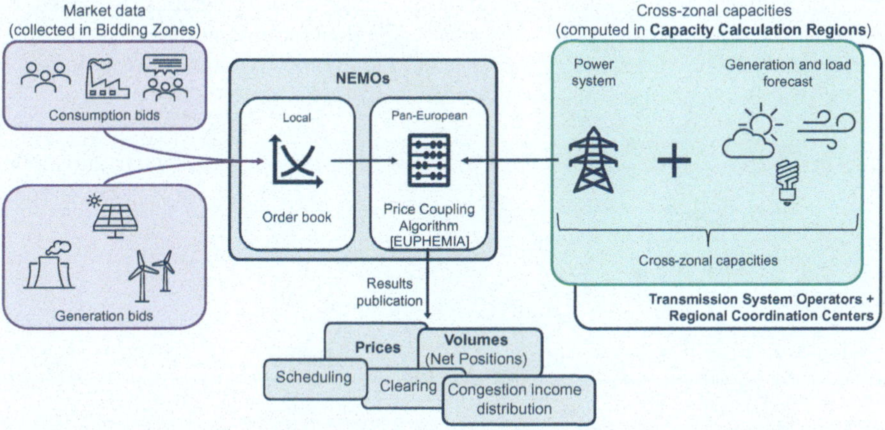

Fig. 3.1 Organization of the European electricity markets. *Source* Author's own illustration based on CACM Guideline

Ideally, CCRs are set-up and defined such that there is no significant impact from one CCR on another in terms of commercial exchanges. This, however, proves to be a challenge, especially in the heavily meshed Continental European power system.

Bidding zone borders involving (at least one) non-EU TSOs are not attributed to any CCR (e.g. bidding zone borders of Switzerland and West Balkan countries). The only exceptions are the Norwegian bidding zone borders after ACER decision 08/2023.[1]

Since Brexit, the former CCRs "Channel" and "Ireland and United Kingdom" were dissolved. Consequently, the SEM bidding zone (comprising the TSOs EirGrid in the Republic of Ireland and SONI in Northern Ireland) was not attributed to any CCR. However, the Celtic Interconnector, a HVDC interconnector, which is being constructed between Ireland and France and is expected to be operational by the end of 2026, will create a new bidding zone border between France and Ireland (FR-SEM). In 2023, all TSOs publicly consulted an amendment to the CCR configuration in light of the Celtic Interconnector, assigning FR-SEM to Core CCR. ACER approved this amendment on 19/03/2024 with decision 04/2024.

Additionally, this decision establishes the new Central Europe (CE) CCR as the merger between the CCRs Core and Italy North, with the initial delivery of a day-ahead capacity calculation methodology (DA CCM).

The current CCR configuration, based on the latest ACER Decision on CCRs, is shown in Fig. 3.2.

Each CCR has established governance and working structures for developing, implementing, and operating the cross-zonal capacity calculation methodologies

[1] Note that while Norway is not an EU Member State, it is a member of the European Economic Area (EEA).

Fig. 3.2 Capacity Calculation Regions. Each bidding zone border is attributed to exactly one CCR. Each CCR is composed of all the bidding zone borders of the same colour. *Source* Author's own illustration based on ACER decision 04/2024 on CCR determination

and processes (a few examples can be found in Sect. 3.2). Some CCRs have established special points of contact for engaging with stakeholders. CCR-specific information can be found on the (ENTSO-E) website and the (JAO) website. ENTSO-E also publishes an interactive Capacity Calculation Regions map.

3.1.2 Timeframes

Cross-zonal capacity calculation is performed for all timeframes for which allocation of cross-zonal capacity is performed: Year-ahead (once per year), month-ahead (once per month), day-ahead (daily), intraday (one to three times per day), and balancing (at least daily).

While day-ahead and intraday cross-zonal capacity calculation are based on common grid models (as also described in Chap. 5) that include forecasts (also referred to as "scenario-based approach"), long-term cross-zonal capacity calculation may be performed on a statistical basis (i.e. without determining cross-zonal capacities using a common grid model as a basis). Day-ahead and intraday cross-zonal capacity calculation processes are relevant in market coupling.

The day-ahead capacity calculation process typically starts in the evening two days before delivery time and ends in the late morning one day before delivery (hence the name *day-ahead*). The intraday cross-zonal capacity calculation process may run several times one-day-ahead of delivery and/or during the day of delivery.

Regardless of timeframes, cross-zonal capacities are to be provided at Market Time Unit (MTU)-granularity. The MTU granularity for day-ahead and intraday capacity calculation is currently 60 min. Note, however, that the cross-zonal computation process itself (i.e. the computations) may run on a different granularity than the granularity of cross-zonal capacity allocation (e.g. capacity allocation might run on 15-min granularity, while cross-zonal capacities are computed on 60-min granularity).

The Transparency Regulation (European Commission, 2013) requires the day-ahead cross-zonal capacities to be published at the latest one hour before SDAC (that is, at 11:00 CET D-1, coinciding with the day-ahead firmness deadline[2]). Intraday cross-zonal capacities must be published 15 min before SIDC gate opening time at the latest.

3.1.3 Inputs to the Capacity Calculation Process

Executing the cross-zonal capacity calculation process requires (Fig. 3.3):

1. a methodology and process to be followed.
2. relevant input data (typically provided by the TSOs of a given CCR).

The CACM Guideline lists several inputs that the capacity calculation methodologies shall address and detail further:

- Individual grid models (IGMs):
  - Network models are essential in a scenario-based capacity calculation process. TSOs provide IGMs for different timeframes, and for the sake of capacity calculation, these IGMs always contain a forecast of the expected market outcome (not a fully realized market result). The market outcome might be expressed as (a set of) net positions and/or reference exchanges.

[2] The approved day-ahead firmness deadline can be found at ENTSO-E (2015).

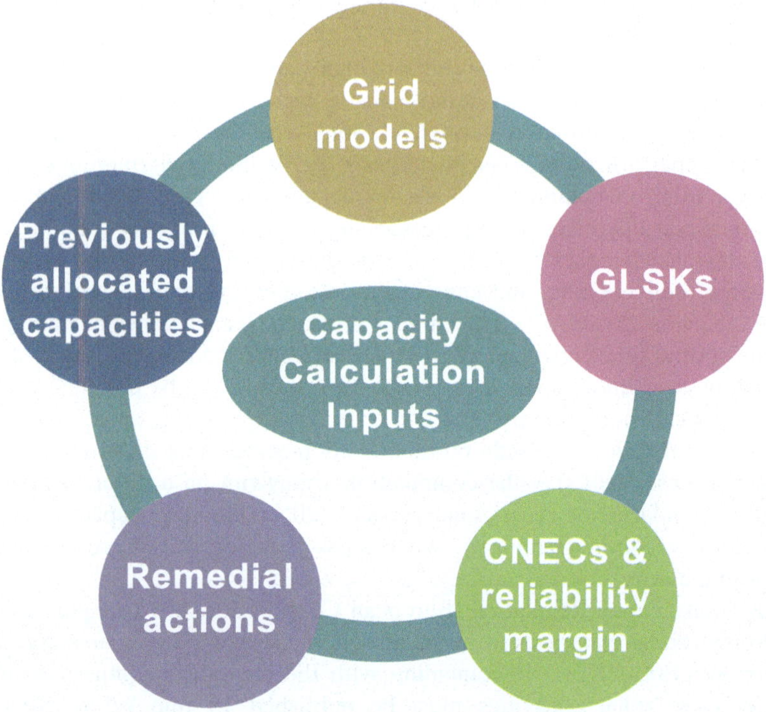

Fig. 3.3 Inputs to cross-zonal capacity calculation.[3] *Source* Author's own illustration based on CACM Guideline, Article 21

Some CCRs use a common forecast, while others rely on the forecasts of individual TSOs.
- IGMs reflect a forecast of the grid topology, power generation, and consumption within the TSO's control area, and flow on HVDC interconnectors. These IGMs are then merged into a common grid models.

- Critical Network Elements (CNEs) under N-1 Contingency (CNECs)
  - Previously referred to as Critical Branches and Critical Outages (CBCOs), CNEs and CNECs can constitute transmission lines (both overhead and underground), transformers and HVDC equipment, such as HVDC lines or back-to-back converters. They are provided with some basic characteristics, such as rated voltage level(s) and the maximum permissible current (which can be dynamic and vary over time, e.g. in the case of dynamic line rating). Within a flow-based approach for capacity calculation (Sect. 6.4), CNECs are typically defined via a name (one per direction of power flow), thermal

[3] CNECs are Critical Network Elements (CNEs) under N-1 Contingency as explained further down. GLSKs are Generation and Load Shift Keys as explained further down.

current or power flow limit (I_{max} / F_{max}), and flow reliability margin (FRM, see next point).

- Reliability margin (RM)
  - To cater for forecast errors, due to issues such as the linear and simplified nature of capacity calculation processes and activations in the balancing timeframe, the available (thermal) capacity may be reduced by a reliability margin (RM). In the various capacity calculation methodologies, RMs might be referred to as flow reliability margin (FRM), total reliability margin (TRM), or similar.
- Generation and Load Shift Keys (GLSKs)
  - GLSKs are used to "transpose" a quasi-nodal power system to a single net position of a bidding zone. GLSKs are a list of factors that translate a change in net position (e.g. increased import or export) to the nodal injections or withdrawals. For example, if a given net position changes by 100 MW, the GLSK "maps" these 100 MW to individual nodes in the grid supplying/withdrawing this additional power (see Fig. 3.4). The sum of the GLSK factors in each bidding zone is equal to 1. It is worth noting that GLSKs themselves do not consider minimum and maximum generation or load limits of generators or loads. Additionally, there is only one GLSK per MTU and bidding zone, regardless of how large the shifts in net positions are. Since GLSKs are linear (their factors remain static), the approximate shifts in net positions are most accurate around the reference point for which they were created, but less accurate with large net position shifts far away from the reference point.
- Remedial Actions (RAs)
  - This input refers to actions that TSOs can take to redirect power flows in the grid. As their naming suggests, RAs are deployed to alleviate congestion on network elements. In cross-zonal capacity calculation, the (coordinated) use of remedial actions is expected to increase cross-zonal capacities.
  - Remedial actions may include topological measures (e.g. opening and closing bus-bar couplers), tapping of phase-shifting transformers, or redispatch (orders to individual power plants or loads to increase or decrease power infeed or withdrawal). Where applicable, the list of remedial actions might be complemented by countertrading. While topological measures and PST (Phase Shifting Transformer) tapping are generally considered non-costly remedial actions, redispatch, and countertrading are regarded as costly remedial actions.
  - It should be noted, however, that remedial actions are not actually dispatched or deployed at the cross-zonal capacity calculation stage; this happens at a later stage when market results are available and firm, and consequently, the actual state of the grid (in terms of power flows and potential overloads) can be assessed.
- Allocation Constraints
  - Allocation constraints are additional constraints used when operational security limits cannot be effectively translated into maximum flows on CNE(C)s.

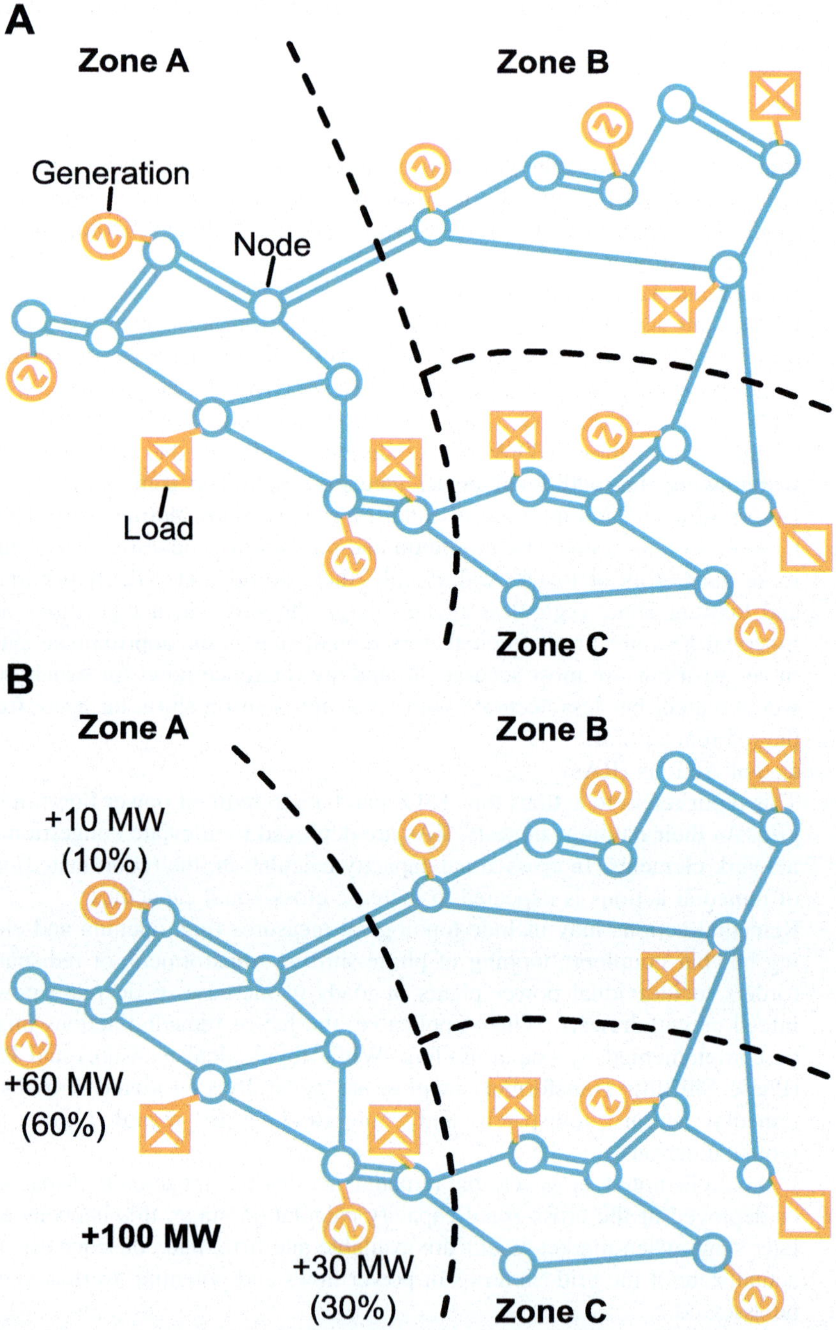

Fig. 3.4 **A** Individual grid models of three bidding zones A, B, and C. **B** Generation and Load Shift Key, showing the contribution of three nodes to a net position increase of 100 MW in Zone A. *Source* Based on author's own illustration

Such constraints are to be respected during cross-zonal capacity allocation to maintain the transmission system within operational security limits.

- The term "allocation constraint" is an umbrella for different designs of such constraints. Two common types of allocation constraints are limitations on net positions and ramping constraints for HVDC interconnectors. Net Position Constraints are used to limit the maximum import and/or export of a certain bidding zone. This limitation can be applicable on either the SDAC level (i.e. a limitation of the overall net position during cross-zonal capacity allocation) or CCR level (i.e. a limitation of the "regional" net position within a certain CCR, e.g. in a CCR applying flow-based capacity calculation and allocation). Ramping constraints provide a maximum gradient for how much power flows on HVDC interconnectors can change from one MTU to the next. As cross-zonal capacity calculation is commonly carried out for each MTU separately without considering preceding or succeeding MTUs, such dedicated ramping constraints are provided as a separate input to the market coupling algorithm, in which consecutive MTUs are coupled.

The above-listed inputs are each to be provided by TSOs. The Coordinated Capacity Calculator (CCC) then combines these inputs and creates "merged" versions containing the inputs of all individual TSOs. The CCC might further receive information on previously allocated capacity (e.g. from long-term allocation) and/or individual grid models from TSOs not part of a given CCR.

3.1.4 Performing Cross-Zonal Capacity Calculation

Generally, when using a scenario-based approach, the process for cross-zonal capacity calculation consists of the following steps (see Fig. 3.5):

1. Input provision: Provide individual grid model and other input data (see Sect. 3.1.3) for capacity calculation.
2. Merging of inputs: Merge individual grid models into a common grid model and merge individual TSO inputs to CCR-wide inputs.
3. Capacity calculation: Compute cross-zonal capacities based on CCR and time frame-specific methodology and perform intermediate steps (e.g. optimization of remedial actions, measures to ensure minimum capacities, etc.).
4. Validation of cross-zonal capacities: Each TSO has the right to validate and potentially adjust cross-zonal capacities.
5. Re-calculation of capacities: As a result of the validation step, cross-zonal capacities might have to be re-computed or adjusted. Furthermore, if applicable, previously allocated capacities (e.g. long-term nominations, cross-zonal capacity allocated for exchanging balancing capacity) are considered.
6. Publication of results and provision to market coupling.

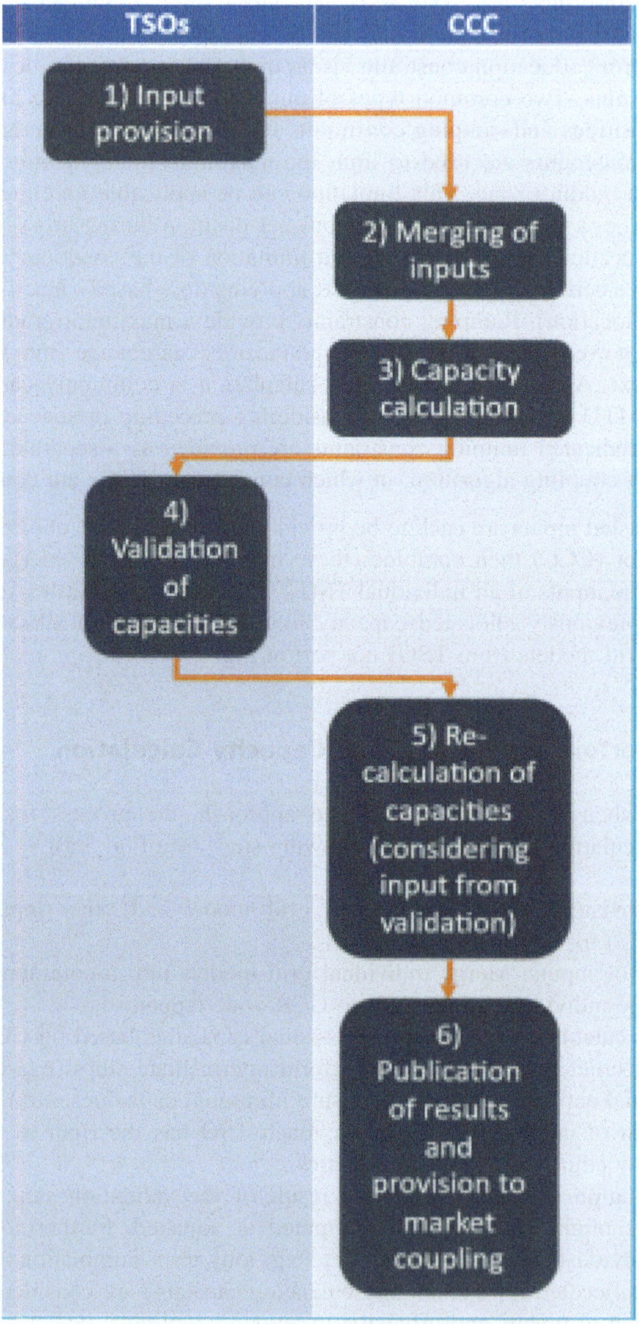

Fig. 3.5 High-level steps in the capacity calculation process. *Source* Author's own illustration based on CACM Guideline, Article 21

Steps 2), 3), 5), and 6) are performed by the Coordinated Capacity Calculator, while steps 1) and 4) are carried out by individual TSOs.

3.1.5 (Coordinated) Net Transfer Capacity Calculation

The coordinated net transfer capacity (cNTC) approach yields cross-zonal capacities as NTC (or Available Transfer Capacity (ATC)) values per oriented bidding zone borders. That is one value for one direction and another for the opposite direction (Fig. 3.6).

In its simplest (and uncoordinated) form, NTCs are computed by each TSO individually before these NTCs are shared with adjacent TSOs. Then, the lower value stemming from either TSO on a given oriented bidding zone border is provided to market coupling. The disadvantage of this approach is that TSOs do not mutually share information on the assumptions taken when computing NTCs and do not coordinate. As a result, NTC levels can be very conservative (and low).

To overcome this drawback, the cNTC approach is based on a common grid model (which is, in turn, based on each TSO's individual grid model). This common grid model serves as an aligned starting point, based on which NTCs are computed. The maximum NTC level is reached when (at least) one critical network element is fully loaded. The (coordinated) application of remedial actions may accompany this approach to increase capacities further. The result is the "NTC domain" (i.e. maximum cross-zonal capacities per oriented bidding zone border).

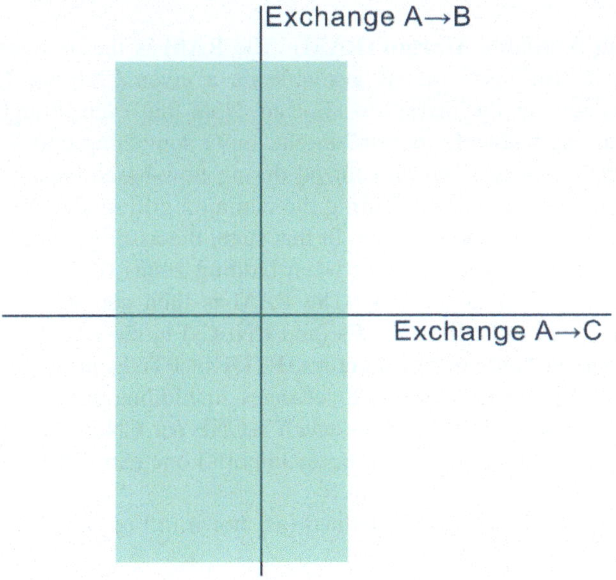

Fig. 3.6 Example of NTC domain. *Source* Based on author's own illustration

Even with cNTC, TSOs consciously decide how much cross-zonal capacity is offered on one oriented bidding zone border versus another (or several other) oriented bidding zone border(s). This decision or choice is particularly relevant for (highly) meshed grid topology, where cross-zonal exchanges on a number of bidding zone borders have an impact on grid elements. Hence, such an approach is sensible in power systems with a radial grid structure (e.g. Italian peninsula) and less suited for meshed power grids (e.g. Central Europe).

The cNTC approach typically favours one (or a few) expected market direction(s) (generally referred to as "corner(s)"). The capacity calculation process is designed to increase cross-zonal capacities for this one (or few) corner(s) and other market directions, which may not be analysed explicitly.

3.1.6 Flow-Based Capacity Calculation

The flow-based approach allows for a more detailed representation of the grid in both the capacity calculation and allocation stage, as it considers several individual grid elements rather than bidding zone borders. As opposed to the cNTC approach, cross-zonal capacities are not determined per oriented bidding zone borders but are described via the flow-based domain. This flow-based domain is a (large) matrix depicting the "solution space", as shown in Fig. 3.7, within which the market can clear. Consequently, cross-zonal capacities are not specific to exchanges on oriented bidding zone borders but to a whole "region".

The flow-based domain is essentially a list of CNEs and CNECs characterized by two main components:

Remaining Available Margin (RAM): The RAM is the amount of transmission capacity (in MW) that is available on a given CNE (or CNEC) and a given direction for the market to allocate. Note that even though the convention "remaining available margin" might imply something that is left over, the RAM is the quantity to be maximized during flow-based capacity calculation. The RAM is determined by shifting the common grid model from its assumed forecast to a "zero-balance" state. In this state, the assumption is that no (commercial) exchanges take place between bidding zones (i.e. the net position of bidding zones is equal to zero). The RAM is then computed based on how much margin is available on CNEs (and CNECs) in the zero-balance state.

Power Transfer Distribution Factors (PTDFs): PTDFs are the sensitivities of individual CNEs (and CNECs) to changes in bidding zone net positions. A single PTDF value describes how much a CNE (or CNEC)'s loading changes when a given bidding zone exports (or imports) one extra MW.

Different kinds of PTDFs can be defined (as shown in Fig. 3.8).

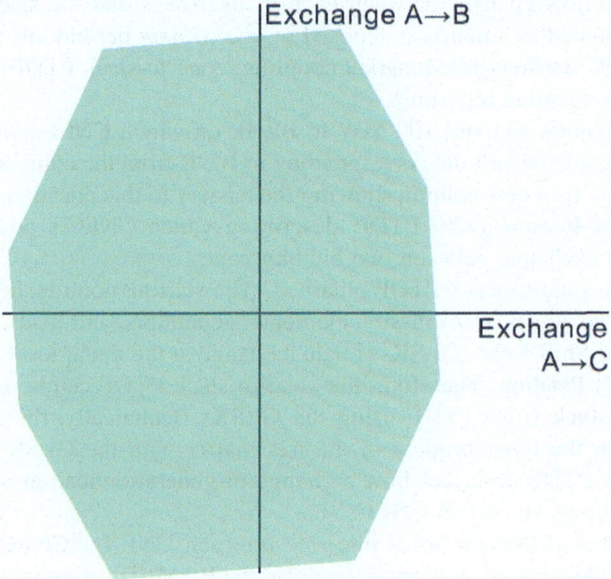

Fig. 3.7 Example of two dimensional projection of a flow-based domain. *Source* Based on author's own illustration

Fig. 3.8 Power Transfer Distribution Factors. **A** Zone-to-slack PTDF. **B** Zone-to-zone PTDF matrix. The column A → B is obtained by subtracting column B from column A of the z2s PTDF. *Source* Based on author's own illustration

A

	BZ A	BZ B	BZ C
CNEC 1	4.9%	4.8%	-3.9%
CNEC 2	4.3%	-24.4%	-11.5%
CNEC 3	14.6%	-2.7%	-10.0%
CNEC 4	0.2%	-2.5%	-1.5%
CNEC n	...	...	...

B

	A→B	B→C	A→C
CNEC 1	0.1%	8.7%	8.8%
CNEC 2	28.7%	-12.9%	15.8%
CNEC 3	17.3%	7.3%	24.6%
CNEC 4	2.7%	-1.0%	1.7%
CNEC n	...	...	...

The **zone-to-slack**[4] (z2s) PTDF describes how the change in Net Position in a bidding zone results in flows on certain CNECs. Because of the model's linearity, the effect of a flow change between two bidding zones is the difference of the

[4] The slack node is the reference point for PTDFs relative to a slack (e.g. node-to-slack or zone-to-slack PTDFs). The slack node is a necessary mathematical construct, but the choice of slack does not influence the result. All PTDFs for the slack itself are zero (flow from slack to slack).

effects of the flow changes between each of the zones and the slack node. This can be represented as a matrix composed of one *column* per bidding zone and one *row* per CNEC. In flow-based market coupling, zone-to-slack PTDFs are provided to the market coupling algorithm.

If BZ A exports an extra 100 MW to BZ B, on which CNECs will this power flow? Subtracting the column corresponding to BZ B from the column representing BZ A will give us a new column showing the answer to this question. This column is called **zone-to-zone** (z2z) PTDF, describing which CNEC's power flows are caused by an exchange between two bidding zones.

How is the zone-to-slack PTDF obtained? The starting point is the Grid Model, composed of its nodes, transmission elements, generators, and loads. The Generation and Load Shift Keys (GLSKs) relate a change at the nodal level to the overall change in Net Position. Therefore, the zone-to-slack PTDF can be obtained from the **node-to-slack** (n2s) PTDF using the GLSKs (technically, the z2s matrix is obtained from the inner product of the n2s matrix with the GLSK matrix). The node-to-slack PTDF describes how a change in generation/load at a certain node translates to flows on certain CNECs.

The flow-based domain holds one *row* for each CNE (or CNEC), one PTDF *column* per bidding zone, and one *column* for the RAM. The number of dimensions of a flow-based domain is given by the number of bidding zones encompassing a CCR. Note that the number of dimensions can be reduced by one because the sum of all net positions within a given CCR must equal zero.

The flow-based approach is especially beneficial in highly meshed power systems. In a cNTC approach, TSOs must split cross-zonal capacities between highly mutually impacted oriented bidding zone borders. In flow-based allocation, "the market" decides where the capacity is allocated.

Chapter 16 provides more information on the transparency of the relevant methodologies, processes, and input and output data.

3.1.7 Ever-Evolving Capacity Calculation Methodologies

With the first round of approved capacity calculation methodologies (CCMs), things have not come to a halt. During the implementation of the CCMs, inconsistencies, and other issues were identified that constituted hurdles for implementation. In some cases, TSOs requested amendments to the CCMs and submitted them to NRAs for approval. So far, amendments to CCMs have been approved in several CCRs already—and this is going to continue.

Another aspect many CCMs had to incorporate after their first approval was the requirements of the 70% min MACZT rule (European Commission, 2019) and (ACER, 2023). By mid-2019, when the Clean Energy Package entered into force, many CCMs had already been approved. Catering to this new rule meant requesting an amendment.

Further changes are already on the horizon. With ACER's proposal for a "CACM 2.0", the 70% rule might also apply in the intraday timeframe. Reconfiguration of CCRs or bidding zones and new (HVDC) interconnectors call for further improvements, and the integration of new features into the CCMs means that there will not be a "steady state" situation.

3.2 Illustrative Examples from Day-Ahead Capacity Calculation in Core CCR

In June 2022, the Core Day-ahead flow-based capacity calculation process went live. At the time of writing, this is the only flow-based process in operation, with intraday flow-based capacity in Core CCR and day-ahead flow-based capacity calculation in Nordic CCR expected to go-live soon. See Chap. 11 for a first analysis of the results after Core FBMC go-live and Chap. 12 for specific information on the Nordic flow-based implementation.

Figure 3.9 provides an overview of the steps of the Core Day-ahead flow-based capacity calculation process, which are further described below.

1. Input File Delivery
   - Each TSO delivers to the Core Capacity Calculation tool (CCCt) a series of input files:
     - Day-2 individual grid models (one per hour)
     - Generation and Load Shift Keys (GLSKs)
     - List of Critical Network Elements and N-1 Contingencies (CNECs) including FRM
     - Available non-costly remedial actions (RA)
     - External Constraints or allocation constraints on the maximum import and/or export net position
   - Other inputs are:
     - Long-Term Allocations (LTAs) are provided by JAO, resulting from long-term allocated cross-zonal capacities in the yearly and monthly LT auctions.
     - Long-Term Nominations (LTNs) are provided by JAO for BZBs with Physical Transmission Rights (PTRs) (as of the end of 2023, used only on HR-SI Croatia—Slovenia)
     - Additionally, the CCC obtains the Day-1 individual grid models from TSOs, not in Core CCR and the Continental European synchronous area.

Fig. 3.9 Steps of the flow-based day-ahead capacity calculation process in Core CCR. *Source* Author's own illustration based on ENTSO-E website

2. Merging and Data Quality Check
   - The CCC checks the IGM data for quality. Afterwards, the IGMs are aligned and merged into one CGM (Common Grid Model). Then, CCCt performs quality checks of the other TSO inputs (GLSKs, CNECs, RAs, etc.)
3. Initial FB computation and CNEC selection
   - The CCCt computes the initial flow-based parameters (PTDF and RAM for each CNEC). Then, a CNEC selection is performed (per timestamp), removing any CNEC with zone-to-zone PTDF lower than 5%[5] on any exchange combination within Core CCR. This means all CNECs not significantly affected by cross-zonal exchanges between *any* two bidding zones within Core are removed from the computation. Approximately 60,000 CNECs are typically provided, of which 15,000 CNECs have PTDF above 5% and are, thus, selected.
4. Non-costly Remedial Action Optimization
   - The CCCt tries to enlarge the flow-based domain by coordinating and optimizing the use of non-costly remedial actions (see 3.1.3 for details).
5. Intermediate FB computation
   - The CCCt computes the intermediate flow-based parameters, now including the selected remedial actions. Additionally, it modifies the previous calculation to (attempt to) satisfy requirements regarding minimum levels of cross-zonal capacities.
   - If needed, the RAM is increased by the Adjustment for Minimum RAM (AMR) to meet the requirement of:
     - 20% of F_{max} for exchanges within Core CCR ('20% minRAM')
     - 70% of F_{max} for cross-zonal exchanges (both within and outside Core CCR) ('70% minMACZT').
     - Note: This second requirement effectively implements the consideration of the 70% rule as introduced by Regulation 2019/943 (Regulation (EU) 2019/943 of the European Parliament and of the Council of 5 June 2019 on the internal market for electricity)
6. Validations
   - The TSOs can validate and, if needed, correct calculated cross-zonal capacities to ensure the grid's operational security. Before reducing cross-zonal capacities, all available remedial actions must be used.
   - At the time of writing, the validation step is performed by TSOs individually or in consortia. A coordinated validation approach is expected to be implemented in the coming years
7. Pre-final FB computation
   - If validation adjustments have been applied during the previous step, the CCCt computes the FB domain again. Otherwise, the result of the intermediate FB computation is retained.

[5] A CNE(C) satisfying a maximum for zone-to-zone PTDF 5% is considered significantly influenced by cross-zonal exchanges.

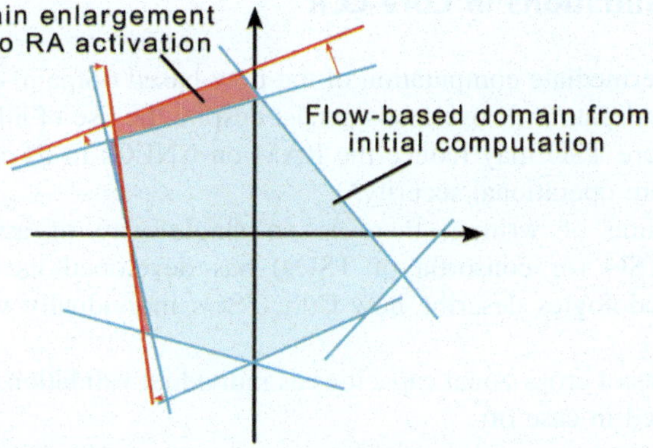

Fig. 3.10 Non-costly remedial action optimization. By applying Remedial Actions (RA), the flow-based domain is enlarged by increasing the RAM on certain CNECs and reducing it on others. *Source* Based on authors' own illustration

8. Final FB computation
   - Upon receipt of both long-term allocated capacities (LTAs) from year-ahead and month-ahead allocation and long-term nominations (LTNs), the final flow-based parameters are computed. The flow-based and LTA domains are shifted to LTN (i.e. to account for already allocated cross-zonal capacity).
9. Publication & DA Market Coupling (SDAC)
   - The final FB and LTA domains are published on the JAO publication tool and provided to SDAC. The SDAC market coupling algorithm uses the Core flow-based domain, LTAs and allocation constraints when performing market coupling.

3.2.1 Non-costly Remedial Action Optimization

After the initial computation of cross-zonal capacities based on forecasts, the Core Capacity Calculation tool tries to enlarge the flow-based domain by attempting to increase the RAM on the CNEC(s) with lowest relative RAM, then repeating the process until the RAM cannot be increased further on any CNEC (Fig. 3.10). This is done by coordinating non-costly remedial actions (such as tapping phase-shifting transformers or applying topological remedial actions). This process is further constrained by monitored network elements that cannot be loaded further and a loop-flow threshold.[6]

Both RCCs (Coreso and TSCNet) perform the non-costly remedial action optimization (NRAO) separately, using different algorithms. The best result per MTU is selected and used in the subsequent computation steps.

[6] The loop-flow threshold is defined in Article 16 of the Core DA CCM (ENTSO-E, 2024).

3.2.2 Validations in Core CCR

After the intermediate computation of the flow-based domain, each Core TSO has the right to validate it. If overloads persist despite the use of all available remedial actions, a Core TSO may reduce the RAM on CNECs in its control area/bidding zone to ensure operational security.

At the time of writing, there is no single way of assessing grid security. Each TSO (or consortia of TSOs) has developed its own methodology. These methodologies describe how Core TSOs individually validate cross-zonal capacities.

Reductions of cross-zonal capacities as individual validation adjustments (IVAs) can be applied in case of:

- Exceptional contingency or forced outages.
- When all costly and non-costly remedial actions are not sufficient to ensure that operational security limits are respected.
- Mistakes in input data resulting in overestimation of cross-zonal capacities.
- Potential need to cover reactive flows.

The size of the IVAs and the frequency with which they are used differ between TSOs. Figure 3.11 shows applied IVAs relative to Fmax for Core in 2022.

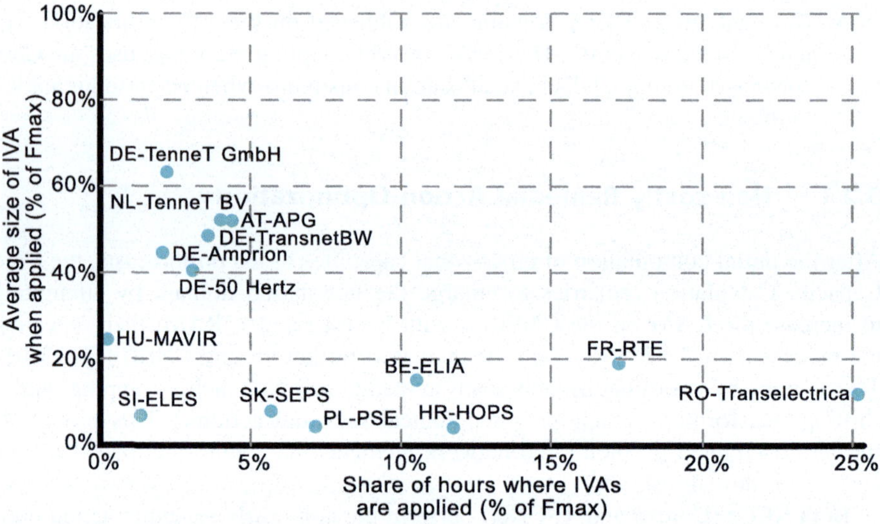

Fig. 3.11 Use of Individual Validation Adjustments by Core TSOs between June and December 2022. *Source* Author's own illustration based on ACER Monitoring Report 2023

3.2.3 Evolved Flow-Based in Core CCR

In recent years, high-voltage direct current (HVDC) interconnectors have been used more frequently as subsea connections between regions (e.g. Germany to Norway) and to connect asynchronous AC grids (e.g. Continental Europe and Nordic). An example of an HVDC interconnector between two bidding zones within Core is ALEGrO, connecting Belgium and Germany, commissioned in 2020.

The flow-based approach inherently considers that changing the power flow on one network element in a synchronous AC network results in the modification of flows in the whole grid (progressively smaller changes for lines further away). On the other hand, an HVDC interconnector is not part of an AC network, as it is connected to two AC/DC converter stations. Because of this, the power flow on an HVDC line must be set explicitly.

The Evolved Flow-Based (EFB) approach is one way to include these characteristics into the flow-based process. This approach essentially builds on modelling the two converter stations as two *virtual* bidding zones, such that they work as Bidding Zones for the capacity calculation and allocation algorithms (i.e. two additional columns in the zone-to-slack PTDF matrix). Note, however, that no market bids can be placed in the virtual bidding zones.

An exchange from zone A to zone B over an HVDC interconnector is expressed as a trade from A to the virtual hub representing the sending end of the HVDC interconnector plus a trade from the virtual hub representing the receiving end of the interconnector to zone B. In the case of ALEGrO, a trade from Belgium to Germany is modelled as BE → AL_BE + AL_BE → AL_DE + AL_DE → DE (with the bidding zones containing "AL" being the virtual bidding zones).

The combined net positions of the two virtual bidding zones must be equal to zero, such that the net positions of the two virtual bidding zones have equal magnitude and opposite sign.

EFB is included in market coupling by sending the virtual bidding zones to the SDAC algorithm and setting the transmission capacity between the virtual bidding zone and the actual bidding zone to the physical capacity of the interconnector. If needed, losses (both at the AC/DC converter stations and on the interconnector itself) are considered during market coupling, while they are ignored during capacity calculation. Similarly, ramping constraints (i.e. the maximum change in power flow between consecutive MTUs) are communicated to SDAC but disregarded during capacity calculation.

3.2.4 Advanced Hybrid Coupling in Core CCR

In cross-zonal capacity calculation, exchanges on bidding zone borders not attributed to the CCR in question but in the same synchronous area (so-called "external bidding zone borders") need to be modelled to account for the impact of these exchanges on the cross-zonal capacities within the CCR in question. There are two common modelling approaches:

Standard hybrid coupling (SHC) entails including the forecasted exchanges with external bidding zones as fixed inputs (as generators/loads for the forecasted imports/exports). This approach is currently implemented in the Core day-ahead flow-based capacity calculation process.

Advanced hybrid coupling (AHC) goes one step further. This approach treats the external Core bidding zone borders similarly to internal ones. The converter station at the Core-end of an HVDC interconnector is modelled as a virtual bidding zone. The net position of this virtual bidding zone represents the flow over the interconnector. The flow-based capacity calculation and allocation processes include the virtual bidding zone. Note that for external bidding zone borders, two (independently computed) sets of cross-zonal capacities are provided to SDAC: First, the cross-zonal capacities from the "native" CCR of the external bidding zone border (e.g. for NL-DK1 this would be Hansa CCR, providing NTCs), Second, the extended flow-based domain with PTDFs for every CNECs for the virtual bidding zone (e.g. for NL-DK1, the virtual bidding zone in the Netherlands in case of Core flow-based or the virtual bidding zone in DK1 in case of Nordic flow-based). Note that AHC only works for bidding zone borders for which cross-zonal capacity calculation is performed in SDAC. For example, AHC is not an option for HVDC interconnectors between Continental Europe and Great Britain, as Great Britain is no longer coupled with SDAC.

3.2.5 cNTC Capacity Calculation in South Western Europe

The CCR South-Western Europe (SWE), comprising the bidding zone borders Portugal-Spain and Spain-France, applies the coordinated NTC approach to calculate cross-zonal capacities (ENTSO-E, 2022). Because of the intrinsic grid topology (radial, not meshed), using a flow-based approach in this CCR would bring about limited benefits, as the influence of bidding zone border exchanges from one on the other is small.

In addition to the inputs listed in Sect. 3.1.2, for the cNTC process in SWE, a few extra details apply:

- The reliability margin is defined per-border as the maximum between a fixed value (200 MW on FR-ES, 100 MW on PT-ES) and a percentage of the Total Transmission Capacity (7.5% on FR-ES, 10% on PT-ES).
- No Allocation Constraints are used.
- A sensitivity analysis (performed once per year) is used to select CNECs significantly influenced by cross-zonal exchanges. Only CNECs with sensitivity above 10% are monitored during the CC process (with few exceptions that must be justified to the SWE NRAs).

The capacity calculation process in SWE (applied both in the day-ahead and intraday timeframes) consists of an iterative computation starting from the coordinated schedule in the common grid model. The coordinated capacity calculator tests several levels of cross-zonal exchange by using Generation and Load Shift Keys and determines if this level of exchange respects operational security limits on all the monitored critical network elements after the occurrence of all the monitored contingencies. Available remedial actions are applied when necessary. At each calculation step, the Remedial Action Optimization monitors the maximum flows, voltage levels, and the maximum voltage phase angle differences defined by the TSOs on all the CNEs.

3.2.6 Allocation Constraints

Allocation constraints constitute additional constraints that the market coupling algorithm must respect (see Sect. 3.1.3). Figure 3.12 depicts the application of two types of allocation constraints in the day-ahead market coupling algorithm.

A. Use of Allocation Constraints in SDAC. The HVDC interconnectors over which ramping constraints apply are highlighted in red.
B. Schematics of Net Position allocation constraints. It is reduced if the Net Position calculated by the CC algorithm is larger than the NP allocation constraint.
C. Schematics of Ramping constraints. If the rate of flow change between consecutive MTUs is larger than the maximum flow difference, it is reduced. The per-MTU requested and realized flows are shown in dashed black and solid red, respectively.

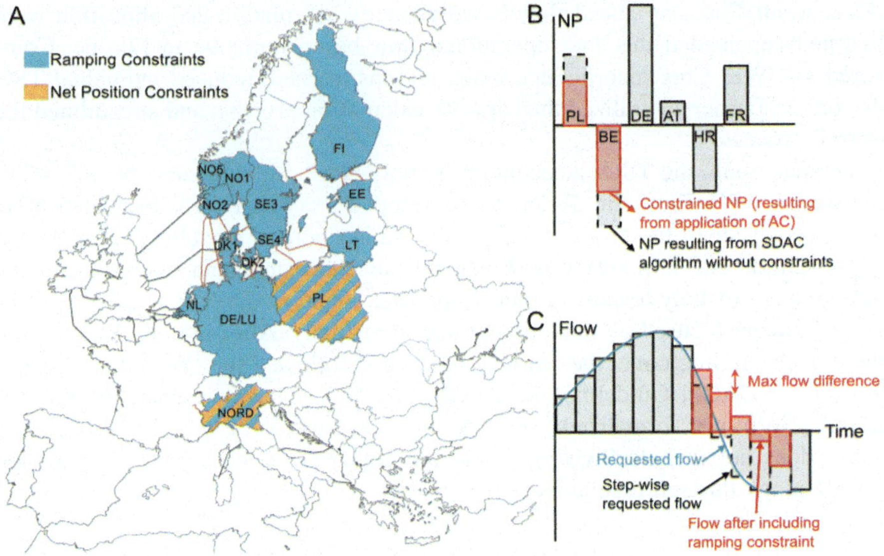

Fig. 3.12 Depicting the application of two types of allocation constraints in the day-ahead market coupling algorithm. *Source* Author's own illustration

Italy North and Poland use net position constraints on the SDAC net position, and the Netherlands uses net position constraints on the CCR net position. Belgium phased out their respective net position constraint in November 2023. The Netherlands is in the process of phasing out its respective net position constraints.[7]

Ramping constraints are used on the majority of HVDC interconnectors, for which cross-zonal capacity allocation is done in SDAC.

3.2.7 History of Coordinated Capacity Calculation

The introduction of the Third Energy Package or the CACM Guideline did not initiate coordination in cross-zonal capacity calculation. Prior to the CACM Guideline's entry into force, several voluntary initiatives existed.

As one example, day-ahead flow-based capacity calculation and allocation was operational in Central-West Europe (CWE). Having gone live in May 2015, the CWE region, formally not a CCR, pioneered the flow-based capacity calculation and allocation. Initially, the TSOs of Belgium, France, the Netherlands, and Germany (except 50 Hertz) worked on implementing CWE flow-based. Discussions with Swissgrid were held about joining, but the exclusion of Switzerland from day-ahead market coupling (not SDAC at that time) wiped the idea off the table. Later, the Austrian TSO, APG, joined the consortium, then still in the common German-Austrian bidding zone border. The introduction of cross-zonal capacity allocation on the German-Austrian bidding zone border in 2018 was implemented in CWE, showing that the system could be extended by additional bidding zone borders. CWE flow-based capacity calculation and allocation was operational until June 2022, when Core day-ahead flow-based capacity calculation and allocation went live and superseded this first operational flow-based initiative in Europe. Compared to CWE, Core incorporated many lessons learned, reduced individual TSO discretion, further centralized the capacity calculation process, and streamlined the overall process.

Further south, the TSOs adjacent to the Italy North bidding zone, namely RTE, Swissgrid, APG, Eles, and Terna, had developed a concept for coordinated NTC computation along the "roof of Italy" (the Italian border, stretching from France to Slovenia). The initiative was driven by the 2006 European blackout and the development of Italy towards a (net) importing country. The NTC capacity calculation concept focused on maximizing imports to Italy. After initially focusing on the day-ahead timeframe, the intraday timeframe followed suit. With the introduction of the CACM Guideline, the coordinated NTC capacity calculation process in Italy North was formalized—both through the creation of the CCR (which formally does not include Swissgrid) and the approval of the capacity calculation methodology through regulators.

[7] See public consultation on 3rd Request for Amendment of the Core Day-Ahead Capacity Calculation Methodology (ENTSO-E, 2024).

Other cooperations and coordinations existed in addition to the examples mentioned above. Poland used to compute its NTC on a profile spanning Germany, the Czech Republic, and Slovakia. The Nordic TSOs have long cooperated when computing their NTCs. Initiatives for explicit day-ahead flow-based processes were launched in Central-East Europe (which merged with CWE to form Core) and South-East Europe.

3.3 What Does the Future Hold for Capacity Calculation Methodologies Across Europe and Its CCRs?

A few key points can be highlighted.

There is a trend to have fewer and, consequently, larger CCRs. The most likely future changes are the attribution of the France-SEM bidding zone border to Core CCR and the merger of the Italy North and Core CCRs (provisionally referred to as "Central Europe" CCR).

Implementing a flow-based capacity calculation approach in areas with a highly meshed grid would be highly beneficial. Important examples are Italy North and countries in the Western Balkans (once joining SDAC).

Improving the consideration of non-EU countries in the capacity calculation methodologies would improve the accuracy and overall quality of the capacity calculation processes. It is not unusual for flows computed in a CCR to be realized in countries outside that CCR. Notable examples are Switzerland (involved in flows between France, Austria, Italy, and Germany) and Serbia and Bosnia Herzegovina (involved in flows between Romania/Bulgaria and Croatia).

Next to geographic extensions, other timeframes than day-ahead gain more traction, namely long-term, intraday, and balancing. Other ongoing implementation processes, such as intraday auctions, long-term flow-based allocation, and the balancing platforms, call for emphasizing other timeframes to deliver cross-zonal capacities to benefit European consumers while maintaining operational security.

Bibliography

ACER. (2023). *ACER's Market Monitoring Report 2023—Cross-zonal capacities and the 70% margin available for cross-zonal electricity trade (MACZT)*. Retrieved November 17, 2024, from https://www.acer.europa.eu/Publications/2023_MMR_MACZT.pdf

ACER. (2024). *ACER Decision No 04-2024 on the amendment to the determination of capacity calculation regions in accordance with CACM Guideline*. Agency for the Cooperation of Energy Regulators (ACER). Retrieved November 17, 2024, from https://www.acer.europa.eu/document/acer-decision-no-04-2024-amendment-determination-capacity-calculation-regions-ccrs

ENTSO-E. (2015a). *All TSOs' proposal for the day-ahead firmness deadline (DAFD) in accordance with Article 69 of the CACM guideline*. Retrieved November 11, 2024, from https://www.acer.europa.eu/sites/default/files/documents/en/Electricity/MARKET-CODES/CAPACITY-ALLOCATION-AND-CONGESTION-MANAGEMENT/8%20DAFD/Action%201a%20-%20DAFD%20approved.pdf

ENTSO-E. (2015b). *Second amendment of the Day-Ahead Capacity Calculation Methodology of the Core Capacity Calculation Region in accordance with Articles 20ff. of the CACM Guideline.* Retrieved November 11, 2024, from https://consultations.entsoe.eu/markets/core_2nd_dafb_ccm/supporting_documents/Core%20DA%20CCM%202nd%20amendment.pdf

ENTSO-E. (2022). *South West Europe TSOs common capacity calculation methodology for the day-ahead and intraday market timeframe in accordance with Article 21 of CACM Guideline.* Retrieved November 11, 2024, from https://eepublicdownloads.entsoe.eu/clean-documents/nc-tasks/SWE%20CCM%20Amendment_January2022_clean-v3.pdf

ENTSO-E. (2024a). *Core DA CC process steps.* Retrieved June 01, 2024, from https://www.entsoe.eu/bites/ccr-core/explained/#fc11

ENTSO-E. (2024b). *Public consultation on 3rd Request for Amendment of the Core Day-Ahead Capacity Calculation Methodology.* Retrieved November 11, 2024, from https://consultations.entsoe.eu/markets/core-tsos-proposal-3rd-amendment-dafbccm/

ENTSO-E. (n.d.). *ENTSO-E Website.* Retrieved November 17, 2024, from https://www.entsoe.eu

European Commission. (2013). Commission Regulation (EU) No 543/2013 of 14 June 2013 on submission and publication of data in electricity markets and amending Annex I to Regulation (EC) No 714/2009. Retrieved November 11, 2024, from https://eur-lex.europa.eu/legal-content/EN/TXT/?uri=CELEX%3A32013R0543

European Commission. (2015). *Commission Regulation (EU) 2015/1222 of 24 July 2015 establishing a guideline on capacity allocation and congestion management.* Retrieved November 11, 2024, from https://eur-lex.europa.eu/legal-content/EN/TXT/?uri=CELEX%3A02015R1222-20210315

European Commission. (2019). *Regulation (EU) 2019/943 of the European Parliament and of the Council on the internal market for electricity.* Retrieved November 11, 2024, from https://eur-lex.europa.eu/legal-content/EN/TXT/?uri=CELEX%3A02019R0943-20240716

JAO. (n.d.). *JAO Website.* Retrieved November 11, 2024, from https://www.jao.eu

Reinhard Kaisinger is a Principal at Magnus Energy. He co-leads Magnus Energy's Technical Advisory practice and is heavily involved in several projects related to electricity market integration. With over a decade of experience in the electricity sector in various roles, he has obtained a wide yet deep understanding of power systems and electricity markets. After obtaining his master's degree in electrical power engineering at KTH, Stockholm, Sweden, he started working on power systems and electricity markets at Vattenfall in Sweden. At E-Control, the Austrian electricity gas market regulator, he focused on network development (TYNDP, PCI) and the implementation of CACM Guideline and System Operations Guideline. Then, he worked on the implementation of Core day-ahead flow-based capacity calculation and allocation at APG (the Austrian TSO) both as an expert and project manager before joining Magnus Energy in 2022. He currently serves as convener of the Core day-ahead capacity calculation project team.

Dr. Iacopo Bertelli is a Senior Consultant at Magnus Energy. His field of expertise is the European electricity market design and integration, both conceptually and analytically. Currently, he is supporting the newly established Central Europe CCR as PMO of the day-ahead capacity calculation project team. He has experience in supporting CCR accessions by TSOs, from assessing the high-level requirements to drafting a detailed local implementation plan. He is a physicist by training, having obtained a PhD in experimental quantum physics from the University of Leiden, Leiden, the Netherlands, and TU Delft, Delft, The Netherlands, during which he published 7 articles on high-impact journals.

4 Role of TSOs in the European Market Coupling Operations

Marius Schrade

Abstract

This chapter presents the high-level overview of pan-European procedures of Single Day-Ahead Coupling (SDAC) and the Single Intraday Coupling (SIDC) in which TSOs are directly or indirectly via service providers or NEMOs involved. In general, all procedures are separated in three steps starting from pre-coupling, to coupling and finally to post-coupling processes. Each of these steps are backed by procedures that cover normal, back-up and fallback operations. As these procedures have a tremendous impact on secure market and system operations, they are reviewed and updated on a regular basis. All pan-European market procedures are published (at least in parts) on TSOs and NEMOs websites.

4.1 Introduction

European day-ahead and intraday market coupling processes are the backbone of the functioning of European electricity markets. TSOs fulfil their role of ensuring safe grid operations and maintaining the stability of electricity supply across Europe not only by providing robust grid capacities to the central market coupling algorithms, but also by actively participating and shaping them.

Since market coupling involves many parties, from TSOs and NEMOs which are directly responsible for the operations, to market parties from across Europe, a clear and stable framework is needed that addresses the "clear weather" scenarios when everything is according to plan, but also prescribes steps when there are

M. Schrade (✉)
50Hertz Transmission GmbH, Berlin, Germany
e-mail: marius.schrade@50hertz.com

A. Estermann et al. (eds.), *European Electricity Market Coupling*, Future of Energy,
https://doi.org/10.1007/978-3-031-86315-8_4

failures within the systems, and the ultima ratio on how to decouple certain bidding zones or even the whole European market area for a process step.

The day-to-day robust, reliable, and under normal conditions successfully coupled European electricity markets are highly complex. Ensuring the efficient and reliable operations of market coupling is hence of very high importance, because in case of a failure certain bidding zones (often entire countries) or even all might be decoupled from each other, leading to drastic loss of welfare and security of supply.

For that reason, TSOs and NEMOs have introduced operational procedures for Single Day-Ahead Coupling (SDAC) and Single Intraday Coupling (SIDC), including Intraday Auctions (IDAs). The focus of the operational procedures is to ensure that the processes are robust and in case of hazardous conditions clear instructions for the involved operators exist with regards to timings (e.g. until when does a process needs to be finished), communication channels (e.g. telephone, email), and who must contact whom.

Although this chapter focuses on operations, it touches upon many concepts that are discussed in more detail in other chapters of this book. For example, building upon the previous discussion in Chap. 3 on capacity calculation, the procedures begin with the process of TSOs submitting the calculated capacities to the market coupling central systems. While our focus here is on TSOs, the subsequent chapter (Chap. 5) shifts attention to the role of Regional Coordination Centres (RCCs). For an in-depth understanding of the underlying algorithms and IT systems that facilitate market coupling, Chaps. 6 and 7 explore the auction-based algorithms of day-ahead and intraday markets and the mechanisms of continuous intraday trading, respectively. The utilization of market coupling results for physical and financial scheduled exchanges is comprehensively covered in Chaps. 8 through 10. Finally, Chap. 14 discusses the legal framework of market coupling in which the procedures are also embedded.

NEMO-only procedures are not covered in this chapter due to the fact that many key procedural steps of the Market Coupling Operator (MCO) and between NEMOs are not made available to the public.

This chapter also does not cover secondary procedures, for example, for different CCRs, cross-regional, regional (e.g. Core CCR), bilateral procedures, or any other procedure that does not apply to the whole SDAC area.

4.2 Operational Procedures in Single Day-Ahead Coupling

Table 4.1 provides an overview of the operational procedures within the SDAC framework. The procedures are categorized into Normal, Back-Up, Fallback, Other, and Special categories.

Normal Procedures are the "clear weather" scenarios, meaning if each step is followed and there are no anomalies as outcomes of the market coupling processes,

Table 4.1 Overview on single day-ahead coupling procedures in production relevant to TSOs

Category	Name	Index
Normal	Cross-zonal capacities and allocation constraints submission	SDAC_NOR_01
	Final confirmation of the results	SDAC_NOR_02
	Market coupling results and scheduled exchanges transfer	SDAC_NOR-03
Back-up	Cross-zonal capacities and allocation constraints submission	SDAC_BUP_01
	Final confirmation of the results	SDAC_BUP_02
Fallback	Incident management	SDAC_FAL_01
	Full decoupling	SDAC_FAL_02
	Partial decoupling	SDAC_FAL_03
Others	SDAC procedures Reading instructions	SDAC_OTH_01
	Internal and external communications	SDAC_OTH_02
	Norwegian bidding area change	SDAC_OTH_04
	Modification of maximum clearing price	SDAC_OTH_06
Special	Impact of second auctions	SDAC_SPE_01
	Impact of price limits in Nordic-Baltic reached	SDAC_SPE_02

Source ENTSO-E A

no further actions are needed. The processes described in the normal procedures are usually highly automated with little to no human intervention.

The naming of the normal and back-up procedures follows a logic. With each process step described in the normal process 01 to 03 there are certain risk cases associated with, for which possible actions are prescribed in the back-up procedures 01 to 02. There is no back-up procedure for the transfer of market coupling results and scheduled exchanges.

All SDAC procedures are the guiding process manuals to which all regional or local procedures must be aligned to.

4.2.1 Normal Procedures

The normal procedures deal with the relevant steps for market coupling operations in a situation without any disturbances.

4.2.1.1 Submission of Cross-Zonal Capacities and Allocation Constraints (SDAC_NOR_01)

The cross-zonal capacities (CZCs) and allocation constraint (AC) submission is the first step and hence also the first procedure. After the TSOs have finished their capacity calculation processes this procedure starts with the transfer of the CZCs and ACs from the TSOs pre-coupling Systems to the NEMO pre-coupling modules. It ends when the data is successfully received. Afterwards they are published by the relevant parties, among others, on the ENTSO-E Transparency Platform and the JAO publication tool.

The target time until when CZCs and ACs, sometimes also called network data, data should be sent to the coupling systems is 10:30. After that that the respective back-up procedure (SDAC_BUP_01) will be followed.

There are different kinds of interconnectors:

- Single submission interconnectors: only one TSO sends its information to a single NEMO.
- Double submission interconnectors: both TSOs sent the cross-zonal capacities to the PMB. The CZCs are then matched (cross-checked) by the two counterpart TSOs (prior to sending to the NEMOs) and then by the two counterpart NEMOs.
- Multi-submission interconnectors: interconnectors for which the CZCs are sent to the PMB by the concerned regional NEMOs and cross-checked in PMB, even if the CZCs are provided by a single TSOs.

With more and more capacity calculation done on a regional level, see for example, Core CCR, TSOs move to the direction to build regional tools that provide the CZC in a single flow to the NEMOs.

There are certain risk cases associated with submission of the CZC in a normal process. All of them are tackled in the respective procedure SDAC_BUP_01. For example, CZCs are not available, or cannot be sent before the Target Time, or NEMO pre-coupling modules cannot receive the flows or rejects them.

The procedure is considered finished and in its "final state" if the CZCs and ACs files are well received. Then SDAC_NOR_02 is started.

4.2.1.2 Final Confirmation of the Results (SDAC_NOR_02)

After the day-ahead algorithm EUPHEMIA has determined the market coupling results, the TSOs verify them in their verification systems against the cross-zonal capacities and in case submitted, allocation constraints, to ensure safe system operations. For example, if the global net position of a bidding zone was constrained by an allocation constraint to a certain value, the TSO verification systems check that this number was not exceeded. The TSO verification systems have in total 12 min for submitting the "Final Confirmation of the Market Coupling Results" to the NEMO coupling systems. After this final confirmation of the market coupling results published either as preliminary or not at all or are considered final and the publication is updated. Following the publication, the market coupling results are considered firm.

The same as for the CZC and allocation constraint submission, there are also certain risk cases associated with the final confirmation of the results.

Risk cases that result in triggering BUP_02 are, for example, verification systems on either site either cannot receive and/or send the results or confirmations, or technical failure of the verification.

If a TSO (or any party entitled by the TSOs) rejects the market coupling results an Incident Committee according to FAL_01 is started.

This process is finished when all systems of the relevant parties have sent a global confirmation of the market coupling results. They are now considered as firm and NOR_03 starts.

4.2.1.3 Market Coupling Results and Scheduled Exchanges Transfer (SDAC_NOR_03)

The last step according to the normal procedures is to distribute market coupling results to the respective NEMO clearing and settlement systems and to the Congestion Income Distribution System operated by the Joint Allocation Office (JAO) for TSOs. From a market coupling point of view an issue in this process step is the least impactful as the main coupling has been performed and the results have been confirmed. However, this step is still important as the distribution triggers the post-coupling processes. Here risk cases usually refer to the failure of sending or receiving of involved IT systems for which local back-up procedures were developed to mitigate any failure.

The post-coupling system calculates the Scheduled Exchanges for the relevant Bidding Zones and sends it to the NEMOs clearing and settlement systems and to the Congestion Income Distribution System (CIDS) handled by JAO.

This process step is completed when the Scheduled Exchanges are received by the NEMOs clearing and settlement systems and the CIDS, and when the Market Coupling Results are received by the CIDs.

4.2.1.4 Trading Confirmation and Scheduled Exchanges Notification (SDAC_NOR_04)

This procedure deals with the sending of the trading confirmation from the NEMOs clearing and settlement system to the CCP shipping system and the sending of the Scheduled Exchanges Notification from the CCP shipping systems to the TSO back-end systems.

In the Core CCR, the Core Capacity Calculation Tool (CCCt) is used to send the scheduled exchanges to the NEMOs clearing and settlement systems.

It is completed when the scheduled exchanges notification has been successfully received by the TSO back-end systems according to the notification deadlines of each TSO.

4.2.2 Back-Up Procedures

Two back-up procedures guide the operators on necessary steps when unforeseen difficulties occur during the submission of CZCs and ACs or the confirmation of the final market coupling results. In general, there are two outcomes of the back-up procedures: either the issue is solved, then the subsequent normal procedure is applied or if it cannot be solved by a certain target time, then the fallback procedures apply.

4.2.2.1 Cross-Zonal Capacities and Allocation Constraints Submission (SDAC_BUP_01)

This procedure describes all back-up cases and solutions with regards to problems in the pre-coupling phase, which might impact the timely submission of the CZCs and optional ACs from TSO to the NEMO systems. In principle, the party that either encounters or observes an issue should inform the other party as soon as possible.

In total the back-up procedure details seven risk cases and possible solutions:

- **CZCs and ACs are not available at target time (10:30) due to technical/ calculation issue**: The concerned TSOs or party entitled by the TSO applies its own back-up procedures in order to solve the issue. If the issue cannot be solved by 11:00 an Incident Committee will be launched.
- **TSO pre-coupling systems cannot send the CZCs and ACs at target time to the NEMO pre-coupling system or the respective NEMO system cannot receive it**: The TSO pre-coupling system operator needs to send the files in the back-up mode according to local procedures, and the relevant NEMO uploads it in accordance with these procedures.
- **NEMO pre-coupling module rejects the CZCs and allocation constraints**: The respective TSO and NEMO operators call each other to identify the root cause and if possible the files are sent once again via local back-up modes. If the problem is solved, then normal procedures are followed again. If not solved by 11:00 an Incident Committee will be launched according to procedure.
- **CZCs and AC need to be updated**: An update of capacity data is always possible until target time. However, after 11:00 any update is not possible due to the day-ahead firmness deadline of capacities in accordance with CACM Article 70 (European Commission 2015). If by 11:30 the integration of the updated CZCs and ACs in the NEMO systems fails, then the CZCs and Allocation constraints are set in accordance with local procedures. Either the earlier CZCs are used, the CZCs are put to zero (de facto decoupling) or no capacity allocation is done on the interconnector. Market participants are informed through external communication.
- **Issues regarding the CZCs for the double-sided submission interconnectors:** The main risk for double-sided submission interconnectors is that there are mismatches in terms of CZCs between the relevant NEMOs for the same interconnector. As TSOs might find different solutions depending on their specific situation, these are usually described in regional/local procedures.
- **Network data file rejected by the NEMO central system**: Since the exchanged data files formats are fully standardized any deviation might lead to an error in the generation of the aggregated network file that is used by EUPHEMIA to run the market coupling. If there are errors in the submitted files the TSO pre-coupling systems need to correct the CZC file and provide a new version to all relevant NEMOs, using the regular method of communication. If there is still no valid network data file by 11:00, then an IC will be launched according to SDAC_FAL_01. If by 11:15 there is still no valid network data file provided

to the NEMO central system, the regional parties and the market participants will be informed about the risk of decoupling of the concerned interconnectors from SDAC in accordance with the external messages in procedure OTH_02. If there is no valid network data file by 11:30, partial decoupling procedure is applied (SDAC_FAL_03).

4.2.2.2 Final Confirmation of the Results (SDAC_BUP_02)

This back-up procedure deals with the situation when the market coupling results cannot be confirmed by TSOs due to several reasons, for example, system issues in receiving, sending, or verifying the results. If the final confirmation is not received within the timespan allocated to it (currently 12 min), the relevant NEMO and TSO operators inquire about the confirmation. There is the possibility to apply deemed acceptance during an Incident Committee call if all NEMOs have confirmed the results already and it is not expected that the confirmation can be generated in time. This is to ensure that local issues in the verification step do not spill-over to other parties and the market coupling is ensured.

4.2.3 Fallback Procedures

The fallback procedures are key to ensure that all parties and operators have a clear guide on the next steps in situations under immense pressure, namely, when an incident occurs and even partial or full decoupling of the whole SDAC area need to be performed. All parties work under the principle to do their utmost to uphold market coupling.

4.2.3.1 Incident Management (SDAC_FAL_01)

As soon as an incident occurs that prevents the timely allocation of the CZCs via the implicit allocation process and/or timely publication of the market coupling results, an Incident Committee (IC) is organized. Based on the rotational scheme of the NEMOs, the IC is organized and convened by the central NEMO coordinator and operator respectively. During the IC the relevant participants identify the issue, assess and agree on potential fallback solutions, these being either Full Decoupling (SDAC_FAL_02) or Partial Decoupling (SDAC_FAL_03).

To ensure robustness and predictability of the market coupling process, ICs can only be triggered until a certain time, also called latest time to start an incident committee, depending on the risk of a potential decoupling case.

Another important deadline is the time until the risk of partial decoupling needs to be communicated externally and in line with SDAC_OTH_02 procedure. This communication is very important because JAO needs it in order to activate the Shadow Auction processes, but also so that market participants can prepare their internal procedures to place their bids for the shadow auctions. Finally, if a certain deadline is reached without a solution has been found, the IC needs to declare the decoupling and the relevant messages need to be sent externally.

The IC has the right to perform all necessary measures, including the declaration of partial or full decoupling. Within an IC exist also different roles. So-called "full right participants" to analyse the incident and assess as well as agree on the solution for the incident, which might also be the declaration of a decoupling. Full right participants are the central NEMO system coordinator and operators, the algorithm provider, and the TSOs and/or NEMOs that are either directly responsible or able to contribute to the solution of the incident. In case of risk of decoupling all participants are full right participants. Full right participants must join the IC call within 5 min after the invitation was sent. If this is not the case the party will be contacted, but the IC will start anyway.

To ensure transparency the IC is minute and a report on the incident needs to be prepared that is then also shared with all NEMOs and TSOs. In case of risk of full decoupling or realized decouplings, an ad-hoc SDAC Operational Steering Committee will be organized on the same working day to ensure that also project experts can contribute to the incident analysis and prevent a similar incident in the next day-ahead market coupling session.

4.2.3.2 Full Decoupling (SDAC_FAL_02)

A full decoupling is the most extreme situation that can occur in the market coupling process. It means that it was not possible to allocate the CZCs via the implicit allocation process and all bids are only matched within a bidding zone but there is no exchange between bidding zones.

In general, full decoupling can be distinguished in two cases:

- Case FD1—the Full Decoupling occurs during the market coupling session.
- Case FD2—the Full Decoupling known in advance.

In FD1, the critical issue leading to the Full Decoupling occurs during the Daily Market Coupling session. In this case, the full decoupling will be declared at 14:00 or earlier if the Incident Committee can unanimously agree to do so (risk of full decoupling will be declared at 13:30). The necessary communication steps are described in SDAC_OTH_02.

Regarding case FD2, the critical issue leading to the Full Decoupling is already known in advance because the issue caused the Full Decoupling for the previous market coupling session. In this case, the Full Decoupling could be declared either in the afternoon of the day before, or during the Daily Market Coupling Session until 14:00 if the Incident Committee considers the issue to be too severe to be solved until the deadline of Case FD1.

Depending on the Full Decoupling case, different actions need to be carried out:

- Inform the market once there is a delay in the publication of the results (FD1) with the message ExC_02: Delay in Market Results Publication.
- Inform the market once there is a risk of Full Decoupling (FD1 & FD2) with the message ExC_03b: Risk of Full Decoupling.

- Inform the market once the Full Decoupling is declared (FD1 & FD2): At the respective Full Decoupling deadline, external communications are sent "ExC_04b: Full Decoupling or ExC_05b: Full Decoupling known in advance)" according to SDAC_OTH_02.

TSOs then send a message to Market Participants, informing them that the nomination deadlines are extended. Depending on the border, different fallback solutions may be used: capacity simply is not allocated in day-ahead but released for intraday, there are shadow auctions organized via JAO, or regional coupling is performed.

Final State of the Full Decoupling is reached when the Full Decoupling is officially declared by informing the Market Participants or in case the issue has been solved before the Full Decoupling deadline.

4.2.3.3 Partial Decoupling (SDAC_FAL_03)

A partial Decoupling is a situation where it is not possible to allocate the CZCs via the implicit allocation for one or several areas and/or interconnectors before the relevant partial decoupling deadline.

In addition to the case when the Partial Decoupling is known in Advance (PC3) that is similar to the case of Full Decoupling known in Advance (FD2), for partial decoupling we distinguish between two cases:

- **PD1**: partial decoupling for CZC-related reasons (PD1) with a deadline at 11:30.
- **PD2**: partial decoupling for reasons not related to CZCs (PD2) with a deadline at 12:40.

With regards to PD1 the latest deadline for an Incident Committee is 11:00 at the latest. An Incident Committee shall be triggered to handle the issue according to SDAC_FAL 01. If at 11:15 the CZCs and AC are still not successfully submitted by TSO systems to the central NEMO system, then the risk of partial decoupling is externally communicated via the relevant message and the local fallback allocation processes are activated. At 11:30 if the CZCs and ACs are still missing, the partial decoupling of the agreed interconnectors is declared by the concerned NEMOs in the Incident Committee, according to local procedures. At 12:00 the order books are closed according to the normal procedure. For the remaining coupled areas and/or interconnectors, the normal procedures will be followed, while for the decoupling areas and/or interconnectors the predefined decoupling values will be used for the CZCs. It should be noted that in case the CZCs have been calculated but their value is zero and the file has been received by NEMOs by 11:30, this is not a decoupling case.

With regards to PD2 another reason for a partial decoupling not related to the CZCs is that a NEMO might have local IT problems and hence cannot participate in the regular processes of market coupling. Since in Europe there are several

NEMOs active with the IT infrastructure to run a successful market coupling process, an IT issue of one NEMO can be mitigated by the other NEMOs, especially in multi-NEMO bidding zones because market participants simply can transfer their bids from one NEMO trading system to the other. Therefore, the deadlines for declaring partial decoupling due to non-CZC-related reasons is later and only starts at 12:40 with a communication towards market participants that there is a risk of partial decoupling. However, if there is an issue within a bidding zone where only one NEMO is active, so-called monopoly NEMO, then this bidding zone needs to be decoupled because the market participants do not have the possibility to submit their bids through the trading systems of another NEMO. For remaining coupled areas and/or interconnectors the normal SDAC procedures will be applied, even though the timings are delayed accordingly. For the decoupled areas and/or interconnectors the local procedures are followed accordingly.

A note on fallbacks: Article 44 (EC 2015) mandates that there should be fallback procedures in case of a decoupling event in order "to ensure efficient, transparent, and non-discriminatory capacity allocation". The current standard fallback procedure to be applied in such a situation are the so-called Shadow Auctions. During the shadow auction process explicit auctions for the cross-zonal capacities are organized via the Joint Allocation Office (JAO), and in general replicate the explicit allocation process as performed in Europe before the market coupling implementation and hence the implicit allocation of CZCs. However, in the situations when shadow auctions were organized, they were hardly utilized, leading to criticism by market participants or regulators. Dierenbach et al. (2022) assessed the shadow auction mechanism in light of this criticism and found that the "probability of a trader holding Long Term Transmission Rights (LTTRs) to allocate cross-border capacity in Shadow Auctions (SA) significantly reduced compared to those traders not holding LTTRs". Further, the "LTTR remuneration based on market spreads significantly reduces the incentives to take part in the SA for LTTR holder". Due to the missing economic incentives of LTTR holders to participate in shadow auctions and introduction of intraday auctions in June 2024, it could be discussed whether the intraday auctions are better suited to achieve the goal of an efficient fallback mechanism than the current set-up of the shadow auctions.

4.2.4 Other and Special Procedures

The normal, back-up, and fallback procedures are the most important guidelines for the daily market coupling operations. Nevertheless, there are also additional procedures that regulate special situations and communications. While OTH_01 provides a useful introduction to day-ahead market coupling and how to read the procedures, OTH_02 standardizes the internal and external communication messages to ensure especially during non-clear weather scenarios the efficient and robust operation of market coupling and the needed transparency for TSOs, NEMOs, and market participants.

The modification of maximum clearing price procedure (OTH_06) is a direct translation of a legal requirement into a daily market coupling procedure and that was also discussed quite extensively during the energy crisis in Europe in 2022. The procedure regulates how the maximum clearing prices, currently at −500 €/MWh and 4000 €/MWh, should be increased in accordance with the "harmonized maximum and minimum clearing prices for SDAC" (HMMCP for SDAC) methodology that was amended by an ACER decision in January 2023 (ACER, 2023). Originally, the clearing prices were supposed to be adjusted in 1000 €/MWh in case the clearing price is above 60% of the price limits in at least one market time unit in one bidding zone. This was first reached on 4th April 2022 in France with prices of 2720 €/MWh and 2990 €/MWh and resulted in an adjustment of the maximum bidding price from 3000 €/MWh to 4000 €/MWh. On 17 August 2022, this threshold was again reached by a maximum clearing price of 4000 €/MWh in the Baltics. However, after the April incident a new HMMCP methodology was consulted to incorporate this critical and new situation by, among other measures, adjusting trigger conditions to 2 market time units of price spike over at least 2 days in a rolling 30 days period instead of 1 price spike on 1 day. The new methodology was adopted via the ACER decision on 10th January 2023.

4.2.5 An Example of Partial Decoupling

Thankfully, in over 10 years of operations a full decoupling of the whole SDAC area has not occurred. But partial decouplings triggered by local issues of parties can happen from time to time. One such example is the partial decoupling of the Slovak bidding zone due to an IT and configurational issue at OKTE, the Slovak NEMO, on 10th May 2022 (SDAC 2022). The Market Coupling Steering Committee published a report on the incident on 8th June 2022. The root cause of the issue was the configurational change due to the adjustment of the maximum and minimum prices (see previous section) in the local trading system of OKTE. Unfortunately, this incident happened before the roll-out of the flow-based capacity calculation methodology in the capacity calculation region Core in June 2022 and as such the problem not only affected the Slovak borders but also spilled over to the German-Czech and German-Polish borders. Shadow auctions were run on the affected borders via JAO, while for the remaining SDAC borders the normal procedures were followed. The local auction in Slovakia was completed on the same day at around 13:20. MCSC parties conclude that "procedures were followed correctly, and the communication was performed in line with those, using the agreed messages". Although the incident was managed well, MCSC parties acknowledge that the increasing number of coupled parties with large numbers of different systems and complexity of operations also increases the risk for partial decouplings.

4.3 Operational Procedures in Single Intraday Coupling

SIDC operates as a continuous trading process, unlike SDAC. Trading starts shortly after the day-ahead market closes—beginning at 15:00 CET—and continues until one hour before electricity delivery. The central tool used is XBID, comprising the Shared Orderbook (SOB), Capacity Management Module (CMM), and Shipping Module (SM), see Chap. 7 for an overview on the XBID solution from a service provider point of view.

Due to interactions with the day-ahead market and the need for capacity recalculations, some cross-border lines are not opened for cross-zonal trading until 22:00 CET on the day before delivery (D-1). Throughout the day, TSOs may overwrite capacities or halt cross-border lines without affecting intra-zonal trading. Unlike in SDAC, TSOs cannot actively verify every trade impacting cross-zonal capacities; instead, XBID automatically ensures capacities are not violated when offers are matched.

Because of its continuous processes SIDC lacks a fallback capacity allocation mechanism, which is why maintaining XBID's 24/7 availability is crucial. Despite best efforts, system malfunctions can occur, so TSOs and NEMOs follow back-up and fallback procedures. In serious cases, an Incident Committee and possibly an ad-hoc Operations Subcommittee (OPSCOM) are convened.

Post-coupling in SIDC is similar to that in SDAC, but due to time constraints, some TSOs do not receive explicit cross-border nominations from NEMOs and instead prepare schedules on their behalf using XBID data—a process known as "nomination on behalf". Continuous trading does not result in a single price within bidding zones, so TSOs currently do not collect congestion income in SIDC. To correctly value scarce transmission capacity, TSOs and NEMOs are introducing three intraday auctions (Table 4.2).

Generally, SIDC and SDAC distinguish the same three phases—pre-coupling, coupling, and post-coupling—as well as the same modes: normal, back-up, and fallback. In this section, we focus on the differences between the operational procedures in SDAC and SIDC to avoid repetition.

Unlike the procedures in SDAC Sect. 4.2, there are explicit procedural differences between joint operations and TSO-only operations, indicated by specific prefixes. In addition to the fallback procedure, two additional exceptional procedures are followed when operating in fallback mode (Table 4.2).

There are four key deadlines: the submission of CZCs and allocation constraints, the allocation deadline, the cross-border nomination deadline, and the publication of results. As long as these deadlines are met—whether in normal or back-up mode—coupling proceeds as usual. If any deadline cannot be respected, it is considered an incident, triggering operations to enter fallback mode.

Unlike in SDAC—where systems can be updated and maintained during periods outside the day-ahead coupling process—the continuous trading in SIDC necessitates additional rules for configuration changes and maintenance windows. This requirement leads to an extra category of procedures.

Table 4.2 List of joint and TSO-only procedures in SIDC

Category	Name	Index
Normal	Submission of cross-zonal capacities	XBID_TSO_NOR_01
	Distribution of allocation information	XBID_JOINT_NOR_02
	Nomination	XBID_JOINT_NOR_03
	Nomination on behalf	XBID_JOINT_NOR_04
Back-up	Submission of cross-zonal capacities	XBID_TSO_BUP_01
	Distribution of allocation information	XBID_JOINT_BUP_02
Fallback	Incident management	XBID_JOINT__FAL_01
Others	Procedures reading instructions	XBID_JOINT_OTH_01
	Internal and external communications	XBID_JOINT_OTH_02
	Maintenance window Local TSO system	XBID_JOINT_OTH_04
	System maintenance	XBID_JOINT_OTH_05
	Planned maintenance Window local Shipper system	XBID_JOINT_OTH_06
	Algorithm monitoring	XBID_JOINT_OTH_07
	Transit shipping	XBID_JOINT_OTH_08

(continued)

Table 4.2 (continued)

Category	Name	Index
Configuration	Shipping module configuration	XBID_JOINT_CFG_01
	Capacity management module—balancing group and user management	XBID_TSO_CFG_01
	Capacity management module—master data management	XBID_TSO_CFG_02

Source ENTSO-E B

4.3.1 Cross-Zonal Capacity Submission (XBID_TSO_NOR_01)

Additional to the submission of CZCs to the XBID system, TSOs or the Coordinated Capacity Coordinator submit the already allocated capacities (AAC) from previous timeframes (long-term and day-ahead) to the XBID system for all interconnectors. This is to ensure that only firm capacities are available for the market and no overallocation of capacities could occur that would endanger system operations. The AACs are also crucial for the accurate calculation of ramping constraints on certain interconnectors. Both CZCs and AACs can be continuously updated by sending new files to the XBID system up until 15 min prior the respective capacity management module contract. Especially the updates of CZCs for certain market time units could happen on a more frequent basis than in day-ahead due to a more frequent capacity calculation process and the reaction of TSOs to changes in forecasts or outages.

Depending on local agreements between the TSOs on the border, they can agree who submits the relevant capacity information to the XBID system. XBID needs to integrate the CZC and AAC within 5 min and sends back an acknowledgement to the TSOs. Upon gate opening time the CZCs and AACs are activated.

4.3.2 Distribution of Allocation Information (XBID_JOINT_NOR_02)

After the trade for a specific market time unit closed according to the gate closure time, XBID aggregates the confirmed trades for all borders at the same time to ensure consistent allocation state for the whole region.

This step generates various data flows for TSOs, including implicit allocation results (capacity rights), scheduled exchanges (cross-border nomination information), shipper hub nomination information, and assigned transit hub volumes. Net cross-border allocation information is also used to create physical schedules on some interconnectors. Shipping agents receive similar information, along with financial details such as CCP-shipper clearing information for source and sink areas, and shipper-shipper clearing information for each border and transit area on the allocation path.

All this information must be sent to TSOs' post-coupling systems, shipping systems, and CCP systems within 5 min after gate closure. Due to the tight time-frame and potential high market activity, local procedures address any delays. Not all files need to meet this deadline; separate deadlines can be specified in local arrangements.

4.3.3 Nominations (XBID_JOINT_NOR_03)

This procedure outlines the nomination process by the Scheduling Agent of the Shipping Agent or Central Counterparty (CCP). Each Shipping Agent or CCP can act as or appoint a Scheduling Agent to provide schedules to TSO post-coupling systems. Only nominations without a "nomination on behalf" arrangement are concerned; where such an arrangement exists, shipping systems do not send nominations.

Nominations must be sent to the TSOs' post-coupling systems before the nomination deadline and can be updated until then. The TSO post-coupling system sends an acknowledgment upon receipt. TSOs verify the scheduled exchanges and may send back the nominated net positions. If multiple files are submitted, the one with the highest version number received before the deadline and positively acknowledged by TSOs is considered.

Risk cases include failures in sending nominations automatically, receiving no or negative acknowledgments, discrepancies in cross-border nominations at TSOs, or TSOs being unable to send aligned nominated positions back to the Scheduling Agent. If issues occur during transmission to the TSO's post-coupling system, incident calls are triggered. If problems arise in aligning schedules or receiving nominated positions back, local back-up procedures may apply.

The procedure concludes when the nominations are successfully received by the TSO's post-coupling systems, and confirmation is received by the Scheduling Agent from the TSOs as per local arrangements.

4.3.4 Nomination on Behalf (XBID_JOINT_NOR_04)

This procedure outlines how TSOs transform information received from the central XBID system into nominations within their post-coupling systems on behalf of shipping agents, CCPs, or explicit market participants. Only TSOs can implement "nomination on behalf"; third parties cannot. Each TSO decides locally whether to use this procedure or the standard nomination process (XBID_JOINT_NOR_03—Nomination), based on the type of allocation and nomination solution in place.

The process distinguishes between ex-ante and ex-post nominations. Depending on local market rules, some nominations are mandatory ex-ante, while others can be ex-post. Generally, "nomination on behalf" is applied only to ex-ante nominations.

Unlike the normal nomination process, TSOs' post-coupling systems create the nominations used in the TSO-to-TSO alignment and verification process. Once these nominated positions are aligned, they are sent to the Scheduling Agent of the Shipping Agent or the scheduling agent of CCPs active in the TSO's scheduling area.

The procedure concludes when all required nominations on behalf are created in the TSOs' post-coupling systems.

4.3.5 Cross-Zonal Capacity Submission (XBID_TSO_BUP_01)

Comparable to the SDAC procedures, the back-up procedures provide an overview of possible risk cases in the process steps and offer applicable back-up solutions to the operators. As a general principle, if a back-up procedure cannot resolve the issue in time, operators refer to fallback procedures. If there are service provider-related issues in the central XBID system then also the predefined response time is specified in the fallback procedure (FAL_01).

We can distinguish between three main risk cases during the submission of CZCs and ACs:

- **Case 1: The AACs and CZCs cannot be calculated by TSOs:** Accurate consideration of AACs in the XBID system is crucial to prevent overallocation of capacities. Local trading within the bidding zone can continue. If AACs cannot be delivered before the scheduled Gate Opening of the concerned interconnector(s), the opening will be postponed, and the market will be informed via a predefined message. Once AACs become available, the market will receive another message announcing the delayed Gate Opening.
- **Case 2: TSOs' pre-coupling systems either cannot send the CZCs and AACs or don't receive acknowledgement for the submission:** If the TSO operator detects an IT problem preventing the sending of CZC & AACs from the pre-coupling system, back-up channels like email or manual file upload through the capacity management module's graphical user interface can be used. TSOs can also support each other by having another TSO upload the capacities after an official request. The interconnector may be closed and reopened via messages, while local trading can continue.
- **Case 3: The XBID system cannot receive or integrate the CZCs and AACs files, might send a rejection or does not activate/update the CZCs and AACs:** If the XBID system cannot integrate the electronically sent AAC files—resulting in no AACs appearing due to a central system issue—a ticket is created. If the problem isn't resolved before Gate Opening, cross-border trading won't be possible, but local trading will continue. A market message will announce the postponement of the interconnector opening using a predefined message.

The process proceeds to the next step when the CZCs and AACs are successfully received, integrated, and activated by the XBID system.

4.3.6 Distribution of Allocation Information (XBID_JOINT_BUP_02)

The Shipping Module (SM) and Capacity Management Module (CMM) of the XBID system transfer physical shipping information to TSO post-coupling systems, shipping systems, and CCPs' clearing systems. This information includes capacity rights and nomination data, which TSOs use to create cross-border and hub nominations on behalf of the Shipping Agent.

Risk cases to be anticipated include central system failures (CMM or SM downtime), the XBID system not receiving acknowledgments, or TSO, Shipping, or CCP clearing systems not receiving allocation information, rejecting it, or sending negative acknowledgments. Since CMM or SM downtime affects the entire system, it triggers Incident Management according to procedure FAL_01.

TSO-related risk cases and possible next steps include:

- **Case 3**: If the TSO post-coupling system or shipping system does not receive allocation information in time, an alarm message is triggered, incident management tickets are created, and the helpdesk is contacted.
- **Cases 4 and 5**: If the post-coupling system, shipping system, or CCP clearing system fails to integrate the allocation information, or if XBID doesn't receive acknowledgment for it, the XBID system automatically resends the nomination information files via back-up communication channels after a predefined time.
- **Cases 6 and 7**: If the TSO post-coupling system, shipping system, or CCP clearing system rejects the allocation information or sends a negative acknowledgment, the XBID system displays an error message in the shipping modules.

The process proceeds to the next step once the physical shipping information is successfully received and integrated by the TSO post-coupling system, shipping system, and CCP clearing system.

4.3.7 Incident Management (XBID_JOINT_FAL_01)

As well as in SDAC, there is also an IC in SIDC that.

This fallback procedure outlines the handling of incidents, including the operation of the Incident Committee (IC) and the application of fallback solutions such as closing and reopening interconnectors (XBID_JOINT_EXC_01), markets, delivery areas, or trading services according to local procedures.

An incident requiring an IC occurs when issues cannot be resolved through local back-up procedures and may breach a Single Intraday Market Coupling (SIDC) deadline (e.g. gate closure or opening). Incident management refers to unwanted events in the XBID system, local NEMO or TSO systems connected to XBID, or the communication channels between them.

When such an incident impacts any SIDC processes, an IC is initiated by the IC Single Point of Contact (SPOC). The IC participants assess and agree on potential solutions. Clear guidelines define when an IC should be triggered and outline participants' roles, responsibilities, and the sequence of actions during the IC.

An IC is only triggered for managing critical or major incidents in the XBID system or Transit Shipping Agent Systems. The triggers include XBID system-related incidents and other issues requiring IC attendance.

To request an IC, the incident reporter must first determine the incident's criticality. For critical and major incidents experienced by a NEMO or TSO, the reporter contacts the service provider and logs a ticket in the management tool.

Incident severity levels are defined as:

- **Critical**: Highest severity with severe functional or performance impact, such as users being unable to access essential functionality or services required for critical business processes.
- **Major**: Significant impact affecting non-critical business processes, where users cannot access certain functions or services, leading to errors in non-critical processes.
- **Minor**: Moderate impact on non-critical business processes, where issues can be bypassed with reasonable effort and do not affect critical operations.

Any proposed solution during the IC must be unanimously accepted by all affected parties. If not accepted, the service provider continues investigating solutions. Parties rejecting the solution must clearly explain their reasons. If consensus cannot be reached, the IC SPOC escalates the issue to an ad-hoc Operations Subcommittee (OPSCOM), which may not convene until the next business day.

A special case within the fallback procedure is the case a transit shipping agent's system unexpectedly goes down and cannot fulfil timely nomination obligations and the agent might then request the closure of interconnectors via the IC SPOC. In case a transit shipping agent system is having an unexpected outage and the transit shipping agent can no longer fulfil its obligations for timely nomination the transit shipping agent may request for closure of interconnectors. This is to be done via the IC SPOC. The IC SPOC then requests as soon as possible the relevant TSOs to close the relevant interconnectors by sending messages. In case not all relevant borders are closed after 5 min since the IC invitation was sent, all IC participants should agree on if and when the XBID system trading service will be set to halt until all relevant borders are closed. If XBID System trading service is to be halted. If the respective Shipping Agent has restored its system, the relevant TSOs apply the process of reopening of interconnectors.

Similarly to the transit shipping system incident, the default of a Shipping Agent leads to any party affected by this to request an IC. Immediate actions to effectively isolate the defaulting party by closing relevant borders immediately. If the relevant borders are not closed after 5 min after the IC invitation, the IC participants should agree on if and when the XBID system trading service will be set to a halt until all relevant borders are closed. After the issue is solved, all IC participants apply the process of reopening interconnectors.

4.4 Outlook

Day-ahead and intraday market coupling apply robust but challenging processes due to the tight timing and the high number of parties and systems involved. If market coupling fails, this results in decoupling which causes significant welfare losses and system operation risks. To ensure process robustness and minimize risk of decoupling, TSOs and NEMOs draft, test, implement, review, and update operational procedures on a regular basis.

As Intraday Auctions (IDAs) only go live in June 2024 procedures are neither provided nor explained in this chapter. Moreover, based on the past experiences of the SDAC and SIDC go-lives it should be assumed that IDA procedures will undergo several (ad-hoc) procedural refinements within the first year of operations, especially as IDAs have for the first time been applied at once in all European countries. Hence, a future revision of this chapter or other publications must take a closer look at IDA procedures. This has most likely further relevance as it can be assumed that with the introduction of IDAs SDAC and SIDC procedures will be (further) harmonized by TSOs and NEMOs.

Bibliography

ACER. (2023). *On the nominated electricity market operators proposal for the harmonised maximum and minimum clearing price methodology for the single day-ahead coupling.* Decision No 01/2023 of the European Union Agency for the cooperation of Energy Regulators of 10 January 2023. Available at: https://www.acer.europa.eu/sites/default/files/documents/Individual%20Decisions/ACER%20Decision%2001-2023%20on%20HMMCP%20SDAC.pdf. Accessed: 03 June 2024.

Dierenbach, J., Holmer, D., Müller, D. et al. (2022). Assessment of the Shadow Auction Mechanism as a Fallback Procedure for Single Day-Ahead Coupling. *Z Energiewirtsch, 46*, 231–244. https://doi.org/10.1007/s12398-022-00333-4

ENTSO-E, A for the cooperation of the E. transmission system operators (a) *Annex 2: Operational Procedures, Single Day-ahead Coupling (SDAC).* Available at: https://www.entsoe.eu/network_codes/cacm/implementation/sdac/#publication-of-contracts. Accessed: 02 June 2024.

ENTSO-E, B for the cooperation of the E. transmission system operators (b) *Exhibit 6: Operational Procedures, Single Intraday Coupling (SIDC).* Available at: https://www.entsoe.eu/network_codes/cacm/implementation/sidc/#intraday-operational-agreement. Accessed: 03 June 2024.

European Commission. (2015, July 25). *Commission regulation (EU) 2015/1222 of 24 July 2015 establishing a guideline on capacity allocation and congestion management.* OJ L 197, pp. 24–72. https://eur-lex.europa.eu/legal-content/EN/TXT/?uri=CELEX%3A32015R1222

SDAC, S.D.C. (2022). *SDAC report on the partial decoupling incident of May 10th, 2022, Single Day-Ahead Coupling (SDAC).* Available at: https://eepublicdownloads.blob.core.windows.net/public-cdn-container/clean-documents/Network%20codes%20documents/NC%20CACM/SDAC%202022/220608-SDAC_report_on_the_partial_decoupling_incident_of_May10th_2022_V1_0.pdf. Accessed: 03 June 2024.

Marius Schrade is an Expert in European Market Coupling projects at 50Hertz Transmission GmbH, one of the four transmission system operators in Germany, within the System Operations Concepts and Analytics department since 2019. At 50Hertz, he started his career as project lead for the integration of the German–Czech and German–Polish border into Single Day-Ahead Coupling, which was completed in 2021. Since then, he focuses on cooperation between TSOs and NEMOs in the Market Coupling Steering Committee where he also lead a task force on day-ahead simulation software. Marius holds a Double Degree in European Economics from the Universities of Maastricht in The Netherlands and Louvain-La-Neuve in Belgium.

5 Regional Coordination Centres and Their Role in European Market Coupling

Jakub Władysław Glegoła

Abstract

This chapter describes the significant transformation of Regional Coordination Centres from their initial voluntary inception to their formal establishment in 2022 building upon the mandates of the Regional Security Coordinators. The chapter then goes on to describe the role of Regional Coordination Centres in the EU electricity system. Regional Coordination Centres are now essential in addressing the complex challenges that transcend national borders. The chapter describes them as well positioned to carry out a range of tasks from capacity calculation, coordinated security analysis, and short-term adequacy assessment to outage planning coordination and common grid model creation. The chapter concludes by describing the roles of Regional Coordination Centres in market coupling.

5.1 History and Background

The energy policy objectives of the European Union are geared towards the establishment of a fully integrated internal energy market. Given the high degree of interconnectivity in the European electricity system, the influence of one country's power grid extends beyond its borders, creating an impact on neighbouring nations.

Historically, regional cooperation and coordination among TSOs have been instrumental in fortifying grid stability. This practice of coordination was previously voluntary in nature. However, recognizing the imperative to advance the

Jakub Władysław Glegoła with contributions from Jayaram Anandha and Alexandra Andor.

J. W. Glegoła (✉)
Accenture Industry X, Kronberg im Taunus, Germany

A. Estermann et al. (eds.), *European Electricity Market Coupling*, Future of Energy,
https://doi.org/10.1007/978-3-031-86315-8_5

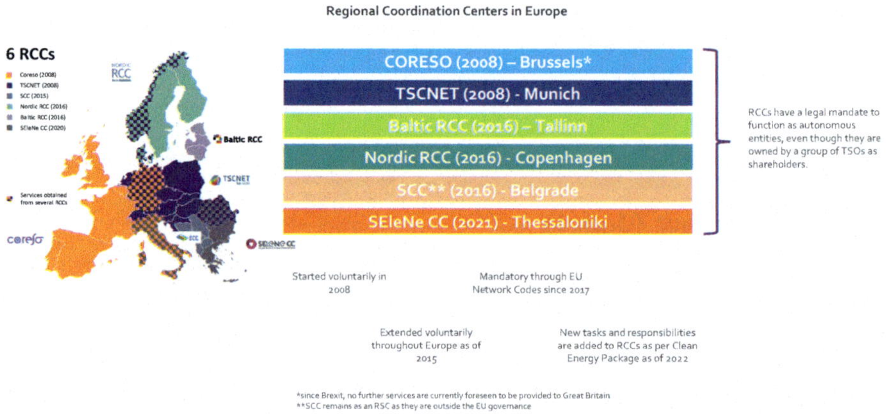

Fig. 5.1 Regional coordination centres in Europe. *Source* Adapted from ENTSO-E (2024)

EU's energy ambitions, the implementation of a mandatory regional coordination framework was identified as essential. This evolution culminated in the establishment of Regional Coordination Centres (RCCs) as of July 2022, building upon the mandates of the Regional Security Coordinators (RSCs), as stipulated by the System Operation Guideline.[1]

RCCs have undergone a significant transformation over the years (see Fig. 5.1) from their initial voluntary inception. They have evolved into a vital legal entity, playing a pivotal role in upholding the security of the European grid and ensuring the efficient operation of the European internal energy market. Beyond their mandated duties outlined in energy packages and network codes, RCCs serve a crucial purpose in regional harmonization, preventing duplication of efforts among Transmission System Operators (TSOs).

The evolution of regional coordination entities has been marked by shifts in scope and nomenclature. Initiatives for Regional Coordination in Europe began as voluntary endeavours long before they became legally obligatory. Initially referred to as joint offices and technical coordination centres, these voluntary initiatives paved the way for the formalization of the concept. During the European Network of Transmission System Operators for Electricity (ENTSO-E) assembly in late 2015 (ENTSO-E, 2015), TSOs endorsed an all TSO Multilateral Agreement for the Regional Security Coordination Initiative (RSCI)—a testament to TSO collaboration. The System Operation Guideline codified the term RSC and defined it as an entity owned or controlled by TSOs. The conceptualization of RCCs originally emerged as Regional Operational Centres (ROCs) in an early draft of the Electricity Regulation 2019/943, as part of the Clean Energy for All Europeans Package, a

[1] Commission Regulation (EU) 2017/1485 of 2 August 2017 establishing a guideline on electricity transmission system operation.

proposal that did not ultimately find its place in the final text. However, the enactment of the Clean Energy Package in 2019 mandated the establishment of RCCs, leading to the transformation of RSCs into RCCs by 1st July 2022 (Commission Regulation (EU) 2019/943).

5.1.1 The Role of RCCs in the EU Electricity System

TSOs are responsible for the secure and coordinated operation of Europe's electricity grid. While TSOs have a pivotal role in managing the electricity system, there are certain activities that TSO cannot effectively perform on their own, due to the cross-border nature of these activities. This is where RCCs step in to bridge the gap and facilitate crucial functions that require collaboration beyond individual TSO capabilities.

Presently, RCCs carry a wide array of responsibilities spanning system operations, system development, and market-related tasks. These include the creation of a common grid model, regional capacity calculation, regional security assessments, evaluations of TSOs' defence and restoration strategies, coordination of outage planning, and thorough post-operation and post-disturbance analyses.

These RCC services and the collaborative framework facilitated by RCCs empower TSOs to make well-informed operational decisions. With their comprehensive regional perspective of the grid, RCCs enhance operational efficiency, curbing the risks of widespread incidents like brownouts or blackouts. This results in reduced costs by optimizing the availability of transmission capacity for market participants. Furthermore, TSO coordination can yield economies of scale—enabling all TSOs to utilize a shared IT system for specific tasks. RCCs are adept at addressing region-specific issues, underscoring the fact that not all challenges require resolution at the EU or ENTSO-E levels.

Activities that RCCs are better positioned to conduct include:

- **Market Operations**: Determining available transmission capacity across borders is a complex task that involves multiple TSOs and requires harmonization. RCCs are instrumental in setting out coordinated actions for capacity calculation to ensure efficient electricity flow and market operation. According to the Electricity Regulation 2019/943, RCCs must carry out the regional sizing of reserve capacity and facilitate the regional procurement of balancing capacity.
- **Operational Security Assessment**: Ensuring the operational security of the electricity grid demands close cooperation among TSOs. RCCs enable deviations from standard procedures for operational security reasons, allowing for flexibility in grid management.
- **Data Facilitation**: Efficient data management is essential for grid operation and market functioning. Presently, there is no comprehensive financial incentive that is applicable for all TSOs to improve their data facilitation tasks. RCCs can play a role in incentivizing and optimizing data exchange among TSOs.

- **Cross-Border Coordination Support**: When activities of RCCs overlap in a System Operation Region (SOR), TSOs decide on the appropriate course of action. RCCs facilitate procedures to consult TSOs, other RCCs, and stakeholders regarding cross-border issues and coordination.

In summary, RCCs are essential in addressing the complex challenges that transcend national borders in the European electricity system. They provide the framework and coordination needed for TSOs to efficiently operate the European electricity grid.

5.1.2 Legal Background

As per Commission Regulation (EU) 2019/943, part of the Clean Energy for all Europeans Package, RSCs have been formally referred to as Regional Coordination Centres (RCCs) since July 2022. The RCCs' responsibilities and tasks are specified in Art. 37(1) of the Regulation 943/2019 (see Chap. 1 for a general introduction to the legal framework).

The Clean Energy for all Europeans Package enabled the working arrangement and governance of RCCs to evolve as an autonomous entity under direct oversight of NRAs and ACER. Now, RCCs need to be more transparent. Just like ENTSO-E has legally mandated reporting requirements, each of the RCCs has multiple reporting requirements and must publish performance indicators of its work. RCCs must develop procedures for consulting TSOs, other RCCs and stakeholders, and NRAs. RCCs must take account of the recommendations of Member States and regional forums and consult with them when needed.

The most noticeable change is the incorporation of RCCs as limited liability companies under the Company Law Directive.[2] Previously the Nordic RCC and Baltic RCC worked as a joint office of its shareholders but since 1st July 2022 both became independent companies. Other RCCs were already created as independent companies. Even though other RCCs are independent companies, shareholders have a key role in adopting statutes, appointing management boards, and other such tasks. By design, TSOs are the main shareholders of RCCs which is meant to cement the tightly-knit relationship between both parties, as they are the recipients of the legally mandated services. This creates an interesting dynamic, as RCCs are expected to act as independent companies, while TSOs remain embedded in their company structures.

In summary, the Network Codes as required by the 2009 3rd Energy Package created the precursor to RCCs with mandates for key tasks. However, it was the Clean Energy for all Europeans Package which really empowered RCCs to play a key role in the EU energy system.

[2] Directive (EU) 2017/1132 of the European Parliament and of the Council of 14 June 2017 relating to certain aspects of company law (codification).

5.2 RCC Tasks

At the ENTSO-E Assembly in late 2015, the TSOs approved and signed an all TSO Multilateral Agreement for Regional Security Coordination Initiative (RSCI), based on TSO cooperation. With this mandatory contractual framework, TSOs across Europe committed to establishing regional coordination services and setting up or appointing Regional Security Coordinators (RSCs) to perform services. This agreement also required that every TSO within the European interconnected area be part of an RSC. Later, the Network Codes not only proposed the creation of the RSCs but also legally mandated RSCs, as of 2022 referred to as RCCs following the Electricity Regulation 2019/943, to offer the following 5 services (see Fig. 5.2):

1. Coordinated Capacity Calculation—CCC
2. Coordinated Security Analysis—CSA
3. Short-Term Adequacy—STA
4. Outage Planning Coordination—OPC
5. Common Grid Model—CGM

5.2.1 Common Grid Model (CGM)

The Common Grid Model Methodology (CGMM) (ENTSO-E, 2016) defines an Individual Grid Model (IGM) as "a data set describing power system characteristics (generation, load and grid topology) and related rules to change these characteristics during capacity calculation, prepared by the responsible TSOs". These IGMs, one for each control area, are then combined into a Common

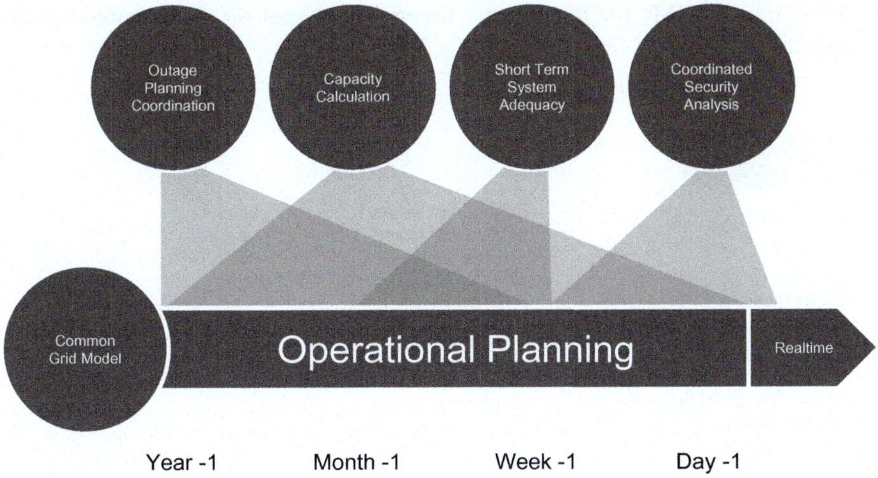

Fig. 5.2 Mandatory services provided by RCCs. *Source* Zimmerman (2019)

Grid Model (CGM) which the CGMM defines as "a Union-wide data set agreed between various TSOs describing the main characteristic of the power system [...]".

The main purpose of the CGM is to provide a common data model representing the power system in the European area, which can be used for performing further analysis through the services performed by the RCCs, in order to ensure a secure power market and security of supply.

The CGMs are created for different time horizons from a near real-time representation of the grid, to one and two days ahead, and all the way up to year-ahead models. The different time horizons have different purposes and provide input data for different services.

When the TSOs build IGMs for the different time horizons, they ensure that the model represents the real power system as closely as possible. The TSOs make sure that the equipment model is up-to-date and incorporate topological changes during the different periods by applying an outage plan.

The TSOs then use forecasts for load, consumption, net position in each bidding zone and the flow on the different DC lines to ensure that the flow in the grid is as accurate as possible.

The RCC service starts when the IGMs are received from the TSOs. The IGMs are imported to the CGM system where the RCC validates that the models follow a set of rules including the convergence behaviour of the IGMs. In case IGMs are missing, invalid or do not converge in terms of power flow calculation, the RCC will follow defined issue resolution and fallback procedures. After having merged the IGMs into a CGM, the model is again validated to ensure that it can be used for the subsequent business processes. Finally, the CGM is made available for the TSOs as well as the services at the RCCs using the CGM as input data.

The following steps are performed for the CGM process (see also Fig. 5.3):

1. TSOs create IGMs for the specific timeframes based on the harmonized data format.

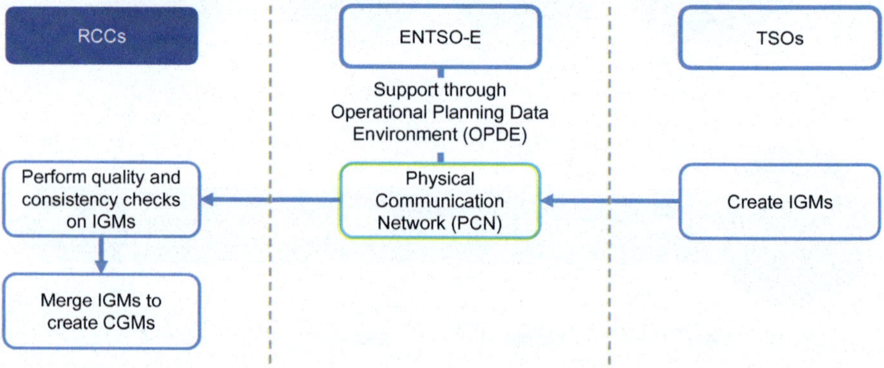

Fig. 5.3 Common grid model process by RCCs. *Source* Own illustration

2. The IGMs are exchanged via the Physical Communication Network (PCN), supported by ENTSO-E's Operational Planning Data Environment (OPDE) platform.
3. RCCs receive the IGMs from TSOs and perform quality and consistency checks.
4. RCCs merge the IGMs to create the CGMs.
5. CGMs are used for performing various RCC services such as OPC, CSA, CC, etc.

5.2.2 Capacity Calculation

RCCs have a dedicated role in the market coupling process, which can be separated into three different phases: pre-coupling, coupling, and post-coupling. Next to their responsibilities as a data intermediary, in the pre-coupling phase RCCs carry out the Capacity Calculation service for one or multiple Capacity Calculation Regions (CCRs), or parts of individual CCRs, depending on which RCC was nominated as the CCC for a respective TSO in a CCR. Capacity Calculation processes produce the foundational input for the market coupling process in the form of maximum cross-border capacity available to the market (see Chap. 3).

In the Capacity Calculation process, with the inputs provided by TSOs, RCCs calculate the amount of capacity available for trading between bidding zones at a specific timeframe, while optimizing and ensuring the grid's security. The Capacity Calculation process aims to maximize the transmission capacity offered to the market. Depending on the regional methodologies, RCCs use Net Transmission Capacities (NTC)[3]-based approach or a Flow-based (FB) methodology to calculate the amount of capacity available for trading.

As per the legal mandates (Network Code FCA[4] and CACM[5]), the CCC service is requested for the following timeframes: day-ahead, intraday, and long-term (yearly and monthly). The recently consulted capacity calculation methodologies in the balancing timeframe (BT) don't strictly foresee a role for RCCs, therefore as of 2023, it remains to be seen if RCCs will be involved.

The CCC service is carried out in three simplified steps: (1) Input data provision and validation; (2) Capacity Calculation; and (3) Capacity provision to the Market Coupling process (See Fig. 5.4).

[3] Net Transfer Capacities and Net Transmission Capacities are used interchangeably.

[4] Commission Regulation (EU) 2016/1719 of 26 September 2016 establishing a guideline on forward capacity allocation.

[5] Commission Regulation (EU) 2015/1222 of 24 July 2015 establishing a guideline on capacity allocation and congestion management.

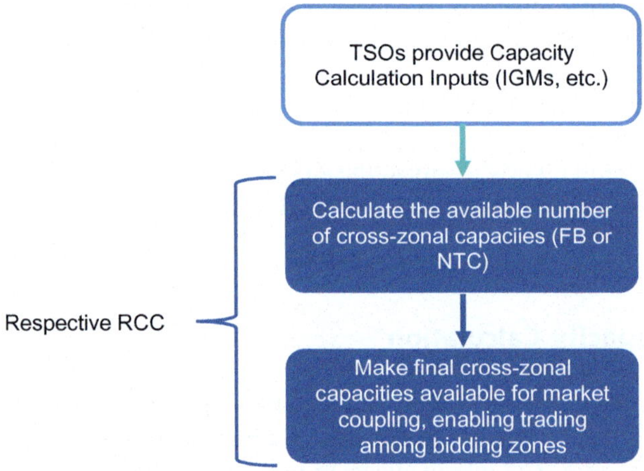

Fig. 5.4 Capacity calculation by RCCs. *Source* Own illustration

5.2.3 Coordinated Security Analysis (CSA)

The Coordinated Security Analysis (CSA) is a critical process aimed at identifying and mitigating operational security risks in the power grid. These analyses are performed for the detection of violations against the operational security of the network for both intraday and day-ahead timeframes. The RCCs perform this analysis on the merged Common Grid Model. In addition, the most efficient and cost-effective coordinated remedial actions are proposed to TSOs to address the identified constraints. All affected TSOs in the region (CCR) are included in the coordination process so that they can evaluate the impact of the proposed remedial action on their grid before agreeing to activate it. Where remedial actions agreed upon within one CCR have a significant impact on physical flows in another CCR, a cross-regional coordination process between these CCRs ensures that the violations in the overlapping zones are addressed.

The CSA service is a game-changer to maintain the security of supply in a meshed European Grid. With the CSA process, the operators in the national control room get an accurate assessment of the grid incidents and events in neighbouring zones.

The System Operation Guideline and its relevant methodology on Regional Operational Security Coordination (ROSC) provides relevant guidelines for the CSA process in each region.[6]

The following shows how CSA is performed (see also Fig. 5.5):

[6] Commission Regulation (EU) 2017/1485 of 2 August 2017 establishing a guideline on electricity transmission system operation; Article 76.1—Common provisions for regional operational security coordination.

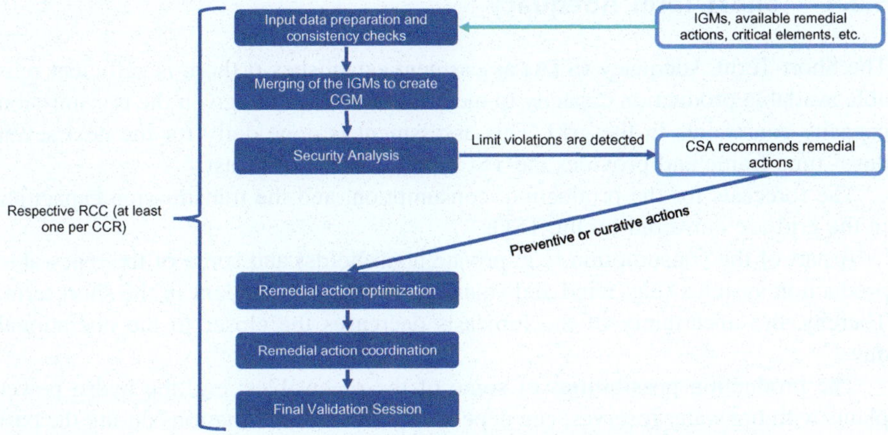

Fig. 5.5 Coordinated security analysis process by RCCs. *Source* Own illustration

1. **Data Exchange (i.e. data provision, consolidation, and validation)**: The CSA process involves the sharing of information related to the power grid, including Common Grid Models (CGMs), contingencies, power transfer corridors (PTC), limits, and system integrity protection schemes (SIPS). This data is essential for conducting a meaningful security analysis.
2. **Security Analysis**: The heart of CSA lies in the security analysis itself. It involves simulating various scenarios, including fault situations where one or more grid elements are connected or disconnected. The focus is on identifying limit violations within the power system.
3. **In Case Security Analysis Yields Limit Violations**:
    (a) **Remedial Action Optimization—Preventive and Curative**: If limit violations are detected, CSA recommends remedial actions. These actions can be either preventive or curative. Preventive actions aim to proactively address potential issues, while curative actions are taken to mitigate ongoing or impending problems in the grid.
    (b) **Remedial Action Coordination—Exchange of Remedial Actions**: CSA involves the exchange of information regarding the suggested remedial actions between relevant parties. Coordination is key, and this information sharing ensures that the necessary measures are taken to enhance grid security.
    (c) **Final Validation Session**: Following the Remedial Action Optimization and Coordination, the results shall follow a final validation by all relevant parties.

By performing these steps, CSA helps TSOs, and other stakeholders, proactively assess and enhance the security of the power grid, ensuring its reliability and stability in the face of various operational challenges.

5.2.4 Short-Term Adequacy

The Short-Term Adequacy (STA) assessment establishes if there is sufficient reliable available production capacity to meet the consumption, given the transmission capacity constraints in the grid. This assessment is done daily for the next seven days' time frame and provides the TSOs with an STA forecast.

The forecasts for the production, consumption, and the transmission capacities in the grid are delivered by the TSOs.

A part of the consumption (e.g. private households) and some of the renewable production systems (e.g. wind and solar) are weather dependent in the short term. Thereby, the uncertainty of the forecasts decreases the closer to the operational day.

The production possibilities of some of the renewables, e.g. the hydro power plants with big water reserves, can depend on the weather situations during the past year. If it has been a dry period, the reservoir capacity can be low in a country or part of a country.

One of the inputs to the adequacy assessment is the capacities in the grid, and the STA assessment is thereby connected to the Outage Planning Coordination (OPC) service and the planned availability of the grid. This cross-service dependency is incorporated into the STA service.

On a European level, the STA assessment is carried out daily as well as for the next seven days. If any adequacy issues are seen, procedures to coordinate and, if possible, reduce the consequences of the lack of adequacy exist.

The STA service is described in System Operation Guideline in Article 81 and Articles 104–107. It is illustrated in Fig. 5.6.

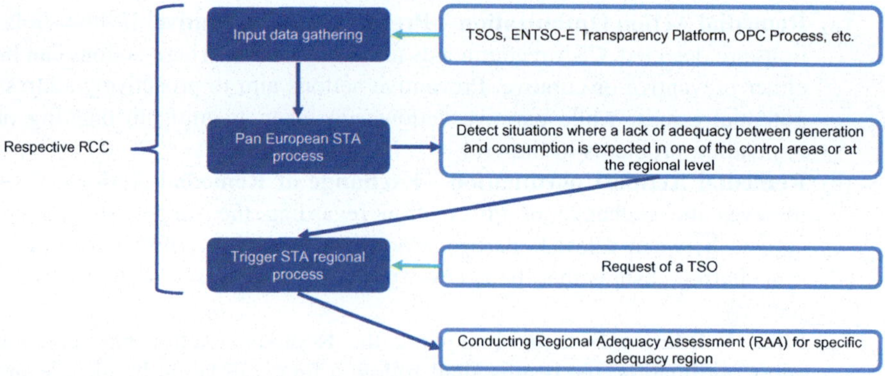

Fig. 5.6 Short-term adequacy process. *Source* Own illustration

5.2.5 Outage Planning Coordination (OPC)

Planning each individual outage is the TSOs responsibility. However, since local actions in one area can have significant and possibly problematic effects on neighbouring areas, there is a strong need to share the information among all impacted stakeholders and to assess the overall compatibility of these outages.

This is the role of the regulated **Outage Planning Coordination (OPC)** service performed by RCCs to preserve the reliability of the interconnected power system.

This OPC service aims to identify Tie-Lines Inconsistencies (TLI) at a pan-European level but also Outage Planning Incompatibilities (OPI) per region. The OPC service then proposes solutions to TSOs to relieve these incompatibilities and coordinates remedial actions with the other RCCs. In terms of solutions, the OPC service can propose at least non-costly remedial actions and adaptations of availability and outages' planning (firstly on grid elements, and secondly on other elements if no solution is available).

During the OPI process, RCCs determine whether the outage planning of TSOs is secure. In case incompatibilities between relevant assets (grid elements, generators, or loads whose availability status has cross-border impact) are detected, RCCs shall propose remedial actions and perform a security assessment to verify whether the grid is secure after the remedial actions have been applied.

This service is requested for the year-ahead and week-ahead timeframes. Updates are done up to one week-ahead based on TSO requests for planning modification or significant changes to the expected operational conditions.

The following steps are performed (see also Fig. 5.7):

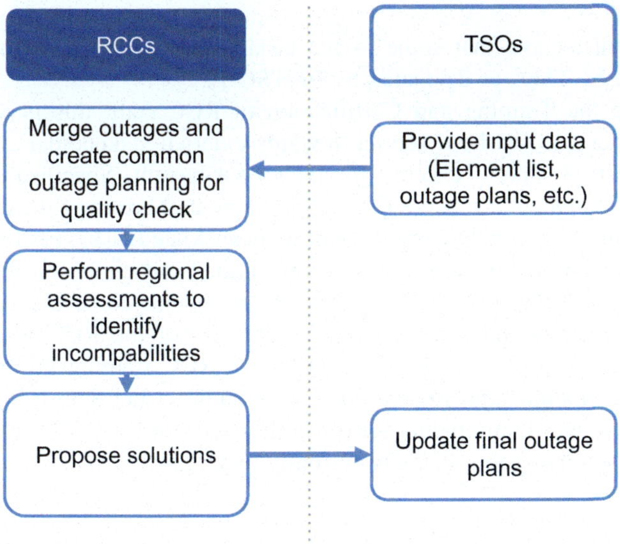

Fig. 5.7 Outage planning coordination process. *Source* Own illustration

1. TSOs provide input data (Element list, Outage plans, etc.).
2. RCCs merge the outages to create a common outage planning snapshot for quality check.
3. RCCs perform regional assessments to identify incompatibilities.
4. RCCs propose solutions (topology changes, change in outage plans, etc.) in order to relieve the identified incompatibilities.
5. Based on the RCC proposals, TSOs update the final outage plans.

5.2.6 Other RCC Tasks and Responsibilities

Besides the five aforementioned tasks as per the Third Energy Package, the Clean Energy Package (CEP) introduced several additional tasks for RCCs to perform:

- Training and certification of RCCs staff.
- Coordination and optimization of regional restoration.
- Post-operation and post-disturbance analysis and reporting.
- Regional sizing of reserve capacity.
- Regional procurement of balancing capacity.
- Optimization of inter-TSOs settlements.
- Identification of regional electricity crisis scenarios.
- Seasonal adequacy assessments.
- Maximum entry capacity available for the participation of foreign capacity in capacity mechanisms.
- Supporting TSOs in the identification of system needs (in the context of TYNDP).

It is important to note that some of the tasks will only be performed by RCCs upon request by TSOs or delegated by ENTSO-E.

Regarding the Training and Certification of RCC staff, this is a unique task as it is not designed to be a service towards external stakeholders. Instead, it is a task that RCCs should execute for their own benefit. Specifically, RCCs are expected to implement a scheme for training their staff ensuring the continued and evolving qualifications necessary to fulfil their mandated tasks. ENTSO-E submitted a proposal to ACER, who subsequently issued a decision on the underlying methodology in 2022. This decision contains provisions on several related topics, including explaining the roles and responsibilities of the RCC training coordinator, setting key aspects to be covered by the RCC training programme, RCC joint training modules and joint training programmes and defining the organization, structure, and requirements for the certification of staff. The implementation of these provisions by the RCCs is currently underway.

The task of supporting the consistency assessment of TSOs' system defence plans and restoration plans was already introduced by Article 6(3) of the Emergency and Restoration Network Code (NC ER[7]). The TSOs define measures and RCCs assess them for consistency within and between neighbouring synchronous areas. These relate to inter-TSO assistance and coordination in emergency state, frequency management procedures, assistance for active power procedure and top-down re-energization strategy. This task is already established.

In 2022, a proposal from ENTSO-E with a decision from ACER was issued also for the RCC task of post-disturbance analysis. This prescribes the involvement of RCCs in the drafting of the ex-post incident reports if a specific regional relevant threshold was met. The implementation of this task by the RCCs has been finalized.

For the tasks of Regional sizing and Procurement of Balancing capacities, ACER approved the methodologies on 19th July 2023 and requested ENTSO-E, TSOs, and RCC to initiate the implementation. The Regional Sizing task requires RCCs to assess short-term availability of sharing reserves and determine the minimum reserve capacity at the relevant SOR, every year. The scope of the regional procurement of balancing capacity by RCCs aligns with the role of TSO in the procurement of balancing capacity. The RCCs' facilitation role in the procurement of balancing capacity at the regional level consists of the calculation of the availability of non-contracted balancing energy bids at the European platforms per type of reserve, direction, and validity period.

The Maximum Entry Capacity task is related to capacity mechanisms, which are (in the current regulatory framework) one instrument that can be introduced in case of adequacy concerns. It is thus tied to the adequacy assessments performed at European (ERAA) or national (NRAA) level. Computing maximum entry capacity available for participation of foreign capacity allows for cross-border collaboration. The underlying methodologies for this task have been drafted, and the implementation by the RCCs is currently underway.

At the time of writing, the following tasks do not yet have a published methodology: supporting restoration, inter-TSO settlement, identification of regional electricity crisis scenarios, seasonal adequacy assessments, and supporting the identification of needs for new infrastructures.

The Clean Energy Package introduced new concepts regarding how RCCs perform their appointed tasks, including coordinated actions, recommendations, and deviation from coordinated actions. For tasks such as CC, OPC, or CSA, the RCC issues coordinated actions that the TSOs in the region implement. However, if a coordinated action would lead to a violation of the operational security limits, the respective TSO shall report the detailed reasons to the RCC and the other TSOs within the region. In turn, the RCC assesses the impact and may propose a different set of coordinated actions.

---

[7] Commission Regulation (EU) 2017/2196 of 24 November 2017 establishing a network code on electricity emergency and restoration.

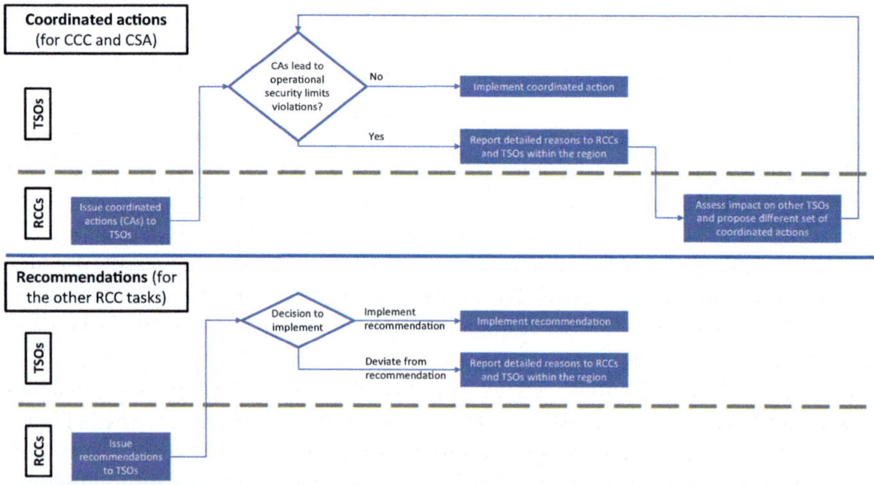

Fig. 5.8 The simplified process of coordinated actions and recommendations issued by RCCs. *Source* Author's own illustration

For the other tasks defined in Article 37(1) of the Electricity Regulation, the RCC issues recommendations to the TSOs. In this case, TSOs decide whether to implement the recommendation. If a TSO decides to deviate from the recommendation, it has to report the detailed reasons to the RCC and the other TSOs within the region.

The process of issuing coordinated actions and recommendations is illustrated in Fig. 5.8. For any coordinated action or recommendation issued by the RCC, one or more TSOs within the region can request a review, after which the RCC shall either confirm or modify the measure.

In the case of capacity calculation, a particular type of coordinated action is defined: the coordinated action to reduce the cross-zonal capacities. RCCs may propose this action if the minimum capacity values would lead to violations of operational security limits. This concept is similar to the "coordinated validation" step from the Core CCR day-ahead capacity calculation methodology, which at the time of writing, is yet to be implemented. However, other capacity calculation methodologies may not have been drafted and processes were not designed to accommodate this step.

5.3 RCCs in Market Coupling

RCCs perform important roles in the pan-European market coupling as well as on system operation and capacity calculation on regional level.

5.3.1 RCCs as CCCs

RCCs have a particular role in European Market Coupling, as they prepare the system-operation-related input (i.e. execute the capacity calculation) to enable the European Market Coupling for both the Day-Ahead Market, as well as the Intraday Market. Most notably, the output generated by RCCs during pre-coupling determines how much cross-border capacity can be made available to the European internal energy market(s) (see Fig. 5.9 for an overview).

Of course, vice-versa, individual RCC services and tasks are dependent on Market Coupling processes. Following the closure of Market Coupling in the Day-Ahead Market, the results are used as input data for the CSA service. With the pending go-live of Intraday Auctions (IDAs) in the EU in June 2024, the IDAs might also be used as input data for CSA in the future.

In their role as CCC, an RCC must interact with TSOs within the CCR, other RCCs, JAO, or with NEMOs. Within a CCR that has more than one RCC nominated as CCC, the involved RCCs can perform the role of CCC either on a rotational basis or by splitting of the tasks. The RCC as CCC also interacts with the TSOs from the CCR throughout the pre-coupling phase, or with the NEMOs when getting closer to the market coupling phase. These interactions mostly relate to the usual flow of information, such as receiving individual inputs from TSOs and providing them with the processed files, or sending input files to NEMOs. Additional interactions are essential in so-called non-happy flows, i.e. when something goes wrong in the process, such as a delay in a process step or a missing/incorrect file. In these situations, the RCCs (Shift Operator in charge) as CCC follow predefined fallback procedures to mitigate the issue and prevent it from

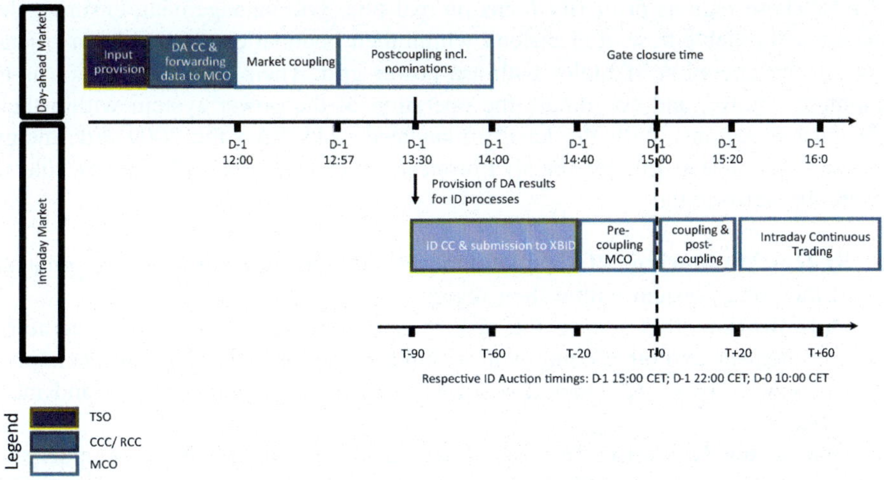

Fig. 5.9 Simplified overview of RCC involvement in market coupling activities. *Source* Author's own illustration

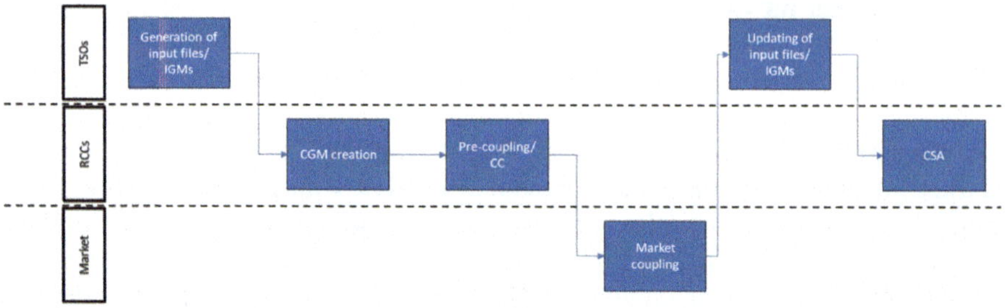

Fig. 5.10 Simplified overview of RCC dependent processes related to market coupling. *Source* Author's own illustration

causing a failure of the whole process. Figure 5.10 shows a simplified overview of the RCC dependent processes related to market coupling.

5.3.2 Role of RCCs Within SOR and CCR

In the context of the European electricity market(s), there exists a distinction between System Operation Regions (SORs) and Capacity Calculation Regions CCRs, each serving a unique role in the efficient functioning of the electricity grid.

5.3.2.1 System Operation Regions

System Operation Regions (SORs) are geographical areas where the daily operational coordination of the electricity system is managed by a TSO or a group of TSOs. These regions primarily focus on real-time grid management, ensuring the secure and reliable flow of electricity within their boundaries. RCCs play a crucial role in the operation of SORs within a power grid. Their primary function is to monitor, control, and coordinate the operation of the power system within their designated regions. With the legally mandated tasks from the 2009 3rd Energy Package and Clean Energy for all Europeans Package, RCCs play a key role in SORs by performing:

- **System Monitoring**: RCCs continuously monitor the status and performance of the power system within their region.
- **Optimization**: RCCs have the task to optimize the operation of generators, transformers, and other equipment within their region. They can suggest adjusting power flows and generation levels to maintain system stability and meet demand.
- **Emergency Response**: In case of contingencies, disturbances, or emergencies, RCCs are responsible for supporting appropriate actions to restore system stability.

- **Communication Hub**: RCCs serve as communication hubs, facilitating coordination and information exchange between different entities within their region. This includes communication with local power utilities, generators, and neighbouring control centres.
- **Coordination with other RCCs**: Since a power grid is often divided into multiple SORs, RCCs collaborate with each other to ensure coordinated operation across regions. This involves sharing information about power transfers between regions and addressing any issues that may impact the overall system.
- **Grid Planning and Analysis**: RCCs are involved in long-term planning and analysis of the power system within their region. They assess future demand trends, plan for infrastructure upgrades, and ensure the reliability and adequacy of the power supply.

In essence, RCCs act as the nerve centre for managing coordination of the power grid within their designated regions, contributing to overall system reliability and efficiency. It is important to mention, however, that RCCs do not take an active role in controlling the grid, but rather as facilitating parties to support stable grid operation.

On the other hand, CCRs encompass areas where capacity calculation methodologies are harmonized to optimize the allocation of cross-border transmission capacity. CCRs facilitate the efficient use of interconnection infrastructure, enabling electricity to flow seamlessly across borders.

5.3.2.2 Capacity Calculation Regions

Capacity Calculation Regions (CCRs) define the geographical areas where coordinated cross-border capacity calculation, capacity allocation, and congestion management are applied (CACM Regulation, Article 2 defines CCRs). At least one RCC shall act as a Capacity Calculation agent for each CCR. RCCs can provide their services to multiple CCRs, which on such a regional level encompass the CCC and CSA services.

In Fig. 5.11 it is easily discernible that, while CCRs have a focus on facilitating market operations and SORs on system operations, they share a relevant geographical connection. RCCs are playing a pivotal role to ensure this connection.

5.4 Outlook

RCCs have come a long way since their voluntary start. Now as a legal entity, they play a key role in maintaining the security of supply; supporting the stable operation of the European grid; and ensuring the functioning of the European internal energy market. Apart from their legally mandated tasks, RCCs also work to ensure harmonization and to avoid duplicity of activities among TSOs. The current focus for the evolution of RCCs is on maintaining and improving the quality and stability of the current services, as well as implementing the remainder of the legally mandated tasks.

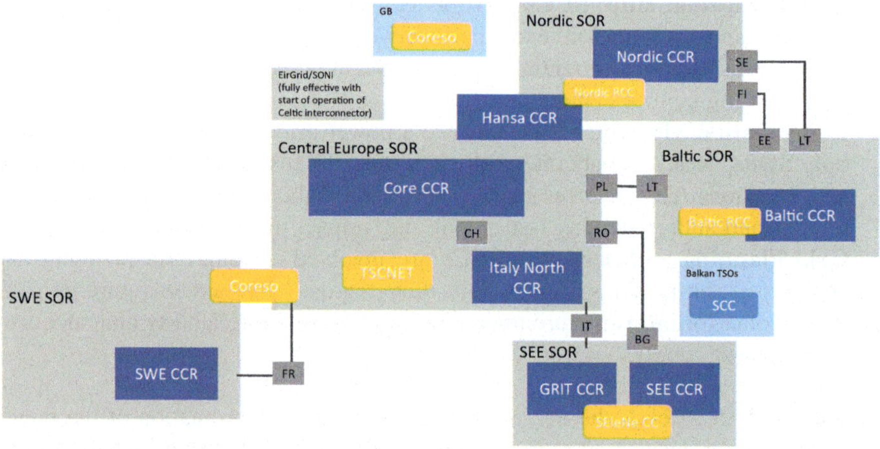

Fig. 5.11 Overview of SORs and CCRs in the EU. *Source* ENTSO-E (2024)

The following are the key RCC themes to look upon in future:

1. **On-Time Implementation of Legal Obligations**: RCC obligations from the 2009 3rd Energy Package are not fully implemented yet; however, obligations for the 4th Energy Package (Clean Energy for all Europeans Package) are already added on top. RCCs have a challenge in hand to develop IT tools, processes, and practices to fulfil obligations as per the legal deadlines.
2. **Navigating a Complex Regulatory Environment**: With periodic updates of Network Codes (CACM 2.0, SO GL, etc.) and capacity calculation methodologies, the introduction of consultations and new reporting obligations, RCCs must ensure ongoing compliance with the complex regulatory environment. They shall also ensure the goal to carry out tasks where their regionalization brings added value compared to tasks performed at national level. However, in the current legal and regulatory framework, RCCs are not the ones "holding the pen" in making proposals for the tasks that they perform. To achieve this, RCCs could proactively engage stakeholders and raise in relevant forums which areas of improvement or added value which could result from their involvement. The introduction of new concepts such as coordinated actions, recommendations, and deviation from coordinated actions further add to the challenge, as existing processes and methodologies would have to be adapted. When RCCs progress to the extent of having their services fully established and running consistently, it might prove to be a worthwhile effort to challenge this paradigm, for instance, in such a way where RCCs themselves might be conceptualizing new tasks to support managing the grid more efficiently.
3. **Data Use Strategy**: Thanks to more than a dozen of legally mandated tasks, RCCs collect an enormous amount of grid and market data. There is a huge potential for leveraging this data to enhance market transparency and develop

insights and intelligence to support different stakeholders such as TSOs, ENTSO-E, and others.

4. **Attracting Talent**: The skills and expertise required in RCCs are scarce in the job market. As a relatively new entity compared with TSOs or other utility companies, RCCs have a real task to build a brand and create new strategies to attract and retain talent.
5. **Streamlining ICT Strategy**: Information and Communications Technology (ICT) Tools and Infrastructures of RCCs are interconnected with multiple stakeholders (including TSOs in Europe, other RCCs, ENTSO-E, Joint Allocation Officer (JAO). It is important for RCCs to streamline the ICT strategy and infrastructure to avoid double and/or redundant investments. RCCs are expected to utilize initiatives like ENTSO-E ICT Committee or other regional IT working platforms to streamline investments and development, thus simultaneously ensuring both the current and future functionalities of each platform and interoperability.
6. **RCC Specializations**: Currently, RCCs are meant to implement all legally mandated services. However, from an operational efficiency point of view, the set-up of all RCCs providing the same services may be challenged. As already mentioned, all RCCs implemented task splitting or rotational principles in service provision in certain regions, due to TSOs being allowed to request services from one or multiple RCCs. Therefore, it remains to be seen if certain unique specializations manifest at individual RCCs, for which they might be considered the preferred service providers for specific tasks in the future.

Bibliography

Commission Regulation (EU). (2015). 2015/122 establishing a guideline on capacity allocation and congestion management. *Official Journal* L197/1, 24.

Commission Regulation (EU). (2016). 2016/1719 establishing a guideline on forward capacity allocation. *Official Journal* L259/1, 42.

Commission Regulation (EU). (2017). 2017/1485 establishing a guideline on electricity transmission system operation. *Official Journal* L220/1, 1.

Commission Regulation (EU). (2019). 2019/943 on the internal market for electricity. *Official Journal* L158/54, 54.

ENTSO-E. (2015). *ENTSO-E assembly meeting minutes*. [online]. Available at: https://eepublicdownloads.azureedge.net/clean-documents/assembly/2015/december/151210_AS_Minutes_vFinal.pdf. Accessed 7 March 2024.

European Network of Transmission System Operators for Electricity. (2016). *All TSOs' proposal for a common grid model methodology in accordance with Article 17 of Commission Regulation (EU) 2015/1222 of 24 July 2015 establishing a guideline on capacity allocation and congestion management*. Available at: https://www.acer.europa.eu/sites/default/files/documents/en/Electricity/MARKET-CODES/CAPACITY-ALLOCATION-AND-CONGESTION-MANAGEMENT/14%20CGM/Action%203%20-%20CGM%20approved%20proposal.pdf. Accessed 15 May 2024.

European Network of Transmission System Operators for Electricity. (2024). *ENTSO-E Annual Work Programme 2024 Edition—ENTSO-E's work on legal mandates.* ENTSO-E Annual Work Programme [online]. Available at: https://eepublicdownloads.blob.core.windows.net/public-cdn-container/clean-documents/General%20ENTSO-E%20documents/ENTSO-E%20Work%20Programme/ENTSO-E_AWP2024_231106.pdf. Accessed 15 May 2024.

Zimmerman, U. (2019). *RSC role in coordinated capacity calculation.* Available at: https://www.google.com/url?sa=i&url=https%3A%2F%2Fwww.energy-community.org%2Fdam%2Fjcr%3A893dc687-0de3-427b-bbce-106fc659ed04%2FECRBREG022019_TSCNET.pdf&psig=AOvVaw0XDog0RDD6U4UWiqhbUaWL&ust=1715852163128000&source=images&cd=vfe&opi=89978449&ved=0CBUQ3YkBahcKEwig5pLLrY-GAxUAAAAAHQAAAAAQBA. Accessed June 2024.

Jakub Władysław Glegoła (until June 2024) is a Team Lead at the Accenture Industry X Division where he focuses on Energy Market and European Regulation topics. He helps energy companies maximize the opportunities provided by market regulations, enabling them to implement new business models and digital capabilities to decarbonize the energy system. Over the past years, he was heavily committed to working with RCCs, both in accompanying their transition from RSCs to RCCs and facilitating the implementation of their overarching responsibilities stemming from various network codes. In his current work, he remains actively engaged with RCCs on CCR level, as well as in facilitating the progression of European Market coupling for various timeframes. Having acquired experience in both Market and System operations on European level, he is dedicated to facilitating the further integration of the European internal market for electricity. He holds a master's degree in Industrial Engineering and Management from the Technical University of Berlin.

Day-Ahead and Intraday Auctions Market Coupling Algorithm

6

Mehdi Madani and Martin Starnberger

Abstract

This chapter presents the high-level functioning of the market coupling algorithm used in the Single Day-Ahead Coupling (SDAC) and for Intraday Auctions (IDA) of the Single Intraday Coupling (SIDC). The fundamentals underlying market coupling calculations, lying at the intersection of power system economics and mathematical optimization, are reviewed and illustrated by examples that do not require any specific technical background. The topics covered include the equivalence between welfare maximization and competitive equilibrium in markets coupled via flow-based or available transfer capacity network representations. Finally, it is briefly highlighted how the energy transition may impact the design and functioning of the market clearing algorithm in the near future in Europe, as well as in both Japan and India.

6.1 Introduction

The fundamentals of market coupling calculations are at the intersection of power system economics and mathematical optimization. They can be traced back to the seminal work of Paul Samuelson (Nobel Memorial Prize in Economic Sciences 1970), who first established the equivalence between the maximization of the welfare in coupled markets (described as a linear optimization problem), and

M. Madani (✉) · M. Starnberger
N-SIDE, Ottignies-Louvain-la-Neuve, Belgium
e-mail: mma@n-side.com

M. Starnberger
e-mail: mst@n-side.com

A. Estermann et al. (eds.), *European Electricity Market Coupling*, Future of Energy,
https://doi.org/10.1007/978-3-031-86315-8_6

Table 6.1 Orders by market participants

Bids	Quantity (MW)	Limit price (€/MWh)
A—buy bid	20	300
B—buy bid	11	200
C—sell bid	15	40
D—sell bid	10	100

the notion of "spatial price equilibrium", thanks to linear optimization duality (Samuelson, 1952).

We cover these elements via examples illustrating the key concepts and which do not require any specific technical background. Some technical elements and references are also provided for professionals willing to have a deeper look into the underlying concepts.

6.2 Welfare Maximization, Competitive Equilibria, and Marginal Pricing Principles

Determining market prices based on the merit order and the intersection of the demand and supply curves is a classic economic principle (Pindyck & Rubinfeld, 2017). We review here how the "marginal pricing principle" is derived from the concept of competitive market equilibrium, equivalent in a precise sense, to welfare maximization (under specific conditions such as the absence of so-called "non-convexities" introduced, for instance, by block orders, and on which we come back in Sect. 6.5)[1] as seen in (Coase, 1946).

This equivalence between welfare maximization and competitive equilibria holds in very general contexts (for instance, with advanced hybrid flow-based market coupling) and is at the core of market clearing algorithms.

We first illustrate the concept based on a simple example. Market orders in the example are given in Table 6.1.

Given these market orders, we can depict the supply and demand curves on Fig. 6.1. The supply curve is here an increasing step curve and the demand curve is a decreasing step curve. We can see that the demand and supply curves cross at the so-called market clearing price; here at a value of 200 €/MWh.

The clearing price (optimal price supporting a competitive equilibrium) is here 200€/MWh because otherwise if the price is below 200€/MWh (for instance, 100€/MWh), the market participant B would prefer that its demand bid be fully

[1] A classic paper in the early general economic literature on marginal pricing and the challenges appearing in the presence of "decreasing returns to scale" (for instance, due to start-up costs or other sources of non-convexities) is (Coase, 1946).

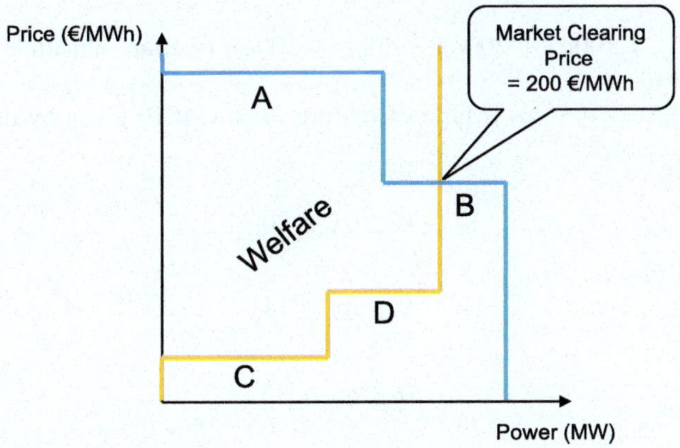

Fig. 6.1 Welfare maximization example. *Source* Author's own illustration

matched instead of being partially accepted, while if the price is above 200€/MWh, it would prefer not to be matched at all. In the context of "inelastic demand" (that is to say, which is price-insensitive), similar considerations would essentially lead to the rule that the last accepted MW of supply needed to meet demand is setting the price (leaving aside considerations where there are so-called "price indeterminacies" as discussed in Sect. 6.7).

Under normal competitive conditions without strategic bidding (assuming no market power, i.e. that bidders truthfully bid their utility/cost), the limit price of demand orders represents the marginal utility of the demand; that is, the marginal value for the buyer of a marginal increase of consumption (or procurement of energy). On the supply side, the limit price corresponds to the marginal cost of supply; that is, the marginal cost increase resulting from a marginal increase of supplied energy.

The welfare maximization problem takes these inputs and finds the matching of bids that maximizes the welfare: the total utility on the demand side minus the costs of production. Thus, the welfare corresponds to the area between the demand curve and the supply curves going from 0 up to the matched power given by the point where the curves cross. Note that in a set-up where the demand is fully price-insensitive, this corresponds to the problem of serving a fixed load at minimum cost.

We can formulate the welfare maximization problem as a so-called linear optimization problem having a linear objective function corresponding to the welfare and linear constraints corresponding to the balance condition (the naming q of the variables stand for "quantity").

$\max_{q_A, q_B, q_C, q_D} 300q_A + 200q_B - 40q_C - 100q_D$ (welfare objective function)
subject to:

$q_A + q_B = q_C + q_D$ [Market Clearing Price (MCP) given by the shadow price]

$$0 \leq q_A \leq 20$$

$$0 \leq q_B \leq 11$$

$$0 \leq q_C \leq 15$$

$$0 \leq q_D \leq 10$$

Optimal Solution:

$$q_A^* = 20 \text{ MWh}, \quad q_B^* = 5 \text{ MWh}, \quad q_C^* = 15 \text{ MWh}, \quad \text{and} \quad q_D^* = 10 \text{ MWh}.$$

When there are hundreds of thousands of market orders in different bidding zones, finding the optimal solution to the welfare maximization requires to solve a large-scale linear optimization problem (or a quadratic problem when so-called interpolated curves, also called piecewise linear bid curves, are in scope), a job done by a dedicated optimization software library called a "solver". Top commercial solvers are, for example, IBM CPLEX, FICO XPRESS, or GUROBI.

If we now drop the balance condition $q_A + q_B - q_C - q_D = 0$, but subtract from the welfare objective the left-hand side of this balance condition multiplied by some price 'p', one gets:

$$\max_{q_A, q_B, q_C, q_D} 300q_A + 200q_B - 40q_C - 100q_D - p(q_A + q_B - q_C - q_D)$$

subject to the remaining constraints:

$$0 \leq q_A \leq 20,$$

$$0 \leq q_B \leq 11,$$

$$0 \leq q_C \leq 15,$$

$$0 \leq q_D \leq 10$$

This directly leads after rearrangement to an individual profit maximization problem per market participant:

$$\begin{aligned}&\max_{0 \leq q_A \leq 20}(300 - p)q_A \text{ for } A;\\&\quad \max_{0 \leq q_B \leq 11}(200 - p)q_B \text{ for } B;\\&\quad \max_{0 \leq q_C \leq 15}(p - 40)q_C \text{ for } C;\\&\quad \max_{0 \leq q_D \leq 10}(p - 100)q_D \text{ for } D\end{aligned}$$

These problems have a direct interpretation. For example, for the market participant C, it consists in choosing the optimal production given that the market price is p and its marginal cost is 40 €/MWh.

In a situation of competitive equilibrium, the matched quantities, which satisfy the power balance conditions by construction, are also optimal for the market participants facing the market price p and seeking to optimize their profits. Thus, there is a competitive equilibrium if demand and supply are equal, and considering this price p as a given, the consumptions and productions are the "best responses" to the price, in the sense of solving the individual profit maximization problems obtained above after rearrangement. If this holds true, it means that given the market prices, no market participant has an incentive to prefer another set of matched quantities than the one centrally determined by the welfare maximization problem.

The problem of finding an optimal price minimizing the sum of these profits obtained as "best response to the price" (neglecting balance conditions during this process) is called in mathematical optimization theory a (Lagrangian) dual of the welfare maximization problem, in which the balance conditions have been "dualized". It takes the form:

$$\mathrm{LD} = \min_p \left[\begin{array}{l} max_{0 \leq q_A \leq 20, 0 \leq q_B \leq 11, 0 \leq q_C \leq 15, 0 \leq q_D \leq 10}(300 - p)q_A + (200 - p)q_B \\ +(p - 40)q_C + (p - 100)q_D \end{array}\right]$$

A fundamental result in mathematical optimization is that, perhaps surprisingly, for an optimal price 'p^*' (that will be the market price), the corresponding sum of these individual profits will be equal to the maximum welfare obtained by solving the initial welfare maximization problem. We say that there is no duality gap, a fact called "strong duality" in mathematical optimization. More importantly, the welfare maximizing solution (given by $q_A^* = 20$ MWh, $q_B^* = 5$ MWh, $q_C^* = 15$ MWh, and$q_D^* = 10$MWh) provides for each market participant an optimal solution to its profit maximization problem if it considers the optimal market price 'p^*' as a given.

A competitive equilibrium can hence be found by optimizing the welfare to obtain quantities satisfying balance conditions and optimizing a dual problem to obtain market prices. Theory then guarantees that, under certain conditions, the matched quantities are optimal for market participants; that is, we reach a competitive equilibrium.[2]

These principles directly extend to set-ups with multiple bidding zones and multiple time periods. When considering multiple bidding zones, specific market participants (Transmission System Operators) offer to transport energy from one location to another thanks to transmission resources such as lines and other critical network elements. Time periods could, for instance, have hourly or quarter-hourly time-resolution. Note that specific line or bid constraints—like ramp constraints on lines or block bids spanning multiple periods—can "couple the periods", in which case it is not possible to solve the problem as a collection of independent sub-problems per period. The next section addresses the topic of market coupling where multiple connected bidding zones are considered.

6.3 Market Coupling: Multiple Bidding Zones, Spatial Price Equilibria, and Congestion Rents

The notions covered in the previous section extend naturally to the context of market coupling, where electricity is traded across multiple locations such as bidding zones. Market coupling refers to the fact that multiple markets are integrated together by considering transmission of electricity between these markets, corresponding in SDAC to a situation where cross-border trades can take place. Note that some countries such as Italy and Sweden contain multiple bidding zones and borders (meaning in our context bidding zone borders).

We first discuss in this section market coupling from an economics standpoint. The fundamental underlying economic concept is the notion of spatial price equilibrium. The algorithmic approaches for calculating market coupling results are like those previously evoked: one needs to solve linear/quadratic optimization problems and their "dual" (which are also linear/quadratic optimization problems) relying on a "solver" (e.g. IBM CPLEX). In practice, however, multiple volumes and/or market prices can be compatible with the requirements, and some additional elements are required to select a single market coupling solution among multiple possible alternatives according to some predefined criterion ensuring uniqueness of the final solution. This topic of indeterminacies is further discussed in Sect. 6.7.

---

[2] A technical exposition of the relation between "strong duality" and competitive equilibrium in a general setup is given, for instance, in the Appendix A in (Torbaghan, et al., 2021). Similar considerations can be tracked back as far as (Rockefeller, 1970).

Table 6.2 Bids by market participants in bidding zones A and B

Bids	Quantity (MW)	Limit price (€/MWh)	Location
A—buy bid	20	300	Area B
B—buy bid	11	200	Area B
C—sell bid	15	40	Area A
D—sell bid	10	100	Area A

Let us consider the same bids as in the simple example of Sect. 6.2, but now located in two separate bidding zones, Area A and Area B, connected by a transmission line (Table 6.2).

We have to adapt the linear optimization problem accordingly. In each bidding zone, the sum of accepted demand minus the sum of accepted supply defines the net position of the bidding zone. Thus, we use here the convention that an importing bidding zone has a positive net position and an exporting bidding area has a negative net position. The flows on the lines between bidding zones are modelled using flow variables which are restricted by the capacity constraints of the lines.[3] To ensure the balancing of the bidding zone, we have to ensure that imported energy minus exported energy (i.e. the sum of the ingoing flows minus the sum of the outgoing flows) equals the net position of the bidding zone.

$$\max_{q_A, q_B, q_C, q_D} 300q_A + 200q_B - 40q_C - 100q_D$$

$$(1) q_A + q_B = \text{NetPos}_{\text{Area}B}$$

$$\left[\text{The shadow price is the MCP in Area } B\right]$$

$$(2) - q_C + q_D = \text{NetPos}_{\text{Area}A}$$

$$\left[\text{The shadow price is the MCP in Area } A\right]$$

$$(3) \text{NetPos}_{\text{Area}A} = \text{flow}_{B \to A} - \text{flow}_{A \to B}$$

$$(4) \text{NetPos}_{\text{Area}B} = \text{flow}_{A \to B} - \text{flow}_{B \to A}$$

$$(5) \text{flow}_{A \to B} \le \text{CAP}_{A \to B}, \text{flow}_{B \to A} \le \text{CAP}_{B \to A}$$

[3] Note that using one flow variable for each direction enables one to model losses on the lines. In our simple case here, we could have instead used a single variable for both directions.

$$0 \leq q_A \leq 20,\ 0 \leq q_B \leq 11,\ 0 \leq q_C \leq 15,\ 0 \leq q_D \leq 10$$

Intuitively, if the transmission capacity is sufficiently large, both coupled markets will behave as a single market and the outcome will be identical to the outcome described in Sect. 6.2. Now suppose the transmission line's capacity is 10 MW (case 2 on Fig. 6.2). In that case, bid A can only be partially accepted, exporting 10 MWh to Area B, leading to a market price of 40 €/MWh in Area A (otherwise, the acceptance of bid A is not optimal for the market participant). For similar reasons, the market price in Area B is 300 €/MWh. The price spread corresponds to the marginal value of transmission capacity: an additional MW of transmission capacity would enable to generate a welfare increase of (300–40) = 260 €. The revenue of the transmission system operator is 10 x (300–40) = 2600€ and is called the "congestion income" (the "congestion rent" corresponds to the congestion incomes minus network costs—for instance, tariffs—included in the market model, which are absent here).

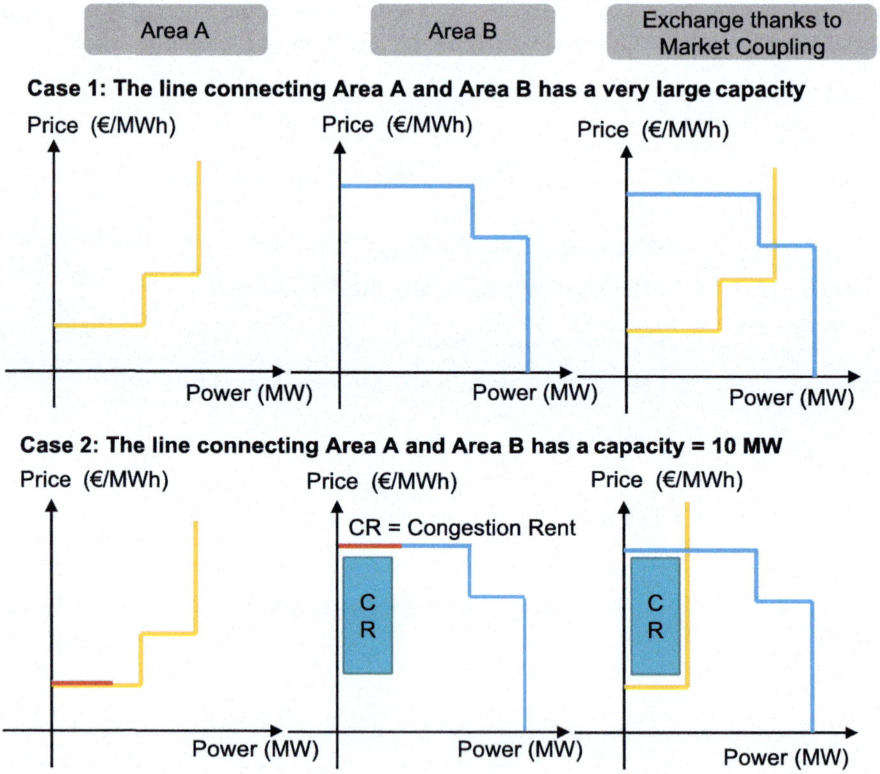

Fig. 6.2 Welfare maximization with multiple bidding zones. *Source* Author's own illustration

In a competitive equilibrium, the usage of the transmission capacity is optimal given the locational marginal prices (i.e. it optimizes the congestion rent or in the language of economists, the profits of the economic agents offering network resources). All these facts are just generalizing the ones presented in Sect. 6.2 and they result from the same underlying principles.

6.4 Network Models

The network constraints in the European grid are modelled using a hybrid version of the so-called ATC model and a flow-based network model. The ATC model corresponds to the previous section where bidding zones are connected with lines having capacity constraints; the flows on these lines have to be at most the available transfer capacity (ATC). This model allows to include other types of constraints such as ramping constraints which limit the change of flow on a line from one period to the next, or losses and tariffs.

The other basic type is the flow-based network model, where the transfer between bidding zones is constrained by linear constraints for critical network elements (CNE) using power transfer distribution factors (PTDF) and remaining available margin (RAM). Critical network elements have limited capacity that has to be considered in the market clearing process, the remaining available margin. The flow through such a critical network element is approximated by multiplying the net exports of the bidding zones with the power transfer distribution factor of the bidding zone for the specific critical network element. Thus, given bidding zones indexed by "bz" (denoting by 'BZ' the set of all bidding zones in the "balancing area"), and with "cnec" indexing the critical network elements,[4] the following equations have to be satisfied (one constraint per "cnec"):

$$\sum_{bz \in BZ} NEX_{bz} PTDF_{bz,cnec} \leq RAM_{cnec} \quad \forall cnec \in CNEC.$$

An important difference to ATC models is that flows on the network can appear non-intuitive to market participants. While a market participant may assume that energy will always flow from bidding zones with a smaller market price to bidding zones with a higher market price in order to increase welfare, this is not necessarily the case in flow-based network models. In flow-based network models, the flows between bidding zones cannot be understood by looking at the individual bilateral trades. Instead, the flow from a bidding zone with high prices to a bidding zone with low prices might enable another flow from a bidding zone with low prices to

[4] CNEC stands for "critical network elements and contingency", as in practice, a so-called "N-1" criterion is used to manage contingencies.

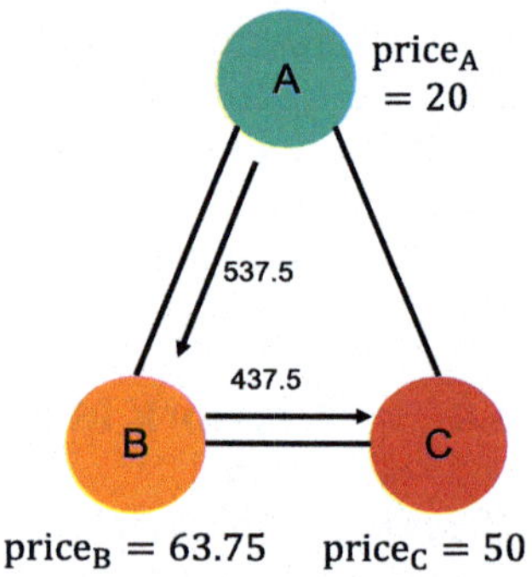

Fig. 6.3 Flows between bidding zones. *Source* Author's own illustration

a bidding zone with high prices leading to an overall welfare increase. A simple example is illustrated in Fig. 6.3.

In the hybrid version that is used in Europe, there can be multiple flow-based areas each containing multiple bidding zones. For the net exports considered in the flow-based constraints only the exchanges within the flow-based area are considered. The bidding zones cannot only be connected to bidding zones within the same flow-based area but also to bidding zones in other flow-based areas or bidding zones outside of flow-based areas. Such connections and connections between bidding zones outside of flow-based areas can be modelled using the constraints available in the ATC model.

On another hand, network domains given to the day-ahead market clearing process must satisfy a "simultaneous feasibility test" condition to ensure that the congestion incomes collected in the day-ahead are large enough to cover payments to holders of Long-Term Transmission Rights (Financial Transmission Rights whose payoff depends on the price spread between two locations). Euphemia implements an LTA Inclusion mechanism[5] to ensure that such simultaneous feasibility test conditions are satisfied (N-SIDE, 2022).[6]

[5] Another (less scalable) LTA Inclusion mechanism based on so-called "virtual branches" was previously part of capacity calculation processes. The Extended LTA Inclusion mechanism implemented in Euphemia was designed in the frame of the Euphemia Lab to reach sufficient scalability for the introduction of Evolved Flow-based in the Central Western Europe region and incidentally enabled to operate a "tight LTA inclusion" in the CORE capacity calculation region.

[6] For details on how the Extended LTA Inclusion is performed in Euphemia to ensure that these conditions are satisfied in practice, we refer to: (N-SIDE, 2022).

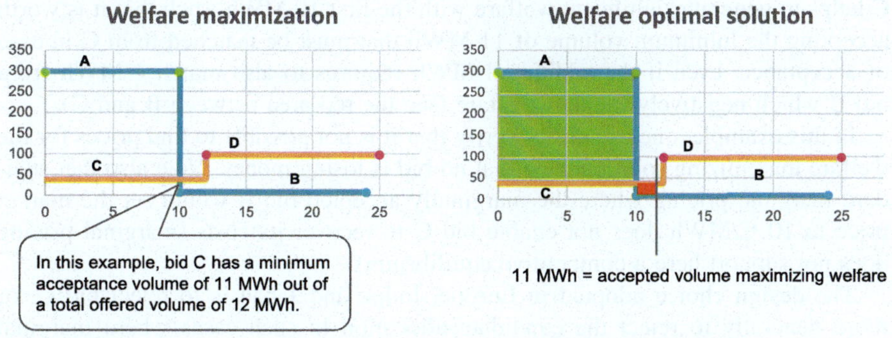

Fig. 6.4 Welfare maximization with block orders. *Source* Author's own illustration

6.5 Bidding Product Overview and Challenges with Advanced Bidding Products (Block Orders and Complex Orders)

Besides classic bid curves, products such as block orders are used to represent technical constraints and cost structures of generation units (or more generally, portfolios of generation assets), and their analogue on the demand side. These advanced products introduce so-called "non-convexities" precluding in general the existence of a competitive equilibrium.

In the example shown on Fig. 6.4, we briefly review why block orders prevent the existence of a competitive equilibrium. The example is like the examples in Sect. 6.2 but we changed prices and volumes of the orders. Additionally, bid C became a so-called block order. We assume that bid C has a minimum acceptance volume of 11 MWh, while offering 12 MWh in total. That means that in every feasible solution 0 MWh of bid C are accepted or any value between 11 and 12 MWh.[7]

If we could accept any value of bid C then the optimal solution with a welfare of 2400 EUR would be at the point where the supply and demand curve meet and where we accept 10 MWh of bid C, which is of course not possible. The optimal allocation is depicted on the right-hand side and has a welfare of 2370 EUR. Bid

[7] This condition leads to a non-convexity as the solution space is not convex anymore. For the solution space to be convex it must contain all convex combinations of any two solutions in the solution space (intuitively, the line segment joining the two feasible points), meaning that if two solutions defined by vectors a and b exist then for each α in [0,1], a solution corresponding to vector α a + (1-α) b exists. The entries of the vectors defining a solution are the values of the variables of the problem. For our example that means that if a solution with bid C being rejected exists and a solution with bid C being fully accepted exists then there has to be at least one solution for each acceptance ratio of bid C, which is not the case. If the objective function of the maximization problem is concave and the solution space is convex then the problem is a convex optimization problem, a class of optimization problems that contains for instance linear optimization discussed before.

C helps to generate significant welfare with the first 10 MWh, such that it is worth accepting the minimum volume of 11 MWh that must be matched from C in case of acceptance, even if the additional MWh requires to also match 1 MWh from bid B which negatively impacts welfare (see the red area between B and C).

In this example, one can easily verify that it is not possible to find prices for the welfare maximizing solution such that no bid is losing money. Note also that standard marginal pricing where the marginally accepted bid B would set the market price to 10 €/MWh does not enable bid C to recover its costs (marginal pricing does not support here a competitive equilibrium).

The design choice adopted in Europe, India, and Japan is to reject bid C, or more generally to reject the candidate allocation in such cases where marginal pricing would lead to block orders which are "paradoxically accepted". In our example, rejecting bid C would lead to accepting bid A fully, accepting 10 MWh of bid D, and 0 MWh of bid B leading to a welfare of 2000 EUR.

Block orders can also cover multiple periods or have a parent–child relationship with other block orders; in this case, the child block order can only be accepted if the parent block order is accepted too.[8] However, block orders are only one example of more complex types of bids that enable the market participants to express their constraints.

We review below the different types of orders in SDAC, leaving aside so-called PUN[9] orders used in Italy which will be phased out as a prerequisite for the introduction of 15 min Market Time Units in SDAC.

6.5.1 Hourly Orders and Simple Orders

The term "hourly order" refers to the simplest type of orders that (a) are fully divisible, and (b) span a single Market Time Unit which as of today is one hour in all bidding zones. Market Time Units will be allowed to span 15 min instead of a full hour after the go-live of the 15 min Market Time Units scheduled in the first quarter of 2025. In the rest of this section, we hence refer more generally to simple orders, sometimes also called curve orders.

[8] Information on various types of block orders available in SDAC can for instance be found in All NEMO Committee. (2020). Euphemia Public Description. Available online at: https://www.nemo-committee.eu/assets/files/euphemia-public-description.pdf (Accessed Friday May 10, 2024), and in HUPX. Linked block orders—Exclusive group block orders. Available online at: https://hupx.hu/uploads/Kereskedes/Keresked%C3%A9si%20rendszer/DAM/Smart%20Blocks%20of%20HUPX%20DAM.pdf (Accessed Friday May 10, 2024).

[9] The National Single Price in Italy (Prezzo Unico Nazionale, PUN) for consumption units is published as the weighted average of the bidding zone prices in Italy in the day-ahead market (Gestore dei Mercati Energetici (2024): Spot Electricity Market, https://www.mercatoelettrico.org/En/Mercati/MercatoElettrico/MPE.aspx (Accessed Friday May 10, 2024).

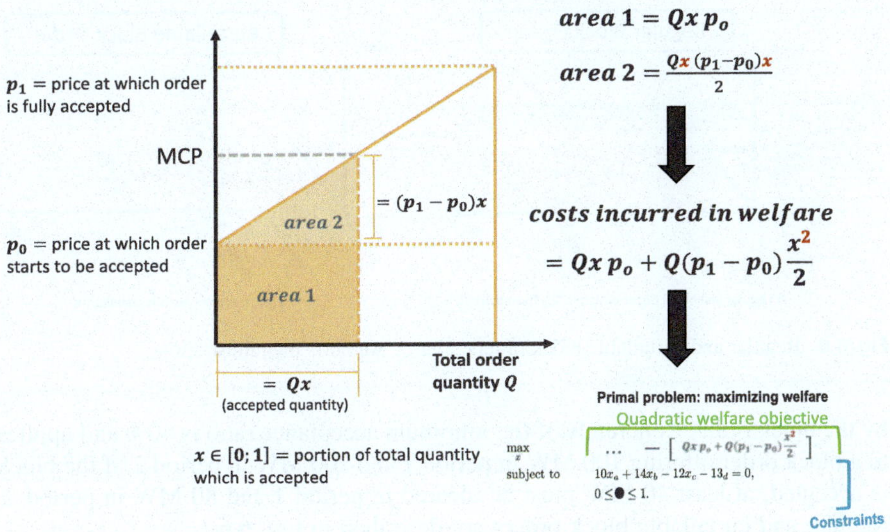

Fig. 6.5 Piecewise linear bid curves lead to quadratic welfare maximization problems. *Source* Author's own illustration

Simple orders denote the simplest bidding products consisting of so-called price-quantity pairs: a given buy or sell quantity, and a limit price. More generally, market participants can submit a demand or supply curve that can consist of multiple steps (stepwise demand or supply curves), or multiple line segments. Examples of stepwise curves can be found on Fig. 6.4 above, while an illustration of a piecewise linear bid curve and why it leads to a (convex) quadratic welfare maximization problem is given on Fig. 6.5.

A demand curve must be non-increasing, while a supply curve must be non-decreasing. These conditions are needed to ensure that the total welfare function to be maximized is a concave function (i.e. that the underlying optimization problem is a convex optimization problem if we relax the binary conditions, that is when binary orders described below are considered divisible during some intermediate algorithmic computations, or when their binary acceptance is considered fixed during specific pricing calculations).

6.5.2 Block Orders

Regular block orders are binary orders with so-called "fill-or-kill" conditions, i.e. fully indivisible orders. The volumes at the different periods which are associated with a regular block order, and which must be either fully cleared or fully rejected, can be period-specific and determine the "profile" of the block order.

Curtailable block orders are semi-indivisible orders in the sense that if they are accepted, a same "minimum acceptance ratio" is enforced at each period spanned

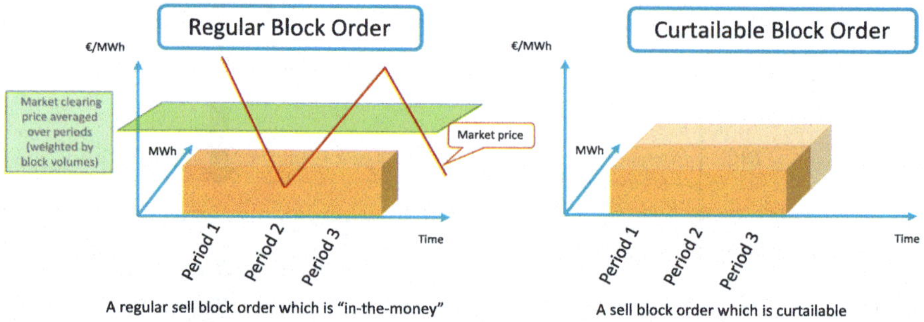

Fig. 6.6 Regular and curtailable block orders. *Source* Author's own illustration

by the block order. Concretely, if the minimum acceptance ratio is 40% and applies to a block order offering 100 MW in period 1 and 200 MW in period 2, if the block is accepted, at least 40 MW must be cleared in period 1 and 80 MW in period 2. Regular and curtailable block orders are described in Fig. 6.6.

Several variants of this base binary order exist in practice: "linked" (a block can be accepted only if the parent is accepted), "exclusive" (only one block in the group can be accepted, or the total acceptance ratios cannot exceed 1), etc. We refer to the Euphemia Public Description (All NEMO Committee, 2020) for a comprehensive description of the block order variants, to EPEX market rules (EPEX Spot, 2023) for technical conditions or restrictions applicable to them, and to a technical document from HUPX (HUPX, 2014) for additional information on linked and exclusive block orders.

As discussed in the beginning of this section, such orders introduce non-convexities usually precluding the existence of a competitive equilibrium supported by uniform prices and the market rules applicable in Europe, Indian, and Japanese markets require to consider block order selections (or more generally non-convex order matchings) such that no block order is paradoxically accepted (i.e. accepted despite being out-of-the-money) according to the resulting marginal prices (The marginal prices are obtained after considering the block order selection as a given, fixing the corresponding binary acceptance variables to the value they take in the considered bid matching, and computing classical marginal prices associated with the resulting "convex" optimization problem).

The next sections discuss how to find solutions for market instances where such rules apply.

6.6 Skeleton of the SDAC Market Clearing Algorithm: Implicit Enumeration (Branch-And-Bound)

Products like block orders or scalable complex orders are modelled using binary variables that can only take the values 0 or 1. These variables express the acceptance of the corresponding order. If the value is 0 then the order is rejected, and if

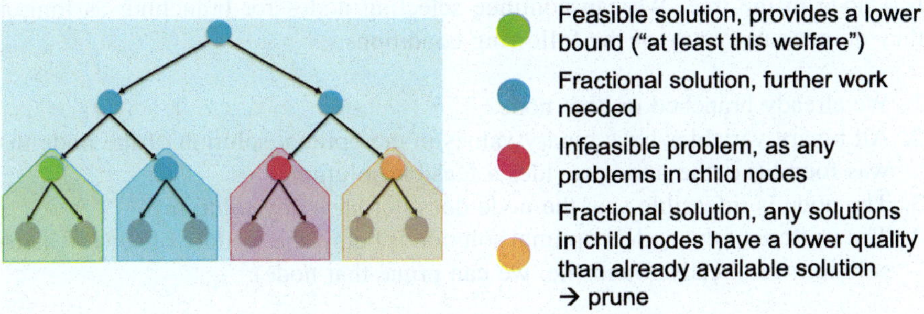

Fig. 6.7 Branch-and-bound method. *Source* Author's own illustration

the value is 1 then the order is fully or partially accepted. The existence of such binary variables in the welfare maximization problems causes non-convexities as we discussed above and requires the usage of mixed-integer programming (MIP) algorithms for solving the welfare maximization problem.

For welfare maximization, introducing new binary variables to the Linear Program (LP) or Quadratic Program (QP) transforms the problems into mixed-integer linear problems (MILP) or mixed-integer quadratic problems (MIQP). Such problems can be solved by combining the branch-and-bound method together with an algorithm for solving quadratic problems like the simplex method.[10] The branch-and-bound method is not a simple enumeration of all binary assignments to the binary variables but uses a tree structure and allows a more efficient search using upper and lower bounds as shown in Fig. 6.7.

Relaxation: The idea behind the branch-and-bound method is to first relax all binary variables. Thus, the binary variables can take fractional values. The optimal objective value to a relaxed maximization problem is always at least as large as the optimal objective value of the original problem and thus provides an upper bound. The reason is that relaxation allows additional solutions while keeping all solutions prior to the relaxation.

Branching: When a binary variable has a fractional assignment in the optimal solution to the relaxation then we can branch on that variable as the solution cannot be accepted for the original problem. That means that we can select one binary variable that has a fractional assignment and create two child problems to the relaxation. In one child, the selected binary variable is set to 0, and in the other one it is set to 1. This branching creates the tree structure, and each sub-problem

[10] While for many specific mixed-integer problems with simple structures dedicated algorithms with better performances do exist, the complex structure of market clearing problems with many different types of constraints leads to the use of generic algorithms that could be applied to any MILP or MIQP. However, there are in general many ways to formulate or approach a specific problem which can have a significant impact on algorithm performances when using such generic algorithms.

is a node in the tree. We can continue selecting nodes for branching as long as they do not satisfy any of the following conditions:

1. We already branched on that node.
2. All binary variables have binary values in the optimal solution of the node that was found (i.e. the node provides a feasible solution).
3. The node is infeasible (i.e. the node does not have any solution).
4. The welfare of the node's optimal solution is below the welfare of some feasible solution we already found (i.e. we can prune that node).

Once all nodes satisfy these conditions the algorithm terminates, and we found an optimal solution or showed that no solution exists. The branch-and-bound method always provides us with a lower bound, which is the maximum objective value of all feasible solutions (assuming that at least one exists), and an upper bound, which is the maximum objective value of all nodes that have no child nodes or have child nodes that were not yet solved. The distance of the upper bound (on the best achievable welfare) to the lower bound (given by the best solution found so far), is the so-called optimality gap. It indicates by how much *at most* the quality of the solution could be improved if the search is further continued, and gives a certificate of near-optimality when this gap is considered small enough. However, this is just an indication, as the upper bound can still improve when the tree is further explored. When no more nodes are available for branching, the optimal solution has been found and the gap is zero.

As not all conditions on a final solution can be modelled efficiently in an MIQP, some conditions can be validated only when finding a feasible solution. Should some condition be violated (e.g. no paradoxically accepted order), the feasible solution can be rejected by adding a new constraint and the node is available for further branching.

Figure 6.8 provides an example of the branch-and-bound method. The root node has an objective value of 13,500 EUR but still contains three relaxed binary variables with fractional assignments (x7, x15, x20). Following the internal branching strategy, the first relaxed binary variable we branch on is x7. We have now two branches with x7 equals zero or x7 equals one.

The internal node selection strategy decides on which node to explore next. However, in both cases there are still some relaxed binary variables with fractional assignments and the branching has to continue.

Following the next branching we end up with the following four cases when exploring the new nodes: (1) all binary variables have binary assignments and we thus obtain a feasible solution, which also provides us with a lower bound of 9500 EUR; (2) some relaxed binary variables still have fractional assignments but we cannot prune this branch and have to continue exploring it as the objective value of 9800 EUR is above our lower bound (this part of the search space could still contain better solutions than the one with a welfare of 9500 EUR); (3) the branch does not contain any feasible solution; (4) the objective value is at 9300 EUR and

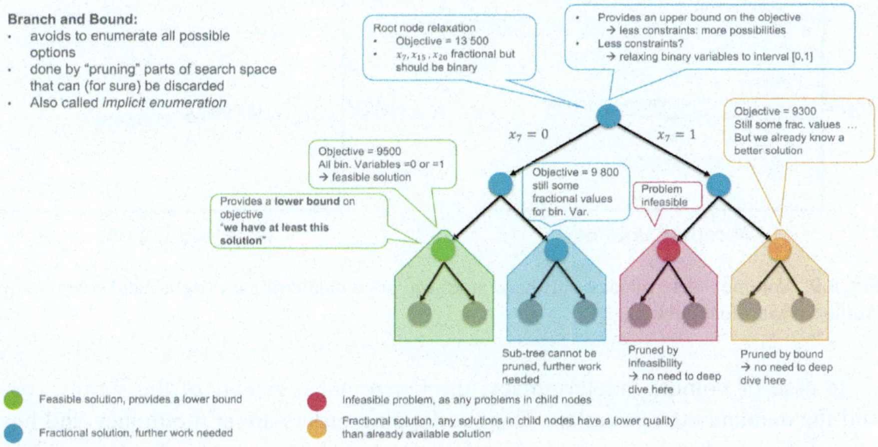

Fig. 6.8 Example of the branch-and-bound method. *Source* Author's own illustration

we can thus prune that branch as the value is below our lower bound of 9500 EUR (no improving solution could be found in that part of the search space).

More technical details on custom algorithms applicable to European-like market rules can be found in the literature, such as (Madani, 2017) and (Müller, 2014). Note however that these references do not deal with PUN orders currently present in Italy. They also do not cover the full set of algorithmic elements present in state-of-the-art production-grade market clearing software for market rules like the ones used in Europe, India, and Japan.

6.7 Managing Volume and Price Indeterminacies

Optimization problems can have multiple optimal solutions, and the convex optimization problems encountered in day-ahead electricity markets may have no solution, a single solution, or infinitely many optimal solutions, as could be observed, for instance, on Fig. 6.9 where the green segment on the left-hand side shows corresponds to a range of possible matchings, with different total cleared volumes but the same total optimal welfare.

Figure 6.9 also displays market clearing problems with step curves that have price indeterminacies, and thus infinitely many possible prices.

This section discusses how such situations can be addressed so as to uniquely determine a single market outcome according to the market rules. Similar rules exist, for instance, for lifting indeterminacies in scheduled exchange calculations (alternative optimal solutions for flows on cross-border lines) which we do not cover in this chapter.

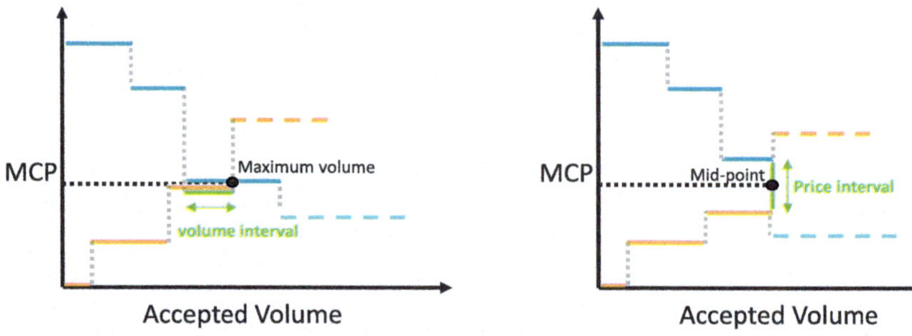

Fig. 6.9 Volume indeterminacy (left-hand side) and price indeterminacy (right-hand side). *Source* Author's own illustration

In case of volume indeterminacy, the horizontal segments of the supply curve and the demand curve overlap. The overlapping orders are at the money and have zero utility in any case, no matter if they are selected or not. In case of price indeterminacies, the vertical segments overlap. Thus, for two distinct market prices the prices of all accepted supply orders are below these market prices and all accepted demand orders have prices above these market prices.

Consequently, those market prices and all market prices in between are valid considering all market participants. In both cases, infinitely many optimal solutions to the welfare optimization problem (or its dual giving the market prices) exist and it follows that secondary objectives beside welfare optimization should be selected to choose the best solution for market participants.

For volume indeterminacies, the maximization of the accepted volume is favoured. A quadratic objective can help here to reach a well-defined volume per bidding area in case of volume indeterminacies in multiple bidding areas. We can select here a strictly convex objective function that is guaranteed to find a unique optimal solution, and which tend to maximize the traded volume.[11]

For price indeterminacies, the distance to the midpoint of the feasible price interval can be selected as optimality criterion. However, other optimality criteria like the smallest price could also be chosen. As for volume problems, a quadratic objective can help to deal with indeterminacies in multiple bidding areas.

While both types of indeterminacies could be avoided by the selection of interpolated curves without horizontal or vertical segments, such restrictions on orders can lead to other challenges to market participants and the algorithm itself. Curve segments with a small but positive slope can, for instance, cause numerical problems and lower precision of solutions.

[11] Having the maximization of the traded volume as a secondary objective is not sufficient to ensure that the market outcome is uniquely defined, because traded volume maximization is technically a "linear objective", which is not sufficient to guarantee that there will be a unique optimal solution. Instead, a "strictly convex" quadratic objective is used, ensuring uniqueness, which in corner cases may not fully maximize the traded volume as a secondary objective.

Another source of indeterminacies are the flows between bidding areas and between NEMOs. These indeterminacies require other dedicated problems with their own objectives that ensure that a unique optimal solution exists for each selection of binary orders.

6.8 Outlook

The design and operation of efficient coupled day-ahead electricity markets, incrementally implemented over the past two decades in Europe, is a formidable achievement that necessitated an impressive degree of coordination among stakeholders, as this entailed the design of methodologies for cross-border capacity calculations, bidding product design, the common market clearing algorithm, fallback procedures, and a wide range of other procedures supporting operations taking place on a daily basis today at a massive scale.

Day-ahead electricity markets around the world will also evolve in the coming years, to cope with the energy transition challenges, and ensure that their design continues to support the functioning of efficient markets. For instance, how to best incorporate in the market design the flexibility provided by new storage technologies, or from smart network resources, constitute interesting topics to further explore.

From an algorithmic perspective, while new challenges may lie ahead, past achievements allow us to look to the future with confidence.

Bibliography

All NEMO Committee. (2020, October). *Euphemia public description.* Retrieved November 03, 2024, from https://www.nemo-committee.eu/assets/files/euphemia-public-description.pdf

Coase, R. H. (1946). The marginal cost controversy. *Economica, 13*(51), 169–182.

EPEX Spot. (2023). *EPEX market rules.* Retrieved May 10, 2024, from https://www.epexspot.com/sites/default/files/download_center_files/EPEX%20SPOT%20Market%20Rules_0_0.zip

HUPX. (2014). *Linked block orders—Exclusive group block orders.* Retrieved May 10, 2024, from https://hupx.hu/uploads/Kereskedes/Kereskedés%C3%A9si%20rendszer/DAM/Smart%20Blocks%20of%20HUPX%20DAM.pdf

Madani, M. (2017). *Revisiting European day-ahead electricity market auctions: MIP models and algorithms.* PhD Thesis. Université catholique de Louvain. Retrieved November 03, 2024, from https://dial.uclouvain.be/pr/boreal/object/boreal%3A183686/datastream/PDF_01/view

Müller, J. C. (2014). *Auctions in exchange trading systems: modeling techniques and algorithms.* Friedrich-Alexander-Universität Erlangen-Nürnberg (FAU). Retrieved November 03, 2024, from https://open.fau.de/items/9c54e6ec-6b2a-43a2-aea4-680170be358d

N-SIDE. (2022). *Extended formulation for LTA inclusion, Description, shadow prices and price formation.* Retrieved from https://www.jao.eu/sites/default/files/2022-03/LTA_Inclusion_Description_202202.pdf

Pindyck, R., & Rubinfeld, D. (2017). *Microeconomics* (9th ed.). Pearson.

Rockefeller, R. T. (1970). *Convex analysis.* Princeton University Press.

Samuelson, P. A. (1952). Spatial price equilibrium and linear programming. *The American Economic Review, 42*(3), 283–303.

Torbaghan, S. S., Madani, M., Sels, P., Virag, A., Le Cadre, H., Kessels, K., & Mou, Y. (2021). Designing day-ahead multi-carrier markets for flexibility: Models and clearing algorithms. *Applied Energy, 285*, 116390.

Dr. Mehdi Madani is a Senior Consultant in power market design and optimization at N-SIDE which he joined in 2018 after a postdoctoral research position at the Johns Hopkins University working on pricing in non-convex electricity markets. He was the project manager of the Euphemia Lab (the R&D program for Euphemia, the European Market Coupling market clearing algorithm) from its inception in 2019 until November 2023 when he moved to the role of expert. Mehdi has deep expertise at the R&D level in market design and solution algorithms for power markets, with a track record of impactful recommendations (bidding product design, pricing, LTA Inclusion, and other TSO requirements, co-optimization of energy and balancing capacity), and, for instance, developed the Extended LTA Inclusion methodology implemented in Euphemia to support the go-live of Evolved Flow-based in 2020. Mehdi holds a PhD in operations research from UCLouvain, a complementarity master in financial risk management from ULiege, and a master in theoretical mathematics also from ULiege. Aside from his consulting activities, since 2022, Mehdi has taught Operations Research at UCLouvain as a guest lecturer at the Ecole Polytechnique de Louvain.

Dr. Martin Starnberger is the Head of the European Energy Market Solutions Team at N-SIDE and former Head of the Optimization Team at N-SIDE. He is an expert on Energy markets and market algorithms. In particular, he has guided the development and evolution of Euphemia (the European day-ahead electricity market algorithm) over eight years in terms of algorithm design, market design, and software architecture through different roles, including tech lead of Euphemia and project manager of the Euphemia Lab (the R&D program for Euphemia). Before his work at N-SIDE, Martin held a researcher position at Vienna University, where he worked on the algorithmic and game theoretical aspects of combinatorial auctions and worked in the financial sector on credit risk modelling. Martin holds a PhD in Computer Science from the University of Vienna and master's degrees in Technical Mathematics and Business Informatics from TU Wien.

Continuous Intraday Market Coupling Algorithm

7

Suzanna Vogeler

Abstract

This chapter begins by describing the background to how existing European intraday markets were finally brought together via a pan-European solution known as Single Intraday Coupling. The chapter introduces the IT-solution (M7 XBID) for the Single Intraday Coupling, which enables continuous intraday trading of energy contracts for the physical delivery of electricity 24 h a day, 365 days a year across Europe. More details on the technical modules of M7 XBID, as well as the data flows including the upstream and downstream systems are provided. The chapter closes with a simplified description of the Single Intraday Coupling continuous trading algorithm.

7.1 Introduction

Single Intraday Coupling is the result of the liberalization of the energy markets in the European Union. Liberalization led to the establishment of energy exchanges and market coupling operators and ultimately single intraday market coupling (see Chapters 1 and 2), as we know it today.

Until 2018, the intraday markets were still in a type of regional coupling state with Germany, France, the Netherlands, and Switzerland coupled via EPEX Spot's intraday trading system (M7 Trading) and Nord Pool Spot coupling the Scandinavian intraday markets with their solution known as ELBAS.

In 2018, the existing European intraday markets were finally brought together via a pan-European solution known as Single Intraday Coupling.

S. Vogeler (✉)
Frankfurt/Main, Germany

A. Estermann et al. (eds.), *European Electricity Market Coupling*, Future of Energy,
https://doi.org/10.1007/978-3-031-86315-8_7

The relationship between Deutsche Börse AG, the TSOs, and the NEMOs for the establishment of a central IT platform and infrastructure for pan-European intraday coupling dates to 2012 when a competitive Request for Proposal (RfP) was issued by the power exchanges APX-ENDEX Power B.V., BELPEX NV, EPEX Spot SE, Nord Pool Spot AS, OMI Polo Español, S.A., and OTE, a.s.

The objective of the RfP was to establish a common cross-border implicit continuous intraday trading solution across Europe, with the allocation of cross-border capacities after the European-wide day-ahead market coupling.

In 2015, Deutsche Börse AG was mandated with the development, maintenance, technical operation, and hosting of the IT-solution (M7 XBID) for the single intraday coupling (SIDC), which was launched in June 2018. It enables continuous intraday trading of energy contracts for the physical delivery of electricity 24 h a day, 365 days a year across 25 EU Member States. This means that an offer submitted in Italy could be matched with a bid submitted in Norway, depending on the bid prices and the available transmission capacity.

With the ever-increasing amount of variable renewable production, pan-European cross-border intraday trading and the delivery of the electricity on the same day gives market participants the opportunity to balance their positions after the closing of the single-day-ahead market (SDAC). It also makes it easier for market participants to allow for unexpected changes in consumption and outages thus limiting their shortfalls or surpluses. Overall, such a mechanism reduces reliance on reserves and the associated costs (ENTSO-E, 2023a, 2023b) and enables a more efficient utilization of the generation resources across Europe (Single Intraday Coupling, 2021).

Since the launch, SIDC has become truly pan-European and, as of early 2024, includes 14 NEMOs: BSP Energy Exchange LLC, Croatian Power Exchange Ltd., EPEX Spot SE, ETPA Holding B.V, Gestore dei Mercati Energetici S.P.A., Hellenic Energy Exchange S.A, HUPX Hungarian Power Exchange Company Limited by Shares, Independent Bulgarian Energy Exchange, Nord Pool European Market Coupling Operator AS, OKTE a.s., OMI—Polo Español S.A., Operatorul Pieţei de Energie Electrică şi de Gaze Naturale "OPCOM" S.A., OTE a.s., and Towarowa Gielda Energii S.A. This geographical expansion occurred in 4 waves from 2018 to 2022 with a fifth wave in 2023 as illustrated in Fig. 7.1.

Since the establishment of the European Energy Exchange (EEX) in Frankfurt in 20 Deutsche Börse AG has helped shape energy trading (Deutsche Börse Group, 2023a, 2023b). In addition to M7 XBID, its M7 product portfolio includes M7 Trading—a trading system for continuous intraday power trading markets, M7 Auction—an auction service for CO2 allowances and Grid Loss as well as M7 Capacity—a capacity management system for intraday cross-border capacity allocation. This is illustrated in Fig. 7.2.

In general, power must be produced and consumed simultaneously and cannot be easily stored. Moreover, the frequency of the transmission grid must remain stable at all times and, thus, consumption must be equal to production at all times. This makes the 24 $\times$ 7 operation, availability and resilience of short-term power markets and the corresponding trading solutions essential. Solutions, such as M7

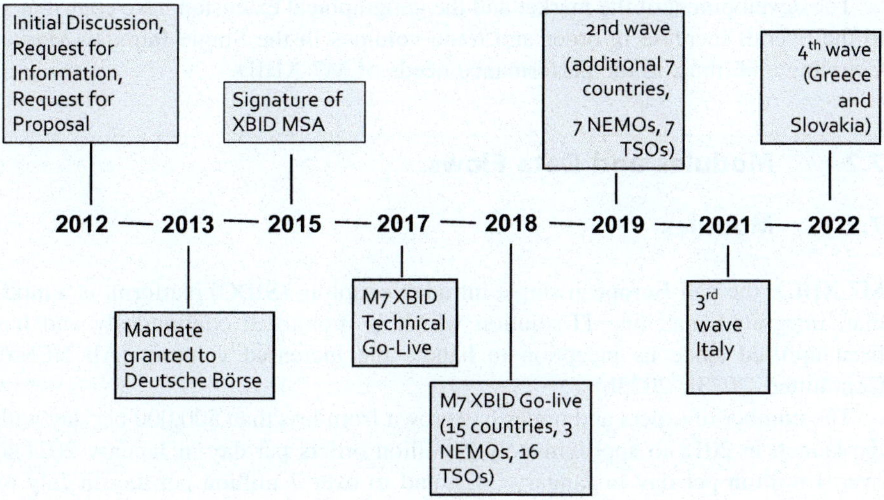

Fig. 7.1 SIDC evolution timeline. *Source* Author's own illustration

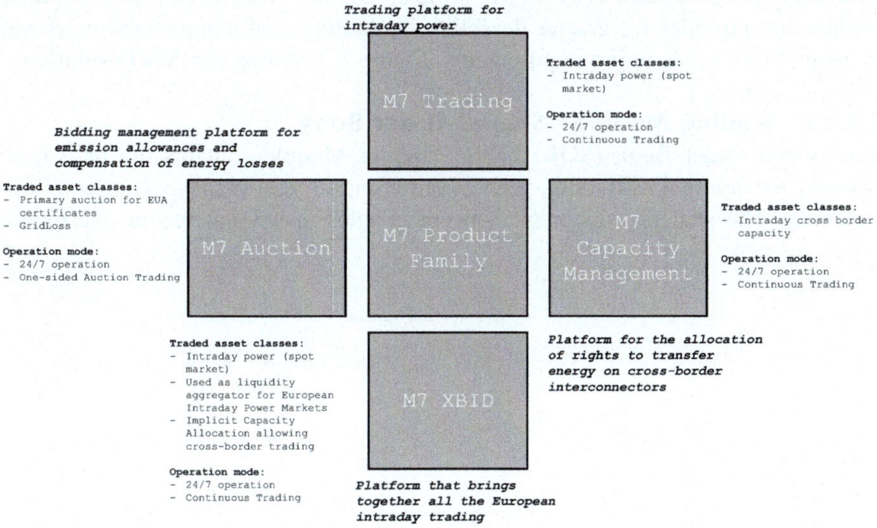

Fig. 7.2 The M7 product family. *Source* (Deutsche Börse Group, 2023a, 2023b)

XBID, help provide electricity where it is needed even across borders by allowing market participants to balance their positions close to real-time. Between mid-June 2018 and the end of February 2023, 242 million trades have been executed on M7 XBID, with an average technical annual availability of 99.97% in 2022 (ENTSO-E, 2023a, 2023b).

The development of the market and the geographical extension have contributed to the overall increase in order and trade volumes in the Single Intraday Market Coupling and, thus, to the performance needs of M7 XBID.

7.2 Modules and Data Flows

7.2.1 Modules

M7 XBID, the pan-European single intraday coupling (SIDC) platform, is a modular, integrated matching IT-solution which is monitored continuously and has been uplifted since its inception to handle the increased volumes (All NEMO Committee, 2023a, 2023b).

The number of orders and trades has grown from less than 500,000 per day with the launch in 2018 to approximately 2 million orders per day in January 2021 to over 4 million per day in January 2023 and to over 7 million per day in July of 2023 as shown in Fig. 7.3.

The solution is based on the Shared Order Book (SOB), the Capacity Management Module (CMM/CMI) and the Shipping Module (SM) as well as the Central Reference Data Module (ENTSO-E, 2023a, 2023b). This performant, modular architecture provides for greater flexibility, scalability, and maintainability as well as adaptability to changing requirements. Figure 7.4 depicts the XBID solution.

7.2.1.1 Trading Module—Shared Order Book

The Shared Order Book (SOB) of the Trading Module contains the base functionality for continuous trading, such as order entry, order management, and order matching. The shared order book concept enables orders entered in one market

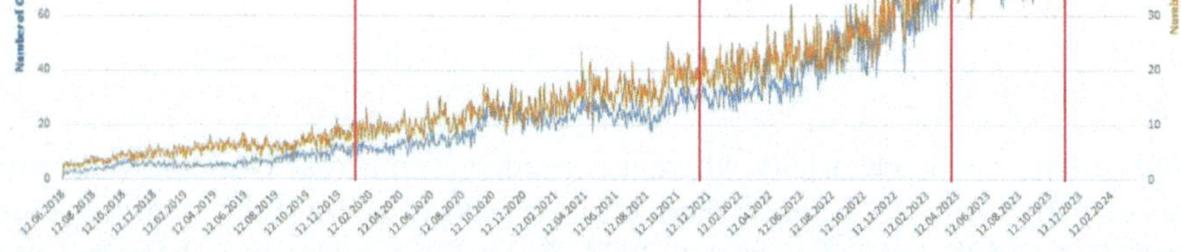

Fig. 7.3 Growth of orders and trades since 2018. *Source* (All NEMO Committee, 2023a, 2023b)

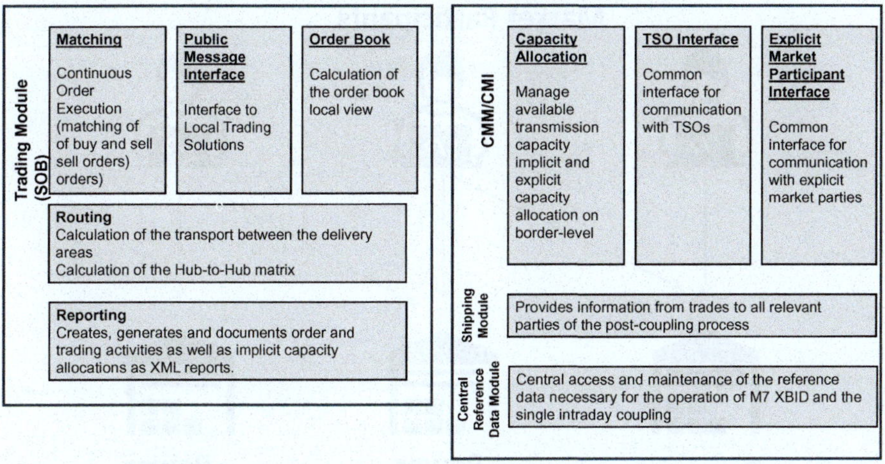

Fig. 7.4 The XBID solution. *Source* Adapted from (ENTSO-E, 2023a, 2023b)

area to be displayed in connected market areas if there is sufficient transmission capacity between them.

The M7 XBID solution enables multiple power exchanges (NEMOs) to connect to the central SOB module as illustrated in Fig. 7.5. Market participants cannot connect directly to the SOB, but instead connect to one or more power exchanges (local trading solutions). Orders are entered in the local trading solutions by the market participants, which in turn connect to the SOB module via the public message interface. Matching of two orders (for a global contract) entered for the same delivery area is performed in the SOB. The trading activities are documented in reports, which are made available to the NEMOs and transmission system operators (TSOs). Global products and contracts refer to those which are set-up in the M7 XBID Solution and which are eligible for matching. Local products and contracts are those which are not set-up in M7 XBID and which are therefore not matched in the M7 XBID solution. They are set-up and matched on the respective local trading solution.

The SOB module maintains a consolidated order book for all global contracts (not local contracts) based on the available transmission capacity between market areas, which is maintained in the Capacity Management Module (CMM). It contains all bid and ask orders with their prices, volumes, delivery area, and contract. The orders are arranged based on their prices; the bid (ask) order with the highest (lowest) price is at the top of the order book as these are the orders that are likely to get matched first (Shinde, 2023).

Orders submitted to the SOB module are either executed immediately or entered in the order book. This means that the order is stored in the system and made visible for other users with respective rights until the order is either deleted or deactivated or a matching order is submitted.

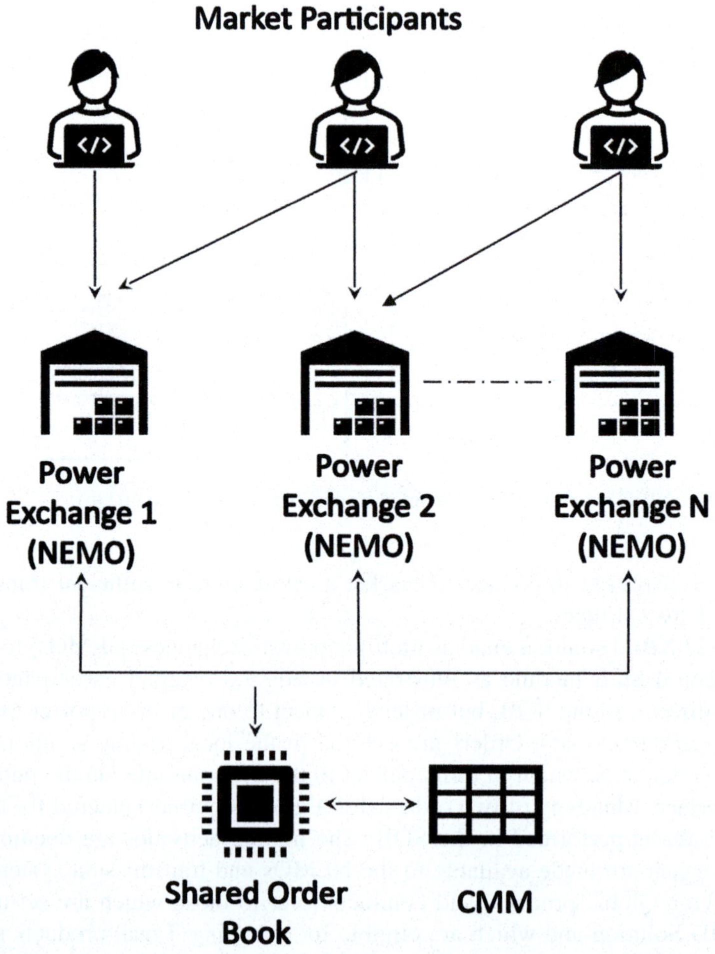

Fig. 7.5 Multiple NEMOs connecting to the central SOB module. *Source* Author's own illustration

For each contract there is one consolidated order book; however, orders in the order book can only be executed against each other if they are entered for delivery areas of the same market area or entered for areas which are connected via cross-border trading. In other words, when market participants submit orders for different delivery areas, they can only be matched if there is enough available transmission capacity between the respective areas (ENTSO-E, 2023a, 2023b).

In a cross-border scenario, the orders that are visible for each delivery area can differ, based on the transmission capacity situation. The Order book data displayed by the SOB module is not simply a collection of all orders entered for a contract, but rather it contains the sum of all orders (local orders and orders from connected delivery areas) that can be executed in the selected delivery area. The selection of the route between delivery areas is referred to as routing and calculated by the routing engine based on the available transmission capacity.

7.2.1.2 Capacity Management Module

The Capacity Management Module (CMM) provides the Transmission System Operators (TSOs) with functionality for managing the transmission capacity between all areas. CMM communicates with the SOB Module via a message interface. An interested reader is invited to see more details about the role of the TSOs in Chapter 4.

M7 XBID allows for both implicit and explicit capacity allocation, which are handled in the same way by M7 XBID.

Implicit capacity allocation or trading refers to the simultaneous allocation of transmission capacity and the corresponding quantity of power, whereas explicit capacity allocation refers to the allocation of only the transmission capacity in a specific direction on a specific interconnector.

In the case of explicit capacity allocation, whereby the two commodities capacity and power are traded in two separate, independent transactions, there is a lack of information about their prices. This may lead to lack of price convergence, less social welfare, more adverse flows and, in general, inefficient use of the interconnectors. Explicit allocations may be performed by Explicit Participants on two borders (as of Q1 2024: French–German and Croatian–Slovenian) by entering direct capacity requests via the M7 XBID CMM GUI or message interface.

The standard procedure on the majority of the M7 XBID borders is implicit capacity allocation during the matching process in the Trading Module. Specifically, this means that when cross-border trades are performed in the trading module, CMM implicitly allocates the required cross-border capacity. In this case, the capacity and the energy are priced together in one transaction. This ensures more efficient use of the interconnectors and minimizes congestion.

7.2.1.3 Shipping Module

The Shipping Module (SM) prepares data required for the nomination and scheduling of power as well as the financial settlement. The Shipping Module receives data about all trades from the SOB module and then enriches it. The enriched information is then provided to parties such as the NEMOs and TSOs and Central Counterparties (CCPs). An interested reader is invited to see more details about Shipping in Chapter 8.

7.2.1.4 Central Reference Data Module

The Central Reference Data Module contains master data such as balancing groups, members, users, products, contracts, etc., which is necessary for the operation of the SIDC.

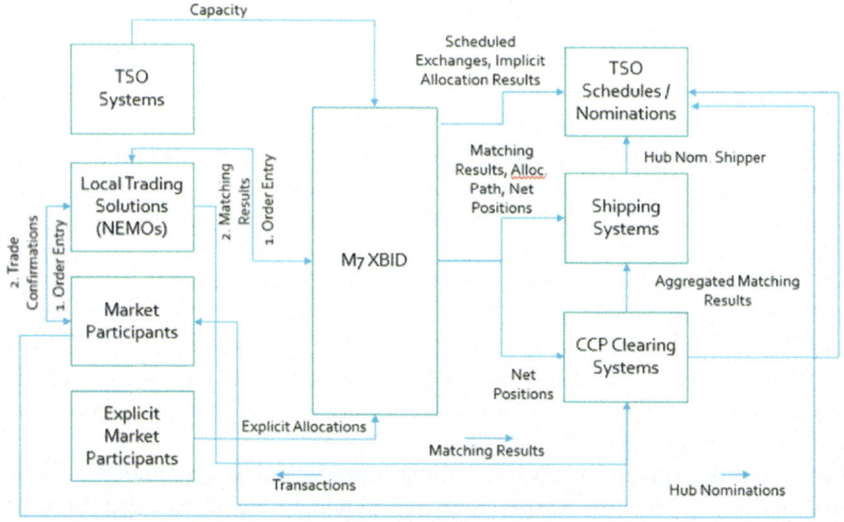

Fig. 7.6 M7 XBID data flows. *Source* Adapted from (ENTSO-E, 2023a, 2023b)

7.2.2 Data Flows

M7 XBID is connected to various upstream systems that provide input data as well as downstream systems to which it provides data resulting from orders, trades, and transmission capacity allocations (ENTSO-E, 2023a, 2023b). An interested reader is invited to see more details on the data flow in Fig. 7.6 below.

7.2.2.1 Upstream Systems

The upstream systems send data to the M7 XBID solution. These upstream systems include the local trading solutions, the TSO systems as well as the explicit participants. The TSO systems provide M7 XBID with the available transmission system capacity on the respective interconnectors. Without this information cross-border trading cannot take place.

The transmission capacity is processed by CMM and made available to the other technical modules.

M7 XBID receives orders from the Market Participants via the Local Trading Solutions, which are operated by the NEMOs. M7 XBID receives explicit allocation requests for transmission capacity from the explicit market participants. These requests are processed by CMM. In addition, the NEMOs, in their role as market administrators, input the master data via the reference data module.

7.2.2.2 Downstream Systems

M7 XBID sends data to the downstream systems as a result of orders, trades, and explicit capacity allocations. Once an order has been received by M7 XBID it is processed by the Trading Module and, if a trade is concluded, the matching results are sent back to the local trading solution, which in turn provides the market participant with a trade confirmation. In the event of an explicit capacity allocation, the explicit participant receives confirmation of the allocation. Data on trades and allocated capacity is enriched by M7 XBID and sent via various files and interfaces to the downstream systems of the TSOs, NEMOs, and Central Counterparties (CCPs) for further processing.

7.3 Trading Model

The European single intraday coupling is based on continuous trading as opposed to the European Single Day-Ahead Coupling which is based on an auction market model. In mid-2024, however, the single intraday coupling continuous trading model will be complemented with Intraday Auctions (IDA). The purpose of these implicit intraday auctions is to harmonize the calculation of the cross-border capacity and also to price it, thereby sending an adequate price signal to the market. An interested reader is invited to see more details about the Single Day-Ahead Coupling as well as IDA in Chapter 6.

In single intraday coupling continuous trading, the market participants can place orders to buy or sell electricity 365 days a year, 24 h a day. The single intraday coupling encompasses the continuous trading of contracts/products for the physical delivery of electricity on day D starting with gate opening on day D-1. The contracts have predefined delivery periods of 15, 30, and 60 min depending on the bidding zone (ENTSO-E, 2023a, 2023b). An interested reader is invited to see more details about the timeline of the markets in Chapter 2.

The market operates continuously, and orders are executed as they match with opposite orders in the order book. Orders are executed on a first-come first-served principle, where the highest buy price and the lowest sell price get served first (Terna, 2021).

The prices in continuous trading are determined by the ongoing supply and demand in the integrated European market. They fluctuate frequently as orders are matched in real-time.

Such continuous matching provides the market with real-time price and trade information, allowing the market participants to see the current state of the market and make decisions accordingly. This ability to balance positions shortly before delivery is beneficial not only for the market participants but also for the power systems. It reduces the need for reserves and the associated costs while allowing enough time for carrying out system operation processes and for ensuring system security (EPEX, 2023). This is especially important in the power markets, where prices can be influenced by factors such as supply availability, fuel costs, and transmission constraints as well as renewable energy, which can be difficult to predict.

Continuous trading can, however, lead to more price volatility than in an auction, as prices are constantly adjusting based on the supply and demand scenario.

Nevertheless, continuous trading promotes liquidity by providing opportunities for buyers and sellers to transact. This liquidity is essential for the efficient functioning of the integrated European market. By being able to adjust their positions in real-time, market participants can respond to market volatility and hedge against price fluctuations, thus reducing their exposure to risk.

The continuous matching algorithm for trading in power is more complex than algorithms for trading tangible assets such as marketable securities. During the matching process, the single intraday coupling continuous trading algorithm must take the physical constraints of electricity and the associated infrastructure, such as the transmission grid into consideration. An interested reader is invited to see more details on constraints in Chapter 4.

During the single intraday coupling continuous order matching process, constraints such as execution restrictions, validity restrictions, market area/delivery area, available transmission capacity, and ramping constraints, among others, must be included in the calculation before a trade is concluded.

- Order execution restrictions refer to the conditions placed on the order by the market participant, such as Immediate or Cancel (IOC) and Fill or Kill (FOK). Orders with type IOC are never entered into the order book. They are either executed fully against one other order or cancelled. FOK orders are executed fully against one or multiple other orders or cancelled. FOK orders are also never depicted in the order book (PSE, 2021).
- Validity restrictions refer to when an order should be deleted by the system, such as at contract expiration or at an explicitly defined date and time (PSE, 2021)
- Market areas and delivery areas refer to bidding zones and scheduling areas, respectively. A market area may contain multiple delivery areas between which there is unlimited transmission capacity. The assumption is that transmission capacity between market areas is subject to congestion. An interested reader is invited to see more details about the zonal model in Chapter 1.
- Available transmission capacity refers to the available transmission capacity between two connected delivery areas. An interested reader is invited to see more details about the capacity calculation in Chapter 3.
- Ramping refers to a change in power flow from one time unit to the next. Ramping restrictions limit the allowed net flow variations on consecutive hours on specific interconnectors (Nordpool Group, 2023).

7.4 Outlook

The single intraday coupling has harmonized the European intraday markets, increased the overall market liquidity and allowed market participants to balance their positions shortly before delivery, without large price swings thus, increasing market efficiency. This is exemplified in Fig. 7.7, which depicts the increasing

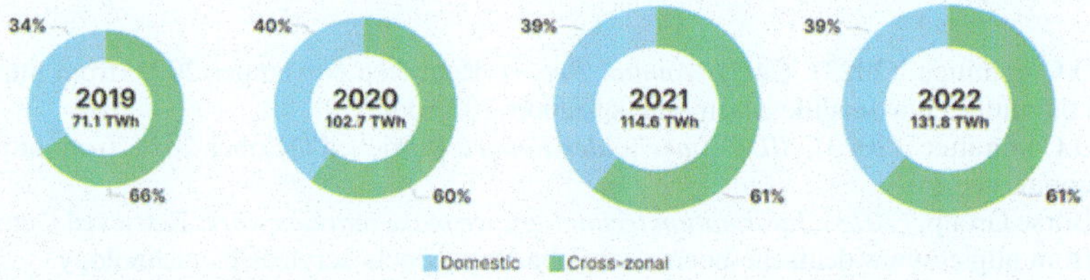

Fig. 7.7 Share of intraday traded volumes according to domestic vs. cross-zonal(border) nature of trades in Europe and yearly continuous intraday traded volumes 2019–2022 (% and TWh). *Source* (European Commission, 2023)

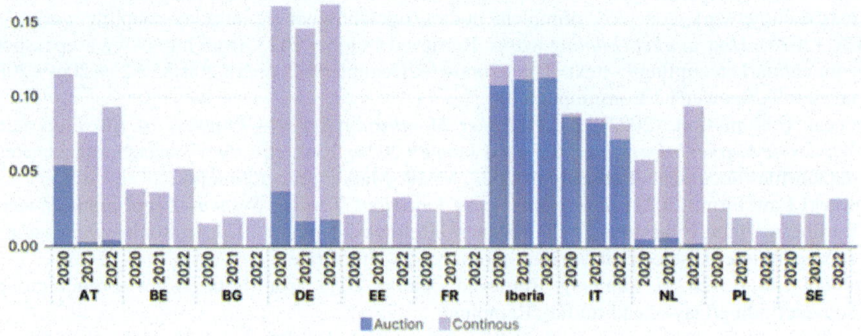

Fig. 7.8 Yearly churn rates in major European intraday markets by type of trade from 2020 to 2022. *Source* (European Commission, 2023)

share of intraday cross-border (cross-zonal) trading in relation to the overall continuous intraday trading volumes in Europe since the inception of SIDC (European Commission, 2023).

The increase in liquidity can be seen via the churn rates, depicted in Fig. 7.8. The churn rates indicate the number of times electricity generated in a market is subsequently traded (Glowacki Law Firm, 2021). The churn rates or factors have increased since 2020, by 6% in 2021 and by 16% in 2022 (European Commission, 2023).

Looking forward, in June 2024, IDA will be implemented to efficiently allocate and price the transmission capacity, thus reflecting their shortage and sending an adequate price signal to the market. It remains to be seen how these auctions will play out and the impact that they will have on the continuous intraday trading.

Bibliography

All NEMO Committee. (2023). *CACM Annual Report.* Retrieved November 2023, from https://nemo-committee.eu/assets/files/cacm-annual-report-2022.pdf

All NEMO Committee. (2023). *SIDC Stakeholder Report.* Retrieved October 2023, from https://nemo-committee.eu

Deutsche Börse Group. (2023). *Information technology, we make markets work.* Retrieved October 2023, from https://www.deutsche-boerse.com/dbg-en/products-services/ps-technology

Deutsche Börse Group. (2023). *Shaping energy trading.* Retrieved October 2023, from https://www.deutsche-boerse.com/dbg-en/products-services/ps-technology/ps-7-market-technology/ps-m7

ENTSO-E. (2023). *ENTSO-E Market Report 2023.* Retrieved November 3, 2023, from https://eepublicdownloads.blob.core.windows.net/public-cdn-container/clean-documents/nc-tasks/ENTSO-E_Market_Report_2023.pdf

ENTSO-E. (2023). *Single Intraday Coupling (SIDC).* Retrieved October 2023, from https://www.entsoe.eu/network_codes/cacm/implementation/sidc/#how-single-intraday-coupling-works

EPEX. (2023). *How market coupling works.* Retrieved October 2023, from https://www.epexspot.com/en/marketcoupling#:~:text=The%20now%2Dcoupled%20area%20is,85%25%20of%20European%20power%20consumption

European Commission. (2023). *2023 Market Monitoring Report—Progress of EU Electricity Wholesale Market Integration.* Retrieved January 2024, from https://www.acer.europa.eu/sites/default/files/documents/Publications/2023_MMR_Market_Integration.pdf

Glowacki Law Firm. (2021). *European Union Emissions Trading System—Legal point of view.* European Union Emissions Trading System. Retrieved January 2024, from https://emissions-euets.com/internal-electricity-market-glossary/954-churn-rate

Nordpool Group. (2023). *Ramping.* Retrieved October 2023, from https://www.nordpoolgroup.com/en/trading/Day-ahead-trading/Ramping/

PSE. (2021). *Single Intraday Coupling (XBID) Information*. Retrieved October 2023, from https://www.pse.pl/documents/20182/30599076/Single+Intraday+Coupling+%28XBID%29IP+-+2021.pdf/

Shinde, P. G. (2023). *A multi-agent model for cross-border trading in the continuous intraday electricity market.* ScienceDirect. Retrieved October 2023, from https://www.sciencedirect.com/science/article/pii/S2352484723008144?via%3Dihub

Single Intraday Coupling. (2021). *Single Intraday Coupling (SIDC) Questions & Answers.* PSE. Retrieved October 2023, from https://www.pse.pl/documents/20182/30599076/SIDC+QA_2021.pdf/

Terna. (2021). *Single Intraday Coupling: What it is and why it is so important for the electric system.* Terna Driving Energy. Retrieved October 2023, from https://lightbox.terna.it/en/insight/dispatching-single-intraday-coupling

Suzanna Vogeler is Vice President of Business Strategy at Deutsche Börse AG. She holds a degree in International Business (1992) from Jacksonville University and a Master in International Management (1996) from ASU Thunderbird School of Global Management. Since 1999, she has worked for the Deutsche Börse Group in various roles and since 2003 has led teams to achieve the successful initiation, definition and launch of several IT solutions with tight market-driven deadlines in the energy commodity sector. She was also Deutsche Börse AG's Project Manag-er for M7 XBID, the solution for the single intraday coupling, and is still one of the Project Executives.

Part III

Physics Meets Finance

Cross-Zonal Shipping and Nominations

8

Jean-Michel Reghem

Abstract

This chapter provides the reader with an overview of physical clearing and settlement processes needed for market coupling. Cross-zonal physical shipping is described in detail, including the roles of Nominated Electricity Market Operators, Central Counter Parties and Shippers. Different shipping arrangements are discussed, including the situation where multiple Nominated Electricity Market Operators are active in the same bidding zone. Examples of shipping and scheduling both in day-ahead and intraday auctions, and intraday continuous trading, are illustrated and explained. Finally, the TSOs matching and scheduling process, which is initiated after each auction or continuous trading gate, is described.

8.1 Introduction

Article 68 of the CACM Guideline (European Union, 2015) deals with clearing and settlement for single day-ahead and intraday coupling. There are two aspects to clearing and settlement of market coupling: physical and financial. This chapter covers the clearing and settlement of market coupling with a focus on the physical clearing and settlement which is better known as "***cross-border and local nomination processes***" towards TSOs systems. Chapter 9 deals with financial clearing and settlement with a focus on Nominated Electricity Market Operator (NEMO) processes. The difference between physical and financial clearing and settlement is briefly illustrated in Sect. 8.2.4.

J.-M. Reghem (✉)
Elia Transmission Belgium, Brussels, Belgium
e-mail: jeanmichel.reghem@elia.be

A. Estermann et al. (eds.), *European Electricity Market Coupling*, Future of Energy,
https://doi.org/10.1007/978-3-031-86315-8_8

To describe the cross-border and local nomination processes, the chapter starts with an introduction to TSO Scheduling, followed by a discussion on the difference between explicit and implicit allocation. For info, actual allocations are available on ENTSO-E's Transparency Platform (ENTSO-E, 2024). The important role of NEMOs, Central Counter Parties (CCPs) and Shippers (or Shipping Agents) are also described in the context of market coupling and nominations.

While monopolistic NEMO arrangements are acceptable in Bidding Zones, Article 45 and 57 in the CACM Guideline provide for the concept of "Multiple NEMO Arrangements" (MNA). NEMO arrangements are important for the cross-zonal shipping and nominations process and are introduced in detail in Sect. 8.2.5. It is also relevant to note that Multiple NEMO Arrangements can generate quite complex shipping arrangements which are illustrated in this section.

For the interested reader, shipping and scheduling in day-ahead auctions, intraday auctions, and intraday continuous trading are illustrated in Sect. 8.3. In addition, the matching and scheduling process of TSOs is also illustrated with some worked examples.

For reference, these topics are covered in Articles 43, 45, 49, 56, 57, 60, 61, 68, and 80(4) of the CACM Guideline.

8.2 Cross-Zonal Shipping

The following section introduces the different elements of cross-zonal shipping including why it is needed, the TSO responsibilities, and the responsibilities of other market stakeholders. The difference between physical and financial shipping is also explained.

8.2.1 The Basis of TSO Scheduling

Each TSO operates its grid operations and schedules with its own nomination and scheduling systems and processes as well with the specific local rules applying to its Balance Responsible Parties (BRP). There are several differences for each TSO, however a principle is common: it is crucial for TSOs to have access to the "balanced net position" of its control area and of each BRP, both in the Day-ahead and Intraday timeframes.

At BRP level, this balanced net position reflects the combination of injection nominations and offtake nominations from power units, industries, and Distribution System Operators (DSOs) that are part of the BRP perimeter. Additionally, it includes local hub nominations, which represent internal commercial trade schedules between BRPs, and cross-zonal nominations (external commercial trade schedules) (Figs. 8.1 and 8.2).

Local hub offtake and injection nominations play a crucial role in enabling the TSOs to monitor the balance status of each BRP and anticipate grid flows.

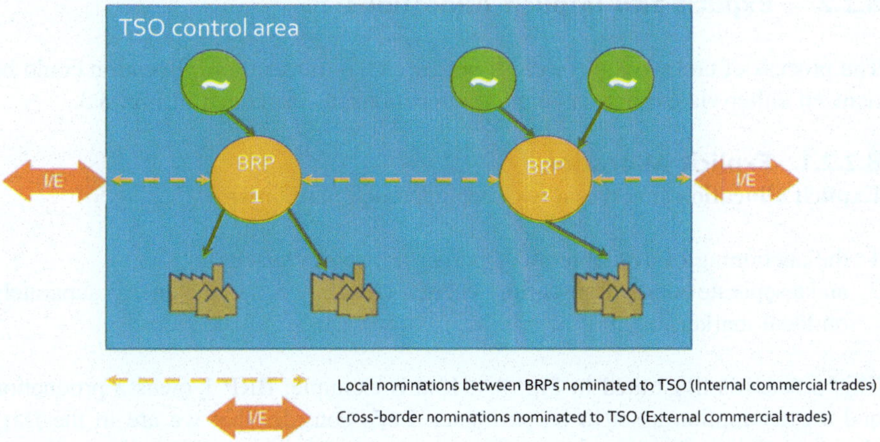

Fig. 8.1 TSO control area and role of BRP. *Source* Author's own illustration

Fig. 8.2 Perimeter of a BRP. *Source* Author's own illustration

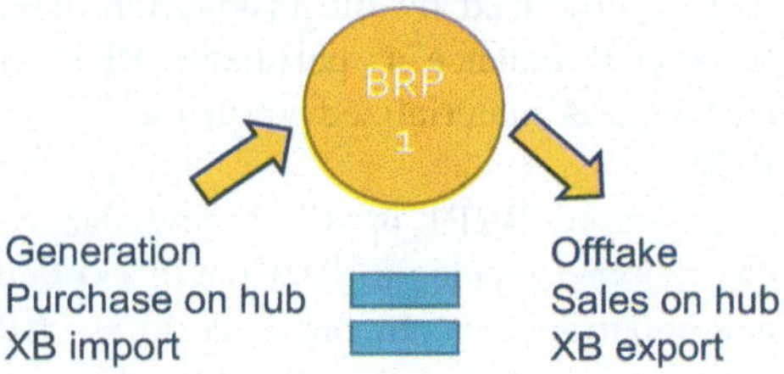

However, it is equally essential not to overlook the significance of cross-zonal nominations, as they are vital for the TSO's scheduling system.

A control area (or balancing area) is a geographical region within an electricity grid managed by one TSO or a group of TSOs. Each control area is responsible for maintaining the balance between electricity generation and consumption within its boundaries. The balanced net position of the control area is the aggregation of all external commercial trade schedules and determines whether the TSO's control area will experience a net import or export position after the Day-ahead and Intraday timeframes. This information becomes invaluable for conducting security analyses, initiating redispatching processes, and ensuring smooth grid operations by resolving potential congestion issues.

In the case of explicit allocation, BRPs are directly responsible for nominating cross-zonal transactions to the TSO. However, when implicit allocation is in effect (as explained in the next section), a specific type of BRP known as ***shipping agent or shipper*** takes on the responsibility of managing the cross-zonal nomination process in collaboration with the TSO.

8.2.2 Explicit and Implicit Allocations

The process of cross-zonal (or cross-border (XB)) transactions allocation could be realized either via explicit or implicit allocations, as explained in Fig. 8.3.

8.2.2.1 Explicit Allocation

Explicit allocation involves a two-step approach per border:

1. the auctioning of available cross-border transport capacity
2. and a separate energy transaction where traders buy or sell the energy separately on local markets.

This process is illustrated in Fig. 8.4. In this example, BRP 1 owns a production and a consumption asset in its perimeter. Let's consider that we are in the Day-ahead timeframe. For a specific imbalance settlement period (ISP), the production asset is expected to produce some energy represented in Fig. 8.4 by a small rectangle. The consumption unit however expects to consume more energy than what will be produced by the production asset (represented by the bigger rectangle). In order to balance its perimeter, BRP1 buys additional energy from BRP2. This exchange is materialized via a local nomination to the TSO between BRP2 and BRP1.

However, BRP1 is still in imbalance for the specific period, represented by the chopped rectangle. BRP1 will **explicitly** import this missing energy from the neighbouring area. In order to do so, BRP1 will first buy a piece of the cross-border capacity between the 2 zones via an auction organized by JAO, the Joint Allocation Office, operating the explicit auctions on behalf of the TSOs. After the

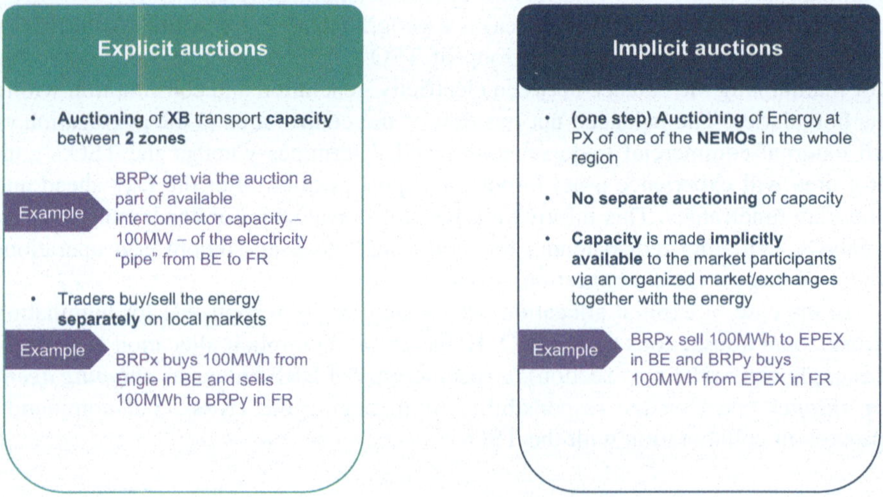

Fig. 8.3 Explicit vs implicit allocations. *Source* Author's own illustration

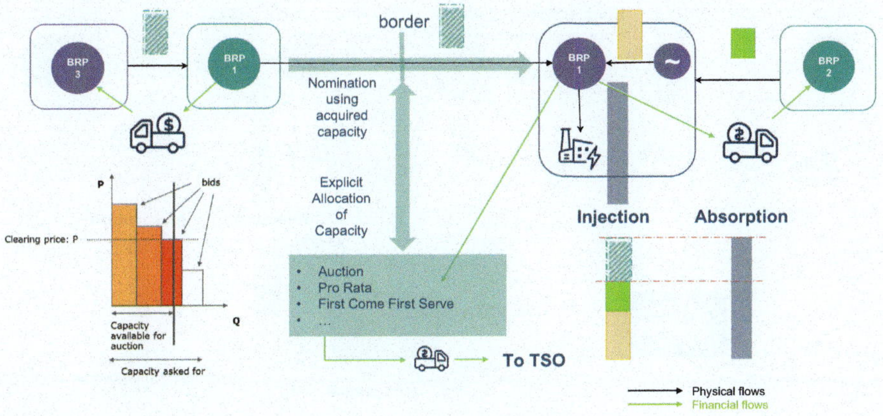

Fig. 8.4 Explicit allocation. *Source* Author's own illustration

result of the auction, if BRP1 bids the requested quantity with a price allowing him to be part of the winners of the auction, it will pay the clearing price and will get in exchange a "right" to use this capacity.

This is not enough however: BRP1 will also have to buy locally from BRP3 the missing energy in the neighbouring area, using another local BRP account for this area. Eventually, BRP1 will nominate cross-border the "chopped rectangle" energy between the 2 zones, using the acquired rights. Once nominated, this additional capacity imported from the neighbour zone will balance the perimeter of BRP1 for this specific period. Besides local and cross-border nominations, BRP1 had to pay BRP2 and BRP3 for the required energy. BRP1 also paid the clearing price to the TSOs on either side of the border, covering the requested capacity.

The explicit cross-border allocation process was then done in two steps: first the acquisition of a right on the capacity, via an auction, and then the purchase of energy in the neighbouring area before its transmission via a cross-border nomination.

8.2.2.2 Implicit Allocation

Implicit allocation, in contrast, streamlines the capacity allocation process into a single step. Energy and capacity available on the borders are auctioned together. There is no separate auctioning of capacity, and it is made implicitly available (along with energy) to the market participants through organized markets via the power exchange of one NEMO.

As illustrated in Fig. 8.5, BRP1 is operating in a day-ahead timeframe as in the previous example: the chopped rectangle energy is still missing in order for BRP1 to be balanced, and it needs to be imported cross-border. This time however, BRP1 will buy this missing energy via an implicit allocation process organized by its NEMO in its area. BRP1 provides a bid requesting to buy energy equal to the

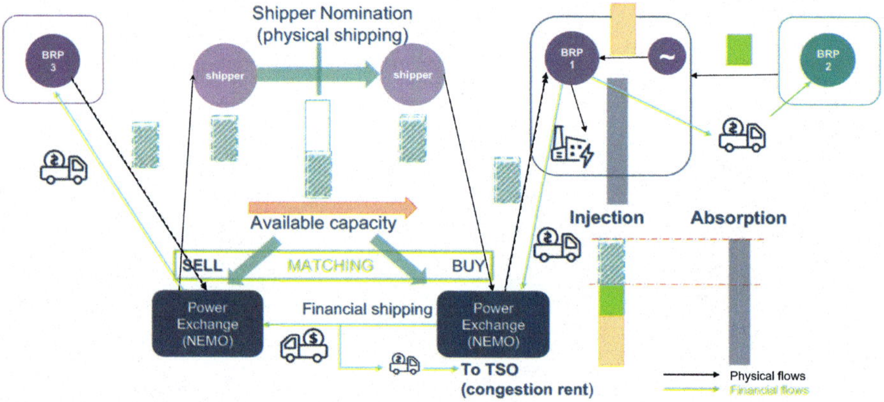

Fig. 8.5 Implicit allocation. *Source* Author's own illustration

chopped area at a specific condition (for instance maximum price it is ready to pay).

The local NEMO is coupled with another NEMO hub in another zone. A NEMO hub can be a separate local market of the same NEMO or a different NEMO entity. BRP3 entered a sell bid (offer) on this other NEMO hub for a compatible amount of energy and price. Both bids are matched by the coupling algorithm.

BRP1 gets its missing energy, pays its NEMO for this energy at the local clearing price defined by the market coupling algorithm. BRP3 will get the local clearing price defined by the market coupling algorithm in exchange for the energy sold to its NEMO. The price spread (equal to the difference of clearing prices in both zones) is transferred to JAO on behalf of the relevant TSO, as congestion rent. Finally, a shipper is transferring the energy between the 2 NEMOs of the 2 zones, via cross-border import nominations and export nominations to both TSOs.

For BRP1 and BRP3, the process is simple and done in one step. They do not know who is the seller or who is the buyer, and BRP1 does not know if the energy is coming from another country or from its own local zone, via another BRP customer of the NEMO. In this example the capacity was allocated implicitly together with the exchange of energy.

Central Counter Parties (CCPs) and Shippers of the involved NEMOs are responsible for transferring the electrical energy (nominations to the TSOs = physical settlement) and the money (financial settlement) between market participants and TSOs, where applicable. NEMOs (and associated sub-roles) are therefore key players in Implicit Allocation schemes.

8.2.3 Market Coupling and Nominations: The Role of NEMOs, CCPs and Shippers

In the CACM Guideline, NEMOs are responsible for the Market Coupling Operator (MCO) function. The CACM Guideline defines the different missions of the NEMO separated into 3 major roles: the NEMO itself, the Central Counter Party (CCP), and the Shipping Agent (SA—or also called Shipper), where the roles of CCP and SA can be operated either by the NEMO (or delegated to a third party entity) or by a TSO (appointed by a member state). In some countries (for instance where NEMOs are monopolistic and regulated entities), this task is assigned to a TSO. CCPs and SAs are the entities involved in the process of local or cross-border nominations, in direct interaction with the TSOs, BRPs, and other SAs, as illustrated in Fig. 8.6.

The mission of the NEMO consists notably of "receiving orders from market participants, having overall responsibility for matching and allocating orders in accordance with the single day-ahead and intraday coupling results, publishing prices and settling and clearing the contracts resulting from the trades according to relevant participant agreements and regulations".

The Central Counter Party (CCP) is the entity with the task of entering into contracts with market participants, by novation of the contracts resulting from the matching process, and of organizing the transfer of net positions resulting from capacity allocation with other central counter parties or shipping agents. In other words, the CCP is then the "BRP of the NEMO", which will nominate locally to the local BRPs the results of the market coupling.

The Shipping Agent (or Shipper) is the entity with the task of transferring net positions between different CCPs. The shipper is a special BRP associated mostly with the NEMO or the CCP, but also sometimes to a TSO of the zone or operating

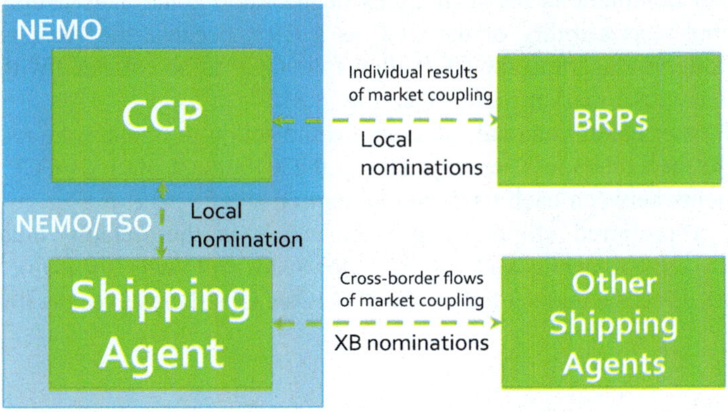

Fig. 8.6 NEMO, CCP, and SA roles, and their interactions with BRPs and other SAs. *Source* Author's own illustration

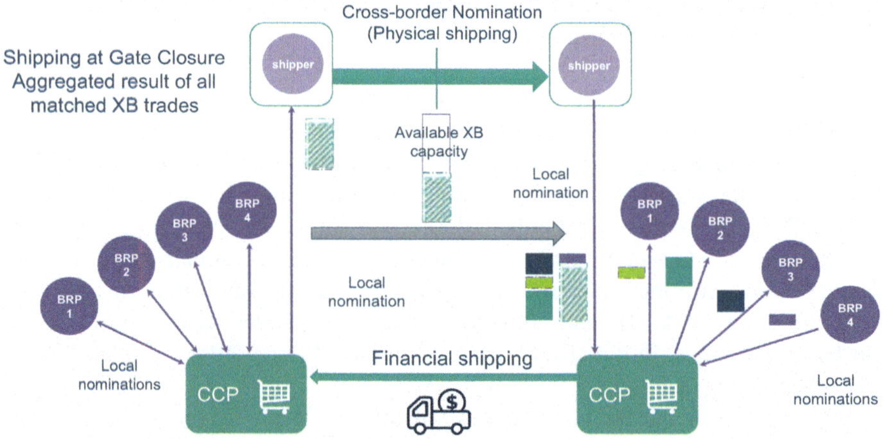

Fig. 8.7 The role of a shipping agent in cross-border nominations between CCPs. *Source Author's Own Illustration*

a border. It enables the transfer of energy in an implicit allocation between CCPs via cross-border nominations to the TSOs, as illustrated in Fig. 8.7.

8.2.4 Difference Between Physical and Financial Settlement

Physical settlement is the action to exchange energy between CCPs of different zones (cross-zonal cross-CCP exchanges) or in the same zone (intra-zonal cross-CCP exchanges). In other words, this is the cross-border nomination to TSO systems of the cross-border energy flows between NEMOs and Bidding Zones or the local nominations between CCPs in the same zone. This action is linked to balancing responsibility of the CCP as a BRP, because the local nomination from the shipping agent to the CCP allows the CCP to be balanced with the local nominations to the market parties (see Fig. 8.8).

On the other hand, financial settlement is the action to transfer the money associated with the exchange of energy. This is the settlement of cross-CCP payments and payments between each CCP and its participants.

Under a preferred shipper approach, which is the default situation (see Sect. 8.2.5), the financial and physical flows are following the same path, but in opposite direction and settled by the NEMOs/CCPs themselves, as illustrated in Fig. 8.9.

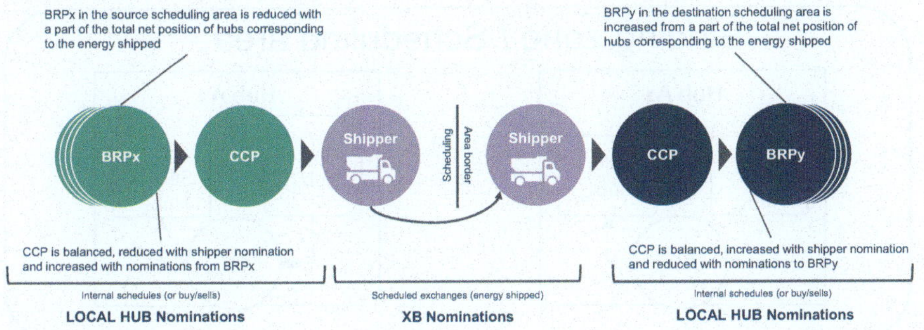

Fig. 8.8 Local and cross-border nominations for the physical settlement of energy. *Source* Author's Own Illustration

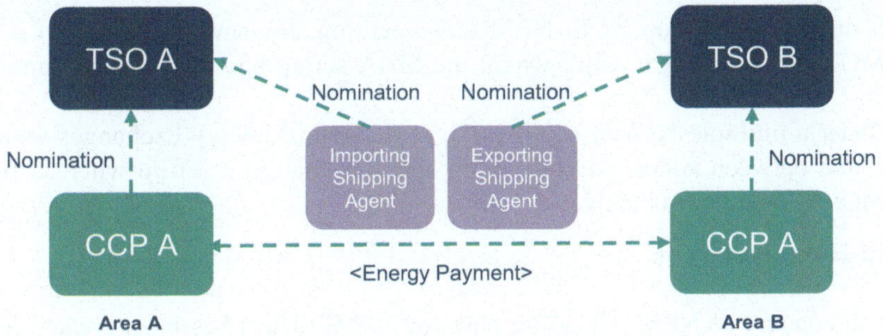

Fig. 8.9 Physical and financial settlement paths. *Source* Author's own illustration

8.2.5 Multiple NEMOs: MNA and Shipping Arrangements

In the CACM Guideline, Articles 45 and 57 introduced the concept of "Multiple NEMO Arrangements" (MNA), requiring TSOs to develop proposals for the day-ahead and intraday timeframes describing how more than one NEMO could be active in the same bidding zone. These arrangements are national, but "MNA countries" in the same regions managed to define common rules allowing the interoperability and the cross-border exchanges, deemed necessary for the cross-border nominations in day-ahead and intraday timeframes.

Not all bidding zones are in MNA configuration. The so-called "monopolistic NEMO areas" where a single NEMO is active and regulated by a local national regulator, had also to be connected to the neighbouring bidding zone in the MNA set-up. The process to introduce MNA within the different bidding zones but also on bidding zone borders in each region and on borders between regions has been in progress since 2016 and is expected to be completed in Single Day-Ahead Coupling (SDAC) and Single Intraday Coupling (SIDC) in 2025 based on the current configuration. However, it is useful to note that this process could be reopened

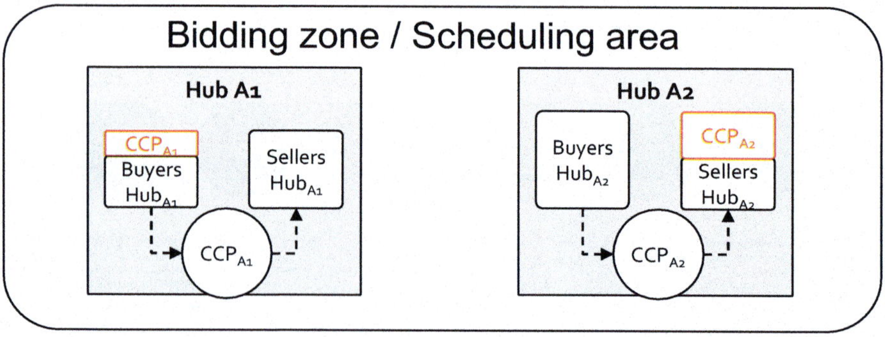

Fig. 8.10 Intra-hub settlement in MNA set-up: concept of NEMO hub. *Source* JM Reghem/B. Lematayer/Elia/RTE

each time a country opens to NEMO competition, for new borders or if a new NEMO enters a MNA country where the MNA set-up has not been implemented yet.

Under a multiple NEMO context, the settlement of energy exchanges within a zone and between zones becomes more complex than in a set-up where a single NEMO/CCP/Shipper is active:

Intra-Hub Settlement:

- The concept of NEMO Trading Hub (or NEMO Hub) has been created in the different local MNA implementations. A NEMO Trading Hub means, for the day-ahead and/or intraday timeframe(s), the place where a NEMO collects the bids from the members of the power exchange it operates, on a bidding zone level (or, if applicable, scheduling area). Figure 8.10 illustrates this concept.
- Locally, the settlement between market participants of a Trading NEMO Hub and CCP/NEMO of that Trading Hub is unchanged and principles remain the same (whether one or more NEMOs are active in a BZ/SA): NEMOs act as central counterparty (CCP) to market participants for all matched orders on their hubs (financial settlement) and nominate volumes of matched orders per BRP (local hub nominations) to TSOs (physical settlement).
- Even though the TSO must integrate more than one NEMO/CCP active in the bidding zone, the situation is also quite transparent for the system operator.
- For each timeframe, each NEMO Hub has a net position which is the position covered by matched bids on other NEMO Hubs, within the same zone (Intra-Zonal) and/or cross-border (Cross-zonal)

Intra-Zonal Settlement:

- MNA introduces this new situation, when there is more than one NEMO Hub in the same zone. Consequently, there is a need to organize the settlement of

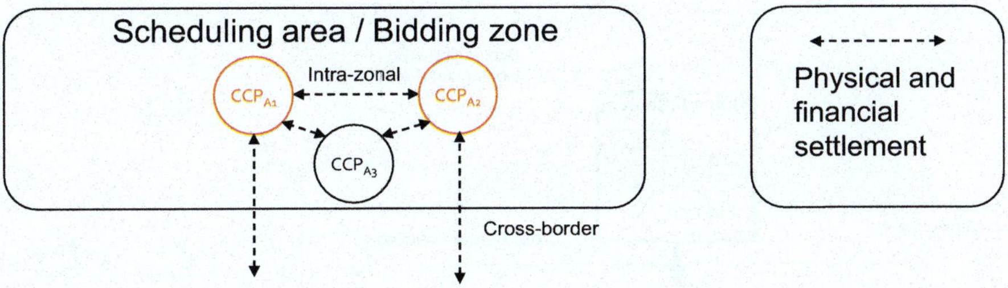

Fig. 8.11 Intra-zonal physical and financial settlement. *Source* JM Reghem/B. Lematayer/Elia/RTE

energy exchanges between NEMO Hubs within this zone (Bidding Zone or Scheduling Area).

- **NEMOs** act as CCP to other CCPs within the bidding zone (or Scheduling Area in Germany). These exchanges, illustrated in Fig. 8.11, consist of
  - Physical settlement of volume exchanged between hubs through relevant BRP
  - Financial settlement with other CCPs and/or Shipping Agents
- Each CCP performs physical and financial settlements according to scheduled exchanges resulting from market coupling of the specific timeframe. For the TSO, these exchanges are materialized with local nominations between CCPs and/or between CCPs and Shipping Agents.

Cross-Zonal Settlement

- Cross-zonal exchanges can occur either (a) between bidding zones or (b) between scheduling areas and a bidding zone, or (c) between scheduling areas and other bidding zones. These exchanges involve not only the CCPs, but also the shipping agents, where applicable, as such entities are responsible for the settlement of energy exchanges between NEMO Hubs of adjacent zones.
- In Fig. 8.12, the different actors involved in the financial settlement and the different steps for the physical settlement are illustrated in an example with 2 different NEMO Hubs in 2 adjacent Bidding zones.
- Fig. 8.13 illustrates another example between a Bidding Zone 1 and its single Scheduling Area A containing 3 NEMOs and one single shipping agent SA1, and a Bidding Zone 2 containing 2 Scheduling Areas B and C, each of them containing 3 NEMOs with the same single shipping agent SA2. Cross-zonal exchanges between the 2 scheduling areas of Bidding Zone 2 are performed by SA2, while cross-zonal exchanges between Scheduling Area B (or C) and Bidding Zone 1/Scheduling Area A are performed by SA1 and SA2 on both sides of the interconnectors.

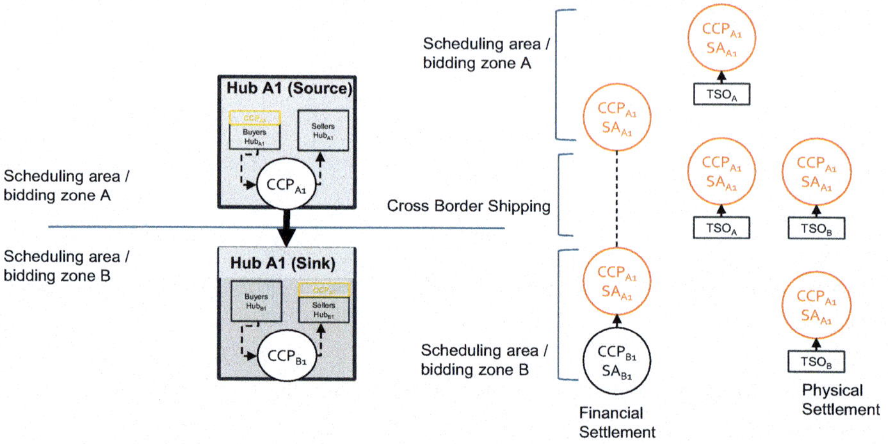

Fig. 8.12 Cross-zonal financial and physical settlement. *Source* JM Reghem/B. Lematayer/Elia/RTE

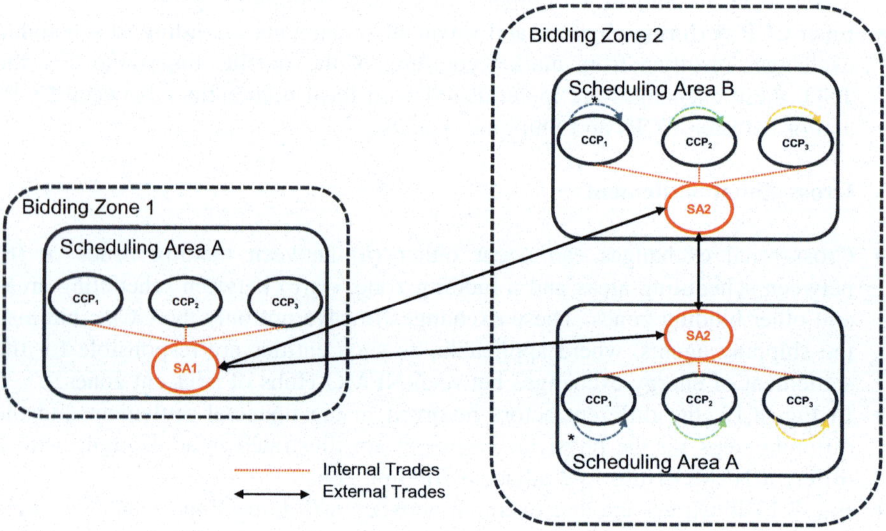

Fig. 8.13 Example of cross-zonal exchanges. *Source* JM Reghem/B. Lematayer/Elia/RTE

- By default, on borders between MNA areas, the "preferred shipper" approach is used. Each NEMO assigns a preferred Shipping Agent (pSA) for its NEMO hub within each Bidding zone. By default, each NEMO selects its own CCP as its own preferred shipper.
  - The preferred shipper performs physical and financial settlement according to scheduled exchanges resulting from coupling.

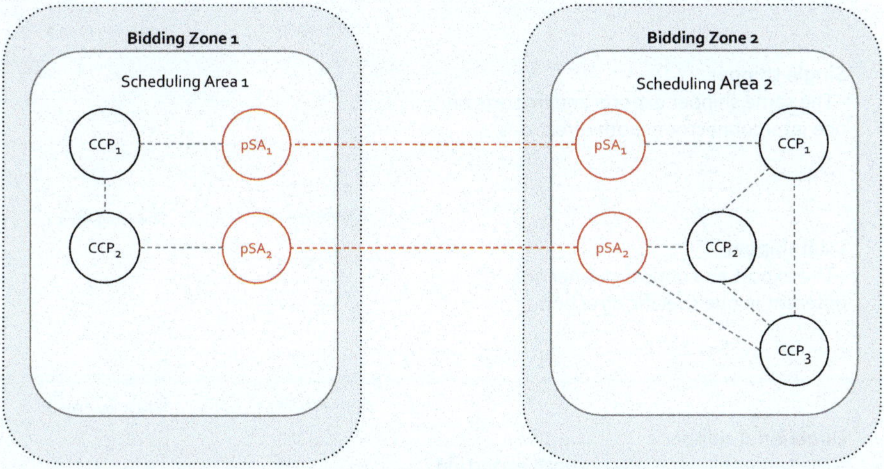

Fig. 8.14 Example of "preferred shipper approach" between 2 MNA zones. *Source* JM Reghem/B. Lematayer/Elia/RTE

- In the intraday continuous markets the preferred shipper of the source zone is responsible for physical settlement in the source and sink zone and between zones.
- The preferred shipper of the source zone performs the financial settlement in the sink zone with the sink CCP.
- With the preferred shipper, the physical and financial settlement follow the same path.
- It is the NEMO/CCP which decides its own shipper, if it is not obliged to use a dedicated single shipper on the border.

This concept is illustrated in Fig. 8.14, where CCP1 and CCP2 are using their own preferred shipper for cross-zonal exchanges, where CCP3 is using the same pSA2 as CCP2, via a specific agreement for cross-zonal exchanges.

On some borders (for example on commercial interconnectors or on borders with a zone of a monopolistic NEMO), other specific shipping arrangements than "preferred shipping approach" can be predefined and must then be followed by NEMOs operating in the zones connected by this interconnector. Figure 8.15 illustrates three of the more common arrangements which could be configured on a border.

On borders between a MNA and a non-MNA area, even if these three situations could be encountered, with a single shipper in the MNA area operating the cross-zonal settlement for all NEMOs, there could also be a hybrid preferred shipper / dual shipper situation where there is a single shipper in the non-MNA zone and multiple preferred shippers in the MNA zone (Fig. 8.16).

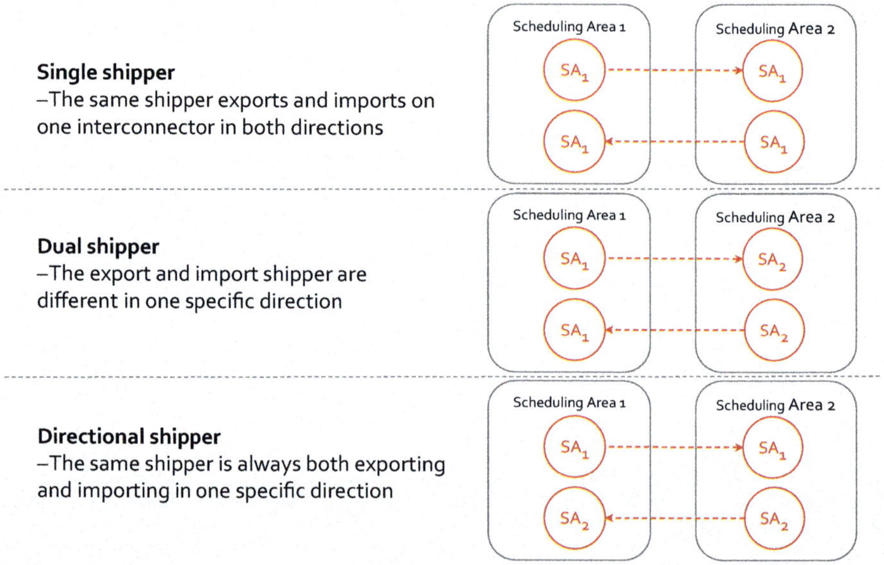

Fig. 8.15 Other shipping arrangements. *Source* Author's own illustration

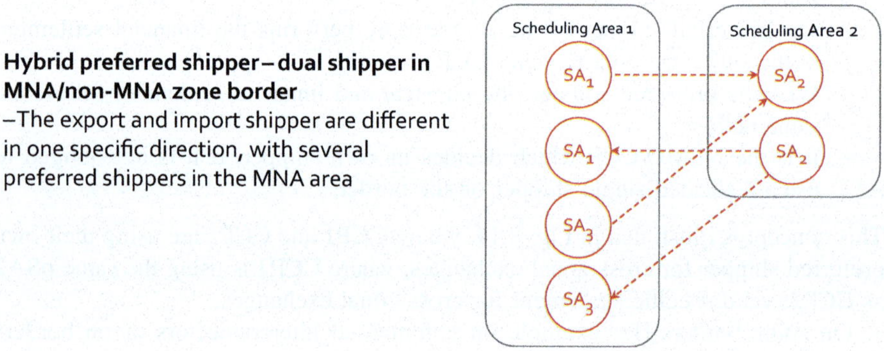

Fig. 8.16 Hybrid preferred shippers: dual shipper. *Source* Author's own illustration

In some of these cases and in accordance with Article 68(6) of the CACM Guideline, the shipping agents are the TSOs operating the zones. For instance, on the border Austria-Hungary, MAVIR (Hungarian TSO) is performing the shipping agent role on the Hungarian side of the border.

8.3 Illustrative Examples

8.3.1 Shipping and Scheduling in Day-Ahead and Intraday Auctions

The outputs from the Euphemia algorithm in the day-ahead market coupling (or intraday auctions) are the price, the net positions and the flows on HVDC interconnectors. To implement the physical and financial settlements, TSOs and NEMOs need to know the commercial flows, which are the scheduled exchanges on each bidding zone/scheduling area borders for each market time unit.

In the simplified example illustrated in Fig. 8.17 and involving four bidding zones, the algorithm has provided for each of the 4 zones a price and a net import/export position for a specific market time unit. Following this, a second step involves the bilateral scheduled exchanges (SEC) process. This process utilizes mathematical optimization to minimize the exchanges between zones (represented by the arrows). The mathematical optimization model for energy flows relies on linear and quadratic coefficients for each border as shown in Eq. 8.1. Linear coefficients emphasize shorter paths and influence flow allocation across interconnectors (prioritization of one interconnector versus another one), while quadratic coefficients manage congestion costs and deter loop flows between regions. When multiple routes share the same length and quadratic cost, the flow is evenly distributed without favouring any specific route, promoting a balanced and efficient energy exchange.

It is important to note that while these SEC represent commercial flows, they do not directly reflect physical flows. The physical limits were considered during the allocation (and the price/net position definition) but not for the scheduled exchanges calculation which is basically used only for the settlement process and will be translated into cross-border nominations made by shipping agents to the TSOs' systems. The only constraint is that the flows must not exceed the Available Transmission Capacity (ATC), which governs the flows on HVDC interconnectors. The purpose of these commercial flows is to establish cross-border schedules that

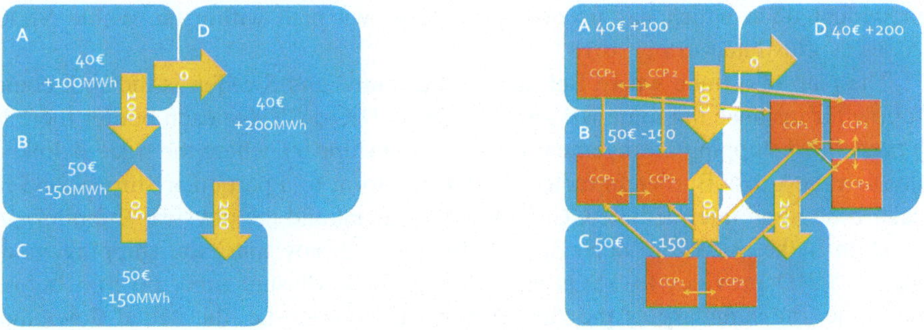

Fig. 8.17 Example of scheduled exchanges calculation in day-ahead coupling. *Source* Author's own illustration

help balance the positions of the bidding zones. Subsequently, security analysis, based on the net positions obtained (and thus not on the commercial flows), will be conducted to forecast the actual physical flows.

$$\text{Scheduled Exchanges Calculation} \\ \left(\sum lc_{i,h}\varphi_{i,h} + \sum qc_{i,h}\varphi_{i,h}^{2}\right) \tag{8.1}$$

With:

- $lc_{i,h}$ = linear cost associated to line i for period h
- $c_{i,h}$ = quadratic cost associated to line i for period h
- $\varphi_{i,h}$ = flow on line i for period h

Once the SEC between bidding zones are determined, they are further divided into SEC between scheduling areas (specifically for Germany which has 4 scheduling areas in its Bidding Zone as of 2024), utilizing fixed pro-rata rules.

Additionally, the algorithm calculates the SEC between NEMO Hubs (NH) as part of the third step in the SEC process. This step takes into account the pre-calculated flows between bidding zones and the net positions of each individual NH following the auction. The algorithm ensures that the flows between NH align with the flows between bidding zones, maintaining the same direction. These calculated flows between NH are crucial for the cross-border nominations made by shipping agents of the NEMOs (in the case of preferred shipper) or the designated shipping agent (in other shipping arrangements) to the border's respective TSOs.

8.3.2 Shipping and Scheduling in Intraday Continuous Trading

Unlike in auctions where the net positions are the output of the implicit cross-border allocation, in continuous trading the output of the allocation is the individual trade which has its own price and own path within the whole SIDC zones as illustrated in Fig. 8.18.

The path of a trade where Belgium is the source and Sweden is the sink could follow for instance the route BE ->FR ->DE ->DK ->SE. On each border there is a predefined shipping arrangement (or preferred shipper), where the trade is linked through the Cross Border Intraday (XBID) platform to a particular shipper and in each zone to an intermediate CCP (for the financial and the physical shipping).

Even if the trades are occurring continuously at any time, the shipping module of XBID provides to TSOs at each trade gate closure time, 60 min before real-time, the summary of the shipping of each border, i.e. the netted flows that each shipper must nominate for the considered market time unit (MTU), being the sum of all individual trades crossing the border and using this shipper for the

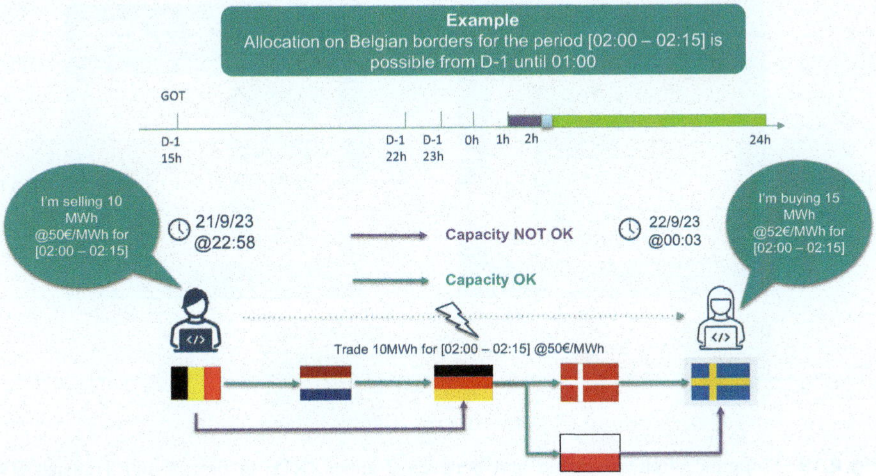

Fig. 8.18 Continuous trading allocation example. *Source* Author's own illustration

considered MTU. On each border, gates are closing either 24, 48 or 96 times per day, depending on the resolution of the SIDC continuous MTU (15, 30 or 60 min) offered for the interconnector and the Operational Time Unit (OTU). Once the TSOs receive this information, the nomination is done immediately by the shipper or by the TSOs (nomination on behalf of the shipper).

The shipping handover process, which depends on the configured preferred shipper or the shipping arrangement on the border, involves:

- the selection of the incoming shipper in a zone,
- the interim CCP,
- and the outgoing shipper to the subsequent zone.

In MNA areas, where multiple NEMOs coexist within each bidding zone and scheduling area, the preferred shipper principle serves as the default shipping arrangement established in the XBID platform. Taking the example of a trade between Spain and Romania, XBID selects the shipper at each border and designates the intermediate CCP within each zone. For instance, if Nordpool/EMCO acts as the CCP in France, it will employ its own shipping agent to nominate to itself as shipper and CCP in Germany (Fig. 8.19).

When borders connect an area with a single active NEMO, typically a single or dual shipper scheme is implemented. In such cases, the handover from a MNA area to a non-MNA area follows a straightforward process utilizing the predefined shipping arrangement of the border. Let's consider an example involving Austria and Hungary. If, in Austria, the involved CCP is EPEX and ECC is the delegated shipping agent, then ECC will handle the shipment to the shipper of HUPX, which is the sole NEMO operating currently in Hungary as shown in Fig. 8.20.

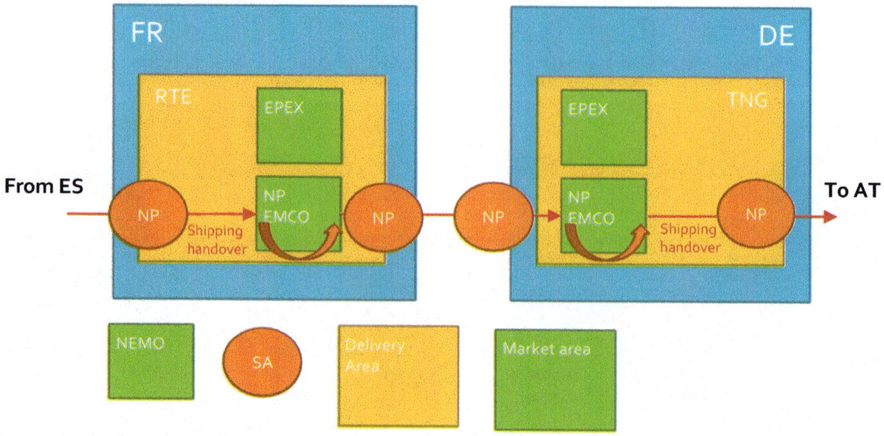

Fig. 8.19 Shipping handover within and between 2 zones (RTE in France and Transnet BW (TNG) in Germany) using preferred shipper approach in continuous trading. *Source* Author's own illustration

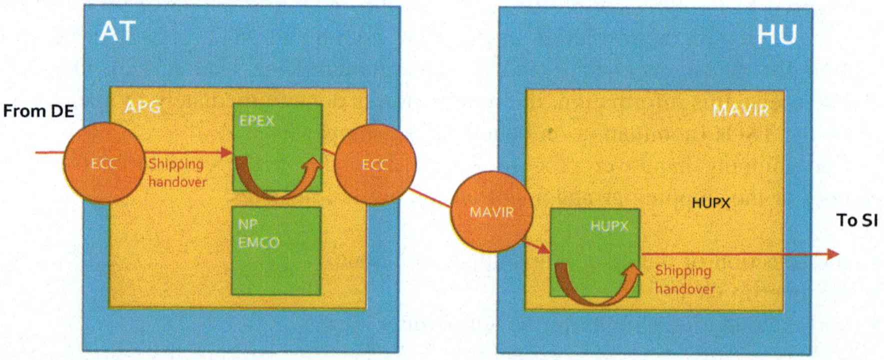

Fig. 8.20 Handover from a MNA area to a non-MNA area in continuous trading. *Source* Author's own illustration

When it comes to the opposite direction, where a trade occurs from a single NEMO area to a multiple NEMO area, the decision-making process becomes more intricate. The single NEMO must determine how to choose among the multiple active NEMOs in the other area. This is where the concept of a transit shipper comes into play. To address this complexity, a rotational basis is established, typically on a three-month cycle between the shipper of NEMOs active in several MNA areas (as of early 2024, EPEX and Nordpool). For instance, let's refer to our previous example, but in the opposite direction as shown in Fig. 8.21. If ECC (delegated shipping agent of EPEX) is designated as the transit shipper, MAVIR will handle the shipping and subsequently hand it over to Austria with ECC as the selected shipper. After three months, when a similar situation arises, MAVIR will

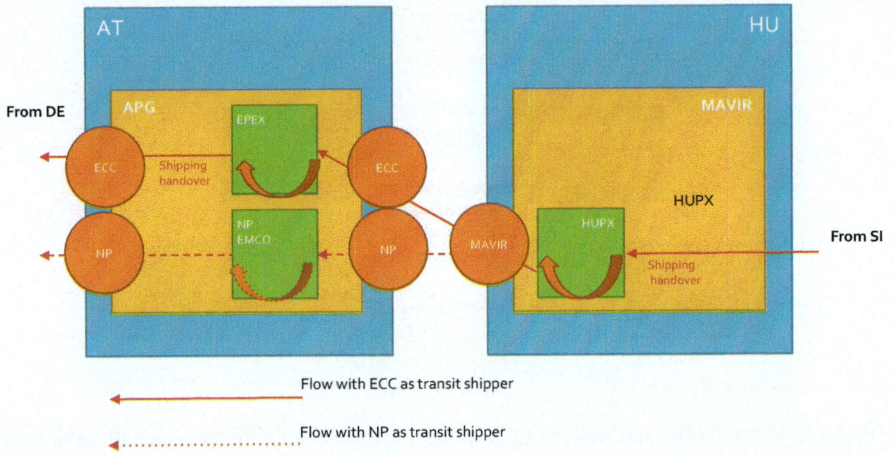

Fig. 8.21 Handover from a non-MNA area to a MNA area in continuous trading. *Source* Author's own illustration

automatically choose NP/EMCO as the shipper for the shipping handover in Austria. This rotational approach ensures fairness and equal opportunities for different NEMOs to act as transit shippers, enabling efficient trade between single NEMO and multiple NEMO areas.

8.3.3 Matching and Scheduling Process of TSOs

After each auction or continuous trading gate, and once the TSOs receive the cross-border nominations from the shippers (or from Balance Responsible Parties in the case of explicit allocations), a matching process is initiated between the neighbouring TSOs. During this process, they exchange their versions of the cross-border nominations and validate that they both have the same results. If there are no issues and the nominations align, the process proceeds smoothly. However, if any discrepancies or problems arise, a manual process between the two operators is undertaken to resolve the issue before the actual energy delivery takes place. This validation and matching procedure ensures the accurate and coordinated transmission of electricity across borders and helps to prevent potential operational challenges or imbalances.

After the matching process is completed and the cross-border nominations are validated, the verified flows are transmitted to the Energy Management Systems (EMS) of the TSOs. These validated flows are then utilized in the creation of the Individual Grid Model (IGM) for subsequent iterations of regional capacity calculations and local security analysis. Additionally, the validated flows are also employed in the security analysis conducted by Regional Coordination Centers (RCCs).

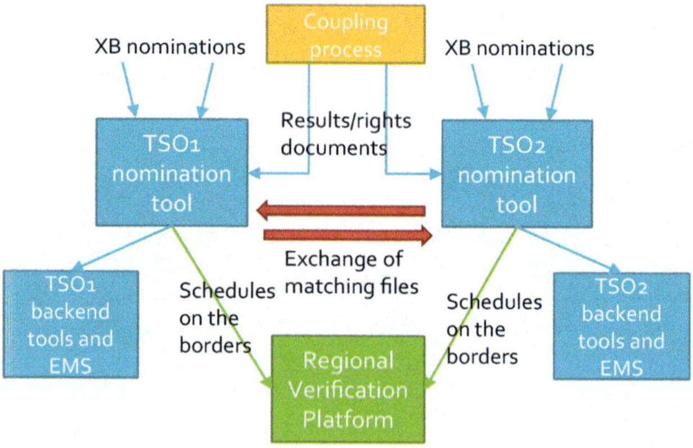

Fig. 8.22 Matching and verification of scheduling process. *Source* Author's own illustration

Once the matching process is completed, the pan-European verification process is initiated. In this process, each TSO sends its matched schedules to its designated regional verification platform (VP), such as the Regional Continental Europe Verification platform (RG CE VP) or verification platforms for Nordic countries, among others. Each VP collects the schedules for each border of the region, performs additional verification checks to ensure accuracy, and then aggregates the information on cross-border flows within its region (Fig. 8.22).

The regional verification platforms then send this aggregated information on cross-border flows to the Pan-European Verification Function (PEVF), where further verification and aggregation takes place for flows between regions. Simultaneously, RCCs are working on merging the IGMs from each TSO into regional Common Grid Models (CGMs). When RCCs combine all the CGMs into one comprehensive grid model for the entire Europe (European Merging Function—EMF), they rely on the output from PEVF to validate that the net position of each area aligns with the scheduled commercial flows (Fig. 8.23).

This pan-European verification and merging process ensures the consistent and reliable management of energy transmission across interconnected regions, maintaining grid stability, and security throughout Europe.

8.4 Outlook

The process of physical shipping, including local and cross-border nominations to the TSOs is becoming more and more important and challenging as the number of nomination gates closer to real-time increase.

In continuous trading, automatic cross-border nomination agreements on behalf of the shippers are implemented more and more by the TSOs. These automatic cross-border nomination agreements use allocation results from the continuous trading platform to speed up the process of collecting schedules on the borders.

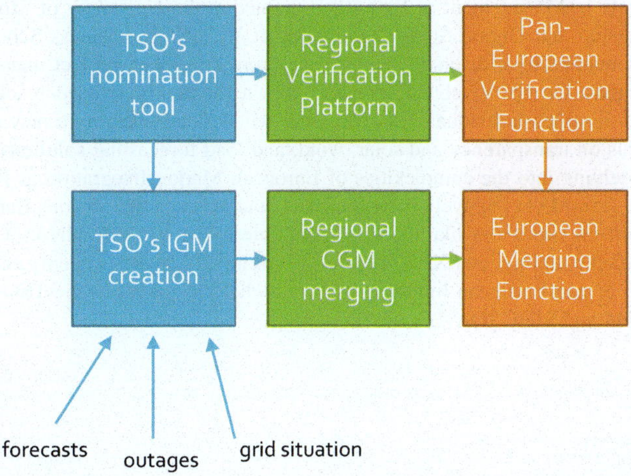

Fig. 8.23 IGM and CGM creation process and verification. *Source* Author's own illustration

Once this is complete, the process of matching and verification of schedules as well as the grid security assessments can commence.

Considering the upcoming projects like the European balancing platforms associated with a specific new balancing timeframe capacity calculation (ACER, 2024), it would be valuable in the future to continue the integration, simplification, and automatization of the matching and verification process. This could be done for instance by moving the bilateral matching, performed border by border by the TSOs, to a central (regional) matching process, performed by the regional verification platforms themselves or the regional capacity calculations platform, using results directly from the XBID platforms.

Bibliography

ACER. (2024, 12). *Balancing Energy Platforms*. Retrieved 12 15, 2024, from ACER - Electricity Balancing: https://www.acer.europa.eu/electricity/market-rules/electricity-balancing/balancing-energy-platforms

ENTSO-E. (2024, 12 13). *ENTSO-E Transparency Platform*. Retrieved 12 15, 2024, from https://transparency.entsoe.eu/dashboard/show

European Union. (2015). Commission Regulation (EU) 2015/1222 of 24 July 2015 establishing a guideline on capacity allocation and congestion management. *Official Journal of the EU*. Retrieved December 15, 2024, from https://eur-lex.europa.eu/legal-content/EN/TXT/?uri=CELEX%3A02015R1222-20210315

Jean-Michel Reghem works at Elia, the Belgian Transmission System Operator, as Manager of the European Market Integration team in the Markets, Grid and Off-shore department. He graduated in 2000 from the Mons Institute of Technology (Faculté Polytechnique de Mons), Belgium, as an electrical engineer, with a specialization in telecommunications. Additionally, Jean-Michel holds a certificate in Management of Innovation, earned at the University of Mons in 2011 and attended Leadership and Digital Strategy programs at VLERICK Business School. His professional career is marked by a diverse range of roles from developer, project manager, to product manager at Acapela Group, a Belgian company known for its text-to-speech solutions, particularly catering to disabled individuals. In 2013, Jean-Michel embarked on a new chapter, joining Elia with a focus on transparency and solar, wind, and load forecasting solutions before shifting gears in 2016 delving into the complexities of European Market Integration as Project and Program Manager. Jean-Michel actively participated in, and at times, led various European projects, leaving his mark on endeavors like the CWE Multiple NEMO (CWE MNA), Core Day-ahead Flow-based market coupling, and XBID projects. Currently, his work is deeply entwined with the implementation of pan-European Intraday Auctions, where he serves as the TSO co-convener of the SIDC QARM and SIDC IDA TG working groups.

Clearing and Settlement

9

Robert Wand

Abstract

This chapter introduces financial clearing and settlement processes in local power spot markets as well as an overview of existing financial clearing and settlement arrangements between NEMOs or their Central Counter Parties (CCP). The first sections describe the generic risk management foundations as well as specific clearing and settlement tasks in the context of European Market Coupling. Thereafter, clearing and settlement arrangements between different NEMOs and CCPs from a practitioner's perspective are described. Finally, the lessons learned and recommendations derived for the evolution of clearing and settlement in European power spot markets are discussed.

9.1 Introduction

This chapter provides an overview of financial clearing and settlement. Readers, who are interested in the specificities of physical shipping, are recommended to look at Chapter 8 of this book.

9.1.1 Power Trading and Risk Management

Large consumers or energy supply firms act as net buyers in power markets. They build and manage a portfolio of transactions, which mirrors their forecasted own or end consumer demand. Securing power demand volumes and hedging prices

R. Wand (✉)
European Commodity Clearing AG, Leipzig, Germany
e-mail: robert.wand@ecc.de

A. Estermann et al. (eds.), *European Electricity Market Coupling*, Future of Energy,
https://doi.org/10.1007/978-3-031-86315-8_9

is usually reached by "structured procurement" of contracts with different delivery time horizons. Delivery horizons can range from long-term derivative contracts with contract duration of several years to short-term power spot transactions. Likewise, power producers aim to hedge their production and secure long-term prices for a share of their produced energy, but also sell short-term power contracts for portfolio optimization or to manage planned or unplanned deviations. In addition to "physical" market players (market participants with physical consumption or production), "financial" market players act as trading companies with the aim to gain profit in long- and short-term power trading. They contribute to liquidity in power markets, which ensures that a counterparty for a transaction can be found at a market price.

Power spot markets play an important role for derivative markets. It is the day-ahead market clearing price or a day-ahead system price,[1] which usually serves as an underlying index against which power futures or option contracts are settled in Europe. Hence, liquid power spot markets are the foundation for functioning derivative markets. Moreover, increasing shares of fluctuating renewable generation elevate the need for short-term portfolio optimization due to deviations in short-term energy production and thus increase the importance of power spot markets further.

In 2022, the SDAC traded volume reached a cleared volume of 1,683 TWh. Each SDAC market coupling session managed an average social welfare of 9.9 B€ per session.[2] Hourly prices strongly fluctuated, ranging from -222.36 to 4,000 €/MWh and even average prices per bidding zone deviated between 25 and 150 €/MWh (All NEMO Committee, 2023). The magnitude of transactions processed in European Market Coupling, the volatility of clearing prices compared to other commodities and the close interaction between different trading venues show the significance of proper risk management for resilient power spot markets.

9.1.2 The Role of Clearing and Settlement

Power trading in liberalized electricity markets entails systemic risk. This risk cannot be eliminated (e.g. through diversification), since it is inherent to the entire market. However, several additional risks, such as counterparty, payment, delivery, currency, legal and concentration risks, also need to be addressed and can be managed in different ways. Bilateral or uncleared broker trading ("over-the-counter",

---

[1] Two examples of day-ahead system prices are the Nordic system price as a recalculation of the SDAC results without transmission constraints in the Nordic region (Nord Pool, 2022) and the National Single Price in Italy (Prezzo Unico Nazionale, PUN) for consumption units. The PUN is published as the weighted average of the bidding zone prices in Italy in the day-ahead market (Gestore dei Mercati Energetici, 2024).

[2] Social welfare, welfare or economic surplus (as used in the CACM Guideline) is determined as the sum of (1) consumer surplus and (2) producer surplus in all coupled bidding zones and (3) the total congestion rent on all coupled bidding zone borders.

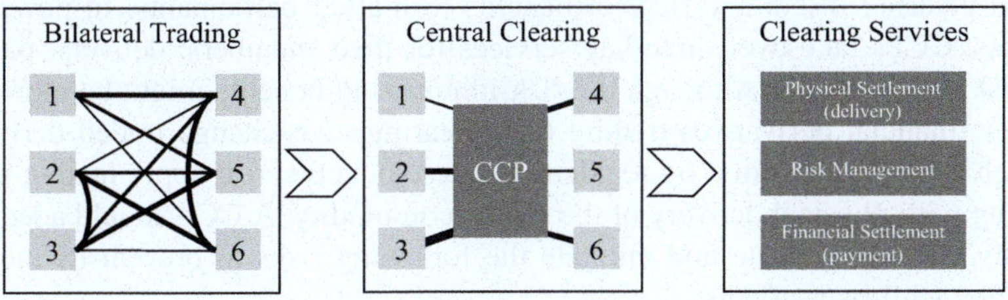

Fig. 9.1 Bilateral trading and central clearing. *Source* Author's own illustration

OTC) requires market participants to manage these risks for each counterparty individually. Despite some level of standardization (e.g. framework agreements), bilateral trading requires constant evaluation and monitoring of a multitude of risks for each counterparty. Potential losses need to be reflected in a market participant's risk bearing ability and result in economic limits for each counterparty. If economic exposure to a risk exceeds a specific limit, mitigation measures might need to be taken.

Alternatively, trades can be matched on the exchange while cleared and settled through a central counterparty (CCP). Each trading participant holds a single membership at the CCP, which becomes the seller to every buyer and the buyer to every seller (see Fig. 9.1).

A market participant's trading volume, trading portfolio structure, counterparties and risk appetite determine the need for central clearing. Also, the type of products determines the resulting exposure to risk. Derivative contracts (usually referred to as contracts with more than two days until expiration) yield higher risks than spot contracts, since the time span between payment and delivery or cash settlement of the derivative product is longer than for spot contracts. Spot contracts refer to products with a duration of up to two days between contract closure and settlement. Products auctioned or continuously traded in the Single Day-Ahead Coupling (SDAC) or Single Intraday Coupling (SIDC) qualify as spot products. Moreover, all products traded and cleared in European Market Coupling are physically settled. This means that the seller of the transaction (or the CCP on behalf of the seller) effectuates the delivery of the traded power contract through nomination with the respective TSO or imbalance settlement operator. Therefore, all participants have to hold valid Balance Responsible Party (BRP) agreements with the relevant TSOs or imbalance settlement operators or need to arrange for delivery by a third party BRP.

9.1.3 Costs and Benefits

Trades cleared and settled through a CCP are agreed on exchange or registered through an exchange for clearing. Whilst this comes at a higher cost (membership and connection, volume fees, deposited collateral to meet margin requirements), central clearing offers a variety of benefits to market participants. In power spot markets, CCPs take over three key services for their members: delivery, payment and risk management. Although the risk-minimizing benefits might be more obvious for financial derivatives trading (and clearing of exchange traded derivatives through a CCP is required by regulation), these benefits also apply to spot market clearing with physical delivery of the traded commodity. A CCP should adequately identify, assess, evaluate and mitigate the following risks to prevent its members from any spill-over effects:

- **Counterparty risk**: The CCP should guarantee the delivery of the traded commodity (in power spot markets for the delivery of power through TSOs). Hence, CCPs should collect margins in the form of eligible collateral from their participants and hold other funds to ensure that all relevant risks in case of non-delivery by the seller (or non-payment by the buyer) are always sufficiently covered. It should also have adequate arrangements to suspend the participant from further trading and clearing without undue delay.
- **Payment risk**: A CCP should provide secure financial settlement solutions through final (i.e. non-disputable) payments to the seller. The CCP should also provide risk management in case the payment agent (e.g. commercial settlement bank) should be unavailable for economic or technical reasons.
- **Physical delivery risk**: In addition to the "commodity value" based on the clearing price of the traded commodity the CCP should also cover potential additional risks stemming from a participant's curtailment by the TSO. In case the CCP does not benefit from alternative mitigation measures (e.g. priority nomination rules, after-market trading), this might result in additional margins to cover resulting imbalances.

Although risk management for its members is one of a CCP's core function, central clearing and settlement yields further benefits and services:

- **Single access to counterparty network**: a multitude of counterparties can be reached through a single CCP membership. The CCP's network effect and standards also reduce the complexity of trade connections, contractual arrangements, delivery and payment solutions and risk management approaches for its members.
- **Non-discrimination and transparency**: all members, who meet the CCP's margin requirements, are allowed to trade, clear and settle independently from their credit worthiness. Margin calculation methodologies, accepted types of collateral and other clearing conditions are transparent.
- **Anonymity** is often appreciated by market participants regarding their trading strategies and positions.
- **Capital efficiency**: various products and positions are netted and can reduce the volume of payments (payment netting). Depending on the number and scope of cleared products, a client might also benefit from cross-margining effects across different market venues, timeframes, locations or asset classes. These

benefits result from different long and short positions, which—depending on their historic correlation of market prices—can be partially offset against each other and result in reduced exposure to risk and margin requirements.

- **Reporting and additional services**: CCPs might offer additional services, such as regulatory reporting, foreign exchange management solutions in case of different settlement currencies or data provision (e.g. clearing prices and volumes).

9.2 Clearing and Settlement in European Market Coupling

9.2.1 Clearing and Settlement: Definition and Set-Up Under the CACM Guideline

Implicit capacity allocation is reached through the simultaneous matching of market participants' bids and offers with available cross-border capacities. Since the CACM Guideline foresees this model as a standard approach (see Chapter 3), cross-border power spot trading is only possible on NEMO-licenced power spot exchanges as central market venues.[3] Market participants must become member at a NEMO marketplace to trade SDAC or SIDC products. According to Article 6(1)(i) of the CACM Guideline, a NEMO "[…] shall be able to provide the necessary clearing and settlement services" as a designation criterion. However, the description of these clearing and settlement services towards the NEMO's members is limited to very few elements mentioned in Article 68(1) and 68(2) of the CACM Guideline (the remainder of Article 68 refers to the exchange of energy between different CCPs or shipping agents):

1. "The central counter parties shall ensure clearing and settlement of all matched orders in a timely manner. The central counter parties shall act as the counter party to market participants for all their trades with regard to the financial rights and obligations arising from these trades.
2. Each central counter party shall maintain anonymity between market participants".

After orders have been matched on exchange the trade is transferred to the CCP, legally either through "novation" of the transaction or as "open offer" (see Fig. 9.2). Novation requires the termination of an existing contract between the buyer and the seller, which is replaced by two contracts with the CCP. The CCP as the new party between the two market participants steps into the contractual chain and inherits all benefits and liabilities of the initial contract. In contrast,

[3] As an exception "explicit capacity allocation", where SIDC capacities are offered separately from continuous trading, is offered for single bidding zone borders in the intraday timeframe.

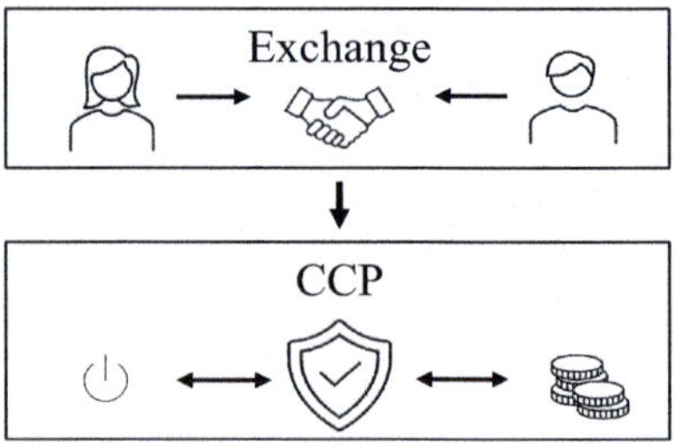

Fig. 9.2 Exchange and clearing transactions. *Source* Author's own illustration

open offer models result in two contracts being directly concluded between the seller and the CCP and the buyer and the CCP after matching of the orders. The CACM Guideline (Article 2 (42)) defines a CCP only with reference to the novation of contracts resulting from the matching process, but practically both concepts occur.

"Clearing" usually refers to all activities from the point in time a commitment for a transaction is made until it is settled. This comprises

- The assumption of counterparty risk as a CCP
- Risk management and mitigation measures (pre-trade limit and member stop processes, monitoring activities)
- Termination and default management

"Settlement" comprises all activities related to

- Netting of payments
- "Physical settlement" through TSO nominations
- "Financial settlement" through central or commercial bank payment transactions
- Reporting, invoicing and support services (the latter often being offered 24/7 for continuous power spot trading)

It becomes obvious that the CACM Guideline does only describe a minor share of clearing and settlement services as NEMO designation criteria. Consequently, the CACM Guideline does not provide further guidance to the NRAs for a joint understanding of the CCP function. Absent further coordination, NRAs in different member states apply deviating legislative or regulatory requirements for the provision of clearing and settlement activities. The application of different NEMO designation criteria to evaluate whether a NEMO effectively operates or procures clearing and settlement services in conjunction with the European passporting principle (Article 4(5)) and NEMO competition (Article 4(6)) have led to partially different understandings of the scope and depths of "clearing and settlement" services. As a result different CCP requirements have been established and coexist —not only between different EU member states, but due to European NEMO passporting also in the same member state or bidding zone.

The possibility to delegate all or part of any task assigned to a NEMO (Article 81) has also resulted in different organizational models for clearing and settlement. Some NEMOs operate all NEMO and CCP activities in the same legal entity. Other NEMOs have procured their CCP services from a third legal entity in accordance with Article 81 of the CACM Guideline. In some cases different NEMOs have also procured CCP services from the same CCP legal entity.

9.2.2 Local and Cross-CCP Clearing and Settlement

The previous chapter focused on clearing and settlement services provided for a NEMO's or CCP's own participants. Coupling these local markets through SDAC or SIDC requires additional interaction between NEMOs, TSOs or their delegates. Irrespective of the type of organization or the level of delegation for CCP services, a distinction should be made between

1. Local clearing and settlement between the participants trading at a certain NEMO or NEMO Trading Hub and
2. Cross-CCP clearing and settlement (shipping) between different NEMO entities or NEMO Trading Hubs.

Figure 9.3 shows some of the characteristics, regulatory foundations and principles of local and cross-CCP clearing and settlement in European Market Coupling.

Legally cross-CCP transactions (sometimes also referred to as "inter-NEMO flows" or "cross-clearing volumes") can occur between different types of entities. As previously described, NEMOs can either provide CCP services on their

Local Clearing (CCP)	**Shipping (CCP or Shipping Agent)**
· Each CCP acts towards its respective participants · Governed by high-level principles of Art. 68(1)-(2) CACM Regulation · Risk management as a core function to ensure safe and anonymous fulfilment of exchange transactions	· Cross-NEMO matching by legal, financial and physical exchange of market coupling transactions · Governed by Art. 68(3)ff. and 77(2) CACM Regulation · CACM Regulation focuses on settlement activities (congestion income, cross-CCP transactions)
· Non-discriminatory treatment of other CCPs compared to a CCP's own members · No contribution of a CCP to other CCPs' default fund or comparable "lines of defense" to prevent spill-over of risks between CCPs	

Fig. 9.3 Characteristics of local clearing and shipping activities

own or delegate these activities to a separate CCP entity. As a third option "shipping agents" can act as a counterparty between different CCPs in accordance with Article 68(6) of the CACM Guideline. This is also reflected in CACM Guideline (Article 2(42)), defining a CCP as an entity "[…] organising the transfer of net positions resulting from capacity allocation with other central counter parties or shipping agents". In practice, if shipping is not done by the NEMO or its CCP, TSOs act as shipping agents on selected bidding zone borders. All of the mentioned entities (NEMOs, delegated CCPs, TSOs in their role as shipping agents) are active as counterparties for the exchange of energy between coupled markets.

After each pan-European matching of SDAC or SIDC transactions (coupling phase) the exchange of energy between different NEMOs or NEMO Trading Hubs (i.e. a specific NEMO's market in a specific bidding zone) has to take place. These activities are also be referred to as "post-coupling activities" and comprise the following tasks for CCPs and/or shipping agents:

1. **Clearing and settlement between local market participants**. This activity is executed by CCPs in accordance with their own clearing rules. Shipping agents may become local market participants at the relevant NEMOs and CCPs, who buy or sell the shipped cross-border volumes at the determined market clearing price.
2. **Clearing and settlement between CCPs** in accordance with SDAC and/or SIDC market coupling results. Where applicable.
3. **Scheduled exchanges** are calculated in accordance with the relevant methodology and determine the volumes, prices, and counterparties for the settlement of cross-CCP transactions. Depending on the involved entities, clearing and settlement can occur between different parties, as shown in Fig. 9.4.
4. **Collection and transfer of congestion income** if market clearing prices between bidding zones differ. Since price differences are needed for the collection of congestion income, it can only result from (i) cross-border flows (i.e. not for intra-zonal cross-CCP handovers) and (ii) SDAC auctions or SIDC intraday auctions (i.e. not from continuous SIDC trades). According to Article 68(8) of the CACM Guideline, CCPs or shipping agents are responsible for the transfer of congestion income to the TSOs or their delegates in accordance with the market coupling results within the defined timings. The distribution of congestion income between TSOs in accordance with the respective congestion income distribution methodology (cf. Chapter 10) is a TSO task instead.

Various configurations for the exchange of energy between different NEMOs or CCPs exist. They can be characterized by the involved entities. Either the same CCP can ship energy between different NEMO Trading Hubs (e.g. two local markets in different bidding zones operated by the same NEMO) or shipping can take place between two different CCPs directly. Alternatively, a shipping agent may act as a third party between CCPs. Figure 9.4 shows the different types of handovers for an exemplified bidding zone border.

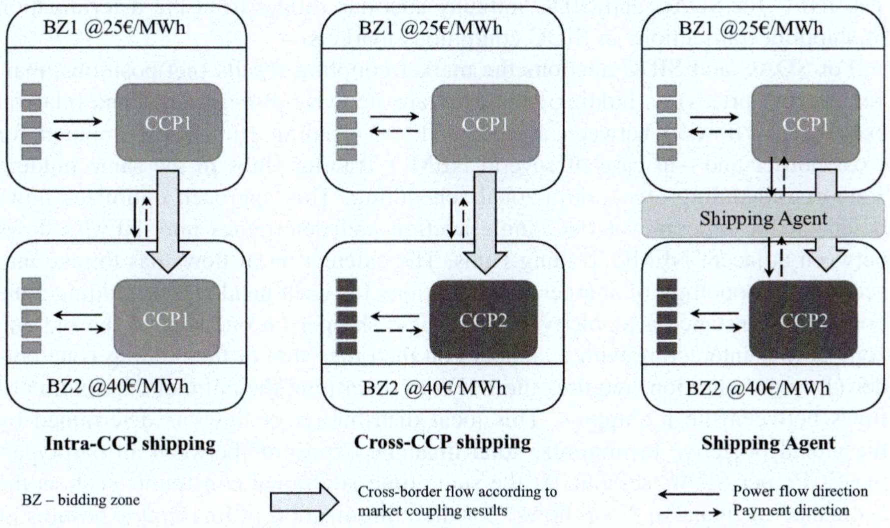

Fig. 9.4 Cross-border shipping configurations. *Source* Author's own illustration

Another differentiation can be made between cross-border and intra-zonal handovers. Whilst the former refers to the exchange of energy across a bidding zone border, the latter describes cross-CCP handovers "inside" a bidding zone. The technical concepts to enable NEMO competition[4] have established solutions for parallel operations of more than one NEMO in the same bidding zone. This also comprises the need for intra-zonal exchange of energy between the NEMOs or CCPs. The combination of the type and the number of involved shipping entities has resulted in different technical terms for shipper configurations or concepts: single shipper, dual shipper, multiple shipper, directional shipper, preferred shipper and other concepts or configurations have been established and are described further in Chapter 8.

9.2.3 Post-coupling: Shipping Path Determination in SDAC and SIDC

The calculation of "inter-NEMO flows" (INFC) determines the volumes, which have to be exchanged between different NEMO Trading Hubs, the applicable clearing prices and the respective shipping entities in the post-coupling phase.

[4] These multi-NEMO arrangement (MNA) concepts have been developed and implemented in accordance with Articles 45 and 57 of the CACM Guideline in EU member states, which are open to NEMO competition and do not apply an exemption in accordance with Article 5.

The INFC for SDAC and SIDC intraday auctions differs from the determination of shipping transactions in SIDC continuous markets.

For SDAC and SIDC auctions the market coupling results (net positions, market clearing prices) on bidding zone level are stepwise disaggregated and bilateral exchanges ("flows") between adjacent NEMO Trading Hubs are computed for cross-border and—in case of several NEMO Trading Hubs in the same bidding zone or scheduling area—intra-zonal scheduling. This approach optimizes flows based on the outcome of one single auction and determines inter-NEMO flows between adjacent NEMO Trading Hubs. The calculation of flows has to take into account the topology of shipper configurations for each handover or bidding zone border. For instance a topology with a single shipper on one side of the bidding zone border interacting with n shippers on the other side of the bidding zone border (1:n configuration) requires the INFC to determine the optimal distribution of flows between the n shippers. This local distribution of flows is determined by the global objective to minimize total financial exposure[5] between all participating CCPs per SDAC session. At the same time additional constraints such as the avoidance of financial "loop flows" for the minimization of financial exposures or active allocation constraints are taken into consideration. The principles, optimization target function, constraints and process steps for the calculation of inter-CCP flows are laid down in the SDAC scheduled exchange calculation methodology and its annexes.[6] A comparable approach applies to the pan-European SIDC intraday auctions. In contrast to the shipping solution in continuous SIDC markets, this model does not necessitate shipping by the same shipper entity along a shipping path to transfer energy between non-adjacent NEMO Trading Hubs.

SIDC continuous markets follow a different transaction model. The shipping module of the XBID System determines a specific price and a set of trades/flows from source to sink for each matched transaction. Depending on the location of source and sink and available cross-border capacity the shipping path can be in the same, two adjacent or across several scheduling areas or bidding zones. This results in a "chain" of shipping trades. Since the initial go-live of the SIDC project the "preferred shipper" principle has been applied. It foresees that each source NEMO assigns its own shipper and ensures delivery until the sink of the transaction. The extension of the SIDC continuous matching to additional EU member states with respective shipper configurations has resulted in a situation that NEMOs could not procure shipping by their "preferred shipper" in all coupled bidding zones.[7] This

[5] Financial exposure is determined as the volume exchanged between two CCPs multiplied by the applicable market clearing price in the bidding zone where the handover is located. Multiple locations of handovers between two CCPs can exist. Optimization is done for the number of market time units of the respective auction.

[6] See ACER (2023), Decision No 10/2023 of 30 May 2023 on the amendment of the methodology for calculating scheduled exchanges resulting from single day-ahead coupling.

[7] The coupling of non-adjacent markets with a national legal monopoly for trading services and designation of a single shipper per country in combination with the reciprocity principle for

caused the need to further develop the SIDC shipping solution and establish "transit shipping" for some combinations of matched SIDC trades. It involves a third party, which takes title to the power and money of the matched SIDC transaction between two NEMOs or CCPs. Transit shipping has been introduced since the SIDC project's "second wave" go-live in 2019.

Further information on the determination of inter-NEMO flows, shipping paths and different shipping topologies in the day-ahead and intraday timeframes is provided in Chapter 8 of this book.

9.3 Market Coupling Arrangements in Practice

9.3.1 Legal Framework for Cross-CCP Clearing and Settlement

A hierarchical system of multilateral agreements between NEMOs and TSOs defines the legal framework for European market coupling. Separate agreements for the day-ahead and intraday timeframes exist. These contracts, separately or jointly concluded between NEMOs and TSOs, are the legal foundations for the practical cooperation between the different market coupling parties and operationalize the described provisions and methodologies laid down in CACM Guideline. Figure 9.5 provides an exemplified overview of these agreements, the contractual parties and governed activities or assets with a focus on All NEMOs' operational agreements for the SDAC and SIDC continuous initiatives. SIDC intraday auctions are governed by a separate set of contracts.

Pan-European Day-Ahead and Intraday Coupling Operations Agreements (SDAC DAOA and SIDC IDOA) represent the highest contractual level for the cooperation between NEMOs and TSOs. These contracts also include a set of operational procedures, which can be further detailed in regional agreements (e.g. the Nordic Day-Ahead Market Coupling Operations Agreement).[8] Due to their multilateral character these procedures usually describe joint NEMO-TSO cooperation tasks. For the post-coupling phase these procedures particularly describe the transfer, reception and confirmation of market coupling results, scheduled exchanges (where applicable), required nominations and the collection and transfer of congestion income (for SDAC and SIDC intraday auctions only). Operational procedures for normal state as well as backup and fallback situations exist.[9] The described multilateral NEMO-TSO cooperation agreements and procedures form the legal

monopolistic and competitive NEMOs, according to Articles 4 and 5 of the CACM Guideline, resulted in the need to find a shipping solution between certain non-adjacent bidding zones.

[8] The European-wide SDAC DAOA and SIDC IDOA agreements and the related operational procedures are publicly available at the websites of ENTSO-E (www.entsoe.eu) and the All NEMO Committee (www.nemo-committee.eu).

[9] Chapter 4 of this book also provides further information on market coupling procedures.

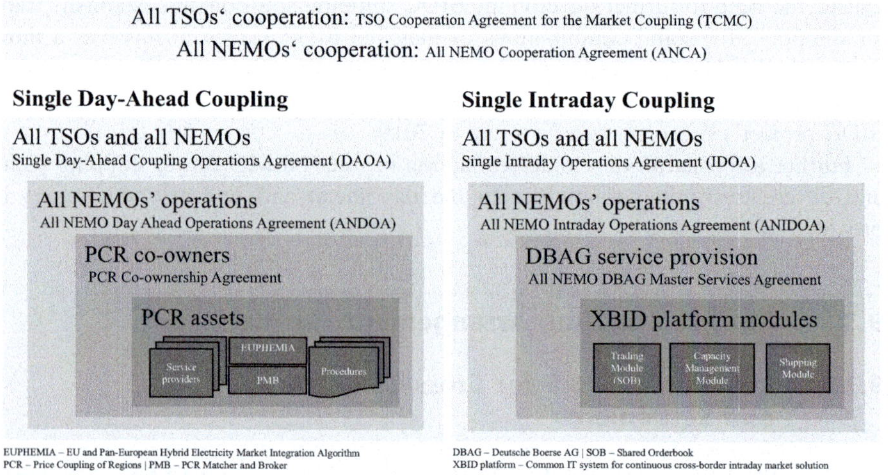

Fig. 9.5 Central agreements and assets for SDAC and SIDC continuous initiatives. *Source* Author's own illustration

foundation for bilateral clearing and settlement agreements, which can be grouped into three categories.

1. **Settlement Link Agreements/Shipping Agreements/Clearing and Settlement Agreements** are bilateral and private contracts, which provide the legal basis for cross-CCP settlement of SDAC and/or SIDC transactions. These agreements exist either between two CCPs or a CCP and a shipping agent. They cover a multitude of aspects related to the main contractual duties for physical and financial settlement of transactions in accordance with the market coupling results, risk management and governance as visualized in Fig. 9.6. Depending on the number and location of bidding zones or bidding zone borders covered, these Settlement Link Agreements can have local to pan-European scope.
2. **Congestion Income Settlement Agreements** provide the contractual basis for the collection and transfer of congestion income between NEMOs, their designated CCP or shipping agents on the one hand and TSOs or the entity appointed by the TSOs on the other hand. Practically, the Joint Allocation Office S.A. has been appointed by the TSOs as the congestion income distribution agent for most bidding zone borders. These contracts are closely linked to the regional Day-Ahead Coupling Operations Agreements and—depending on the region—usually based on generic terms and conditions for all relevant NEMOs, CCPs or shipping agents with individual bilateral participation agreements.
3. **(Shipper) Balance Responsible Party Agreements and/or Imbalance Settlement Agreements** are the legal foundation for the physical settlement of market coupling transactions through nominations to the respective TSO or imbalance settlement operator. These agreements are designed as standard balance responsible party (BRP) agreements and cover admission requirements, nomination

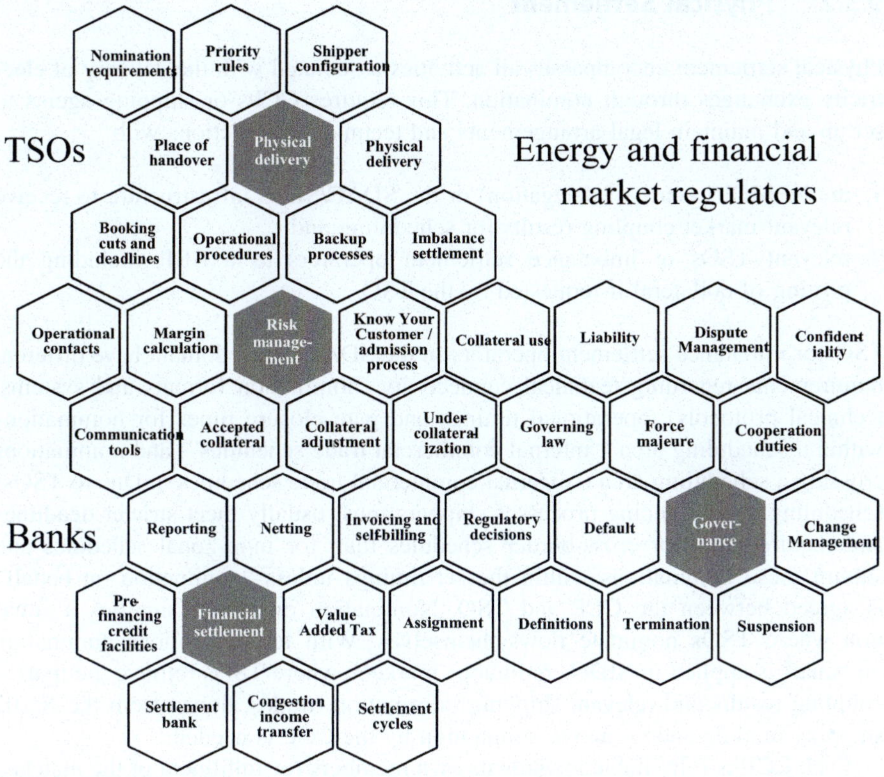

Fig. 9.6 Exemplary topics of shipping agreements and stakeholders. *Source* Author's own illustration

procedures and timings for intra-zonal ("local") and cross-border nominations as well as priority rules, liabilities in case of imbalances due to wrong, late or non-nomination and collateral requirements. Specific TSO rules or side letters for NEMOs or their CCPs might exist.

In addition to the listed contractual framework CCPs adhere to a broader regulatory framework—inter alia determined by the CACM Guideline and other network codes or guidelines, NEMO licence requirements, NRA decisions, Anti-Money-Laundering legislation, tax law and regulations, national specific legal provisions, corporate by-laws and statutes. Also, some CCPs in European market coupling are directly governed by financial market regulation, such as the European Market Infrastructure Regulation (EMIR) and are supervised by European or national financial regulatory authorities as part of the European System of Financial Supervision.

9.3.2 Physical Settlement

Physical settlement encompasses all activities associated with the delivery of electricity exchanges through nomination. This requires CCPs or shipping agents to set up and maintain legal arrangements and technical connections with

1. the NEMO (in case of delegation) or the SDAC/SIDC infrastructure to receive relevant market coupling results for scheduling and
2. relevant TSOs or imbalance settlement operators as a BRP, including the posting of collateral if requested by the TSO.

TSOs or imbalance settlement operators in the SDAC/SIDC domain have different requirements regarding technical connectivity (nomination formats and systems, technical protocols), operational timings and gate closure times for nominations within a scheduling area ("internal commercial trade schedules") and nominations crossing a scheduling area ("external commercial trade schedules"). Due to TSOs' scheduling and balancing processes shippers must usually meet stricter deadlines and requirements for cross-border schedules than for intra-zonal schedules and submit these nominations within the set timings unless "nomination on behalf" is agreed between the CCP and TSO. Nomination on behalf signifies a situation where TSOs nominate flows themselves. With few exceptions nomination on behalf is applied in SIDC continuous markets, where TSOs retrieve the market coupling results and relevant shipping information (trades, flows) from the XBID shipping module and no active nomination by the CCP is needed.

Each CCP is fully liable towards its own members for fulfilment of the matched trades. Hence, a balanced net position (means the sums of all buy and all sell volumes are equal) is key for each CCP. Since this is only possible with reliable buy or sell transactions with its local members and its partner CCPs in the context of SDAC or SIDC, full liability for imbalances due to late, wrong or non-nomination between CCPs is needed. These arrangements are usually stipulated in the bilateral Settlement Link Agreements.

In many scheduling areas TSOs apply special priority nomination rights for CCPs (i.e. their nomination values prevail over the nomination values of market participants in case of nomination value mismatches) or CCPs provide single-sided nominations for their members. However, if two CCPs nominate energy volumes between them, special rules need to be agreed for the provision of priority rules and mismatch management between them. For most scheduling areas these principles are either defined in the national multi-NEMO arrangements or standard BRP rules of the TSO or imbalance settlement operator. Direction dependent priority nomination rules or lesser rules are common arrangements for cross-CCP nominations.

The risk of imbalances due to member suspension by the TSO, missing priority nomination rules or capacity curtailments determine the so-called "delivery risk" for a CCP and the resulting margin requirements to cover this risk ("delivery risk margin"). It reflects the usually higher imbalance prices compared to day-ahead or intraday clearing prices and should cover potential loss. Since a CCP's members have to deposit additional capital as delivery margin, CCPs try to mitigate these risks in constant dialogue with the TSOs. Closely coordinated BRP suspension arrangements with the respective TSOs are sought by the CCPs. Also, CCP-specific priority nomination rules or side letters between individual TSOs and CCPs have been agreed to ensure that no additional delivery risks occur for a CCP and cost-efficient margin requirements for the CCPs' members can be implemented. These arrangements are the result of a constant and long alignment and negotiation between the TSO and CCP(s) in each scheduling area. Still, they have not been implemented in all European power spot markets. An EU-wide mitigation measure has been found on the risk of capacity curtailment in case of implicit capacity allocation. Article 72 of the CACM Guideline stipulates that TSOs will not hold CCPs or shipping agents to account for any imbalances associated with the curtailment of cross-border capacities in the event of force majeure or emergency situations.

9.3.3 Financial Settlement

Financial settlement between CCPs or shipping agents covers payments as well as associated financing, accounting, reporting and invoicing activities for the delivery of cross-CCP electricity exchanges. These arrangements depend on local legal requirements, regulatory needs and CCPs' individual rules and regulations. The decentralized bilateral settlement links between different CCPs in Europe are suited to address these local and individual specificities, but also have proven to be points for discussion and negotiation absent appropriate cost recovery for CCPs or shipping agents.

All SDAC and SIDC products are denominated in euro currency and so are payments between CCPs or shipping agents. For most cross-CCP payments infrastructure solutions from commercial banks are used. The market coupling counterparties' commercial "house banks" might either directly interact with each other or a joint settlement bank might be selected and contractually bound by a trilateral side agreement to the Settlement Link Agreement. In some cases, commercial banks might need to fulfil specific contractual or membership requirements at a CCP. For CCPs, who have access to the real-time gross settlement payment system operated by the Eurosystem (T2), the processing and settlement of cross-CCP payments in central bank money is possible.[10]

Since NEMOs/CCPs or shipping agents settle high financial volumes, netting of settlement amounts is similarly important for risk management and procedural

[10] Only central banks and financial institutions established in the European Economic Area can directly access and submit payment orders in euro to T2. Hence, a CCP must be licenced as a financial institution to be able to access this system as a payment solution.

efficiency. Netting means combining or aggregating different financial obligations to a net obligation with the goal of reducing financial risks and settlement amounts between parties. Practically, netting of cross-CCP transactions and payments take place in different process steps in SDAC and SIDC auctions. Bid and ask offers by a CCP's members are matched and settled by each CCP locally and only the residual net position of the market is settled with other market coupling counterparties. As described further above scheduled exchanges are calculated with the goal of minimizing total financial exposure between CCPs per market coupling session. If more than one settlement link between two CCPs exists (e.g. handovers in different markets or on more than one border), the described optimization also implies that gross payments for the exchange of energy are reduced to one net exposure, which is minimized. In addition CCPs or shipping agents, who are active in both timeframes, aggregate cross-CCP payments from SDAC and SIDC (auctions and continuous) transactions to one single payment per settlement.

The timespan between trading and financially settling a product is called the settlement period or settlement cycle. Financial settlement cycles between CCPs should be as short as possible to minimize the buildup of exposure (risk of unpaid transactions) and reduce associated margin requirements. In case of deviations in financial settlement cycles between CCPs they need to agree which party determines the cycle and arrange pre-financing credit facilities to bridge the gap. Additionally, Settlement Link Agreements also contain provisions on operational timings, technical and economic default in case of late or non-payment, arrangements on bank holidays, invoicing and accounting requirements, reporting needs, bank fees and interest and revenue or cost allocation as well as boilerplate provisions. Also, local requirements regarding tax or customs and the associated need for pre-financing or currency conversion (in case of non-euro tax payments) are covered, where applicable. These local provisions, if not harmonized between countries participating in SDAC or SIDC, can become important external cost drivers for CCPs or shipping agents and increase the cost of trading and clearing.

9.3.4 Congestion Income Collection and Transfer

For SDAC and SIDC intraday auctions, market clearing prices between bidding zones can differ if the demand for commercial cross-border flows cannot be met by sufficient available cross-border capacities (commercial congestion). Congestion income arises as a result of cross-border energy being bought on one side and sold on the other side of a bidding zone border at deviating market clearing prices in these bidding zones. Hence, due to the absence of clearing price differences congestion income does not occur in case of intra-zonal exchange of energy between market coupling counterparties or in continuous SIDC trading. Where congestion income occurs, the shipping configuration also determines the need for contractual arrangements. Where a single shipper is active on both sides of the border it has full liability for the collection and transfer of congestion income. Where more than one shipper exists, one shipping party might collect the entire congestion income or a direction-dependent congestion income collection and transfer to the TSOs (or their congestion income distribution agent) might be agreed. This allows the CCPs involved on a specific bidding zone border to allocate the responsibilities for congestion income collection and transfer, which they have according to Article 68(8) of the CACM Guideline, between them. The principle, which party (importing or exporting shipper) should collect and transfer congestion income, might also be agreed on the level of a Capacity Calculation Region.

Each CCP (or where required shipping agent) collecting and transferring congestion income has a respective agreement with the relevant TSO or congestion income distribution agent. The capacity calculation and allocation methodology also impacts the congestion income collection and transfer. Bidding zone borders with coordinated Net Transfer Capacity (cNTC) capacity calculation methodology have a minimum available capacity of zero (no allocation of commercial transmission capacity in the respective market time unit). Moreover, INFC always calculates flows from the lower price area to the higher price area. This results in a positive (at minimum zero) congestion income payment from the collector to the TSO. The introduction of flow-based capacity calculation in some Capacity Calculation Regions has changed this set-up since non-intuitive flows (from high-price to low-price areas) can occur, which can lead to negative congestion income. This results in the need to cater for special arrangements in the congestion income settlement agreements since the direction of payment could reverse from the TSO or congestion income distribution agent to the CCP or shipping agent in specific cases.[11] Hence, existing congestion income transfer agreements needed to be further developed due to the introduction of plain flow-based capacity calculation and allocation. Today, bi-directional payments between the TSOs and CCPs/shipping agents with subsequent effects on liabilities, payment timings, bank arrangements, invoicing and reporting as well as default management are agreed for regions with flow-based capacity calculation.

9.3.5 Risk Management

Risk Management is an integral part of the cross-CCP clearing and settlement arrangements. It might be counterintuitive that CCPs collateralize their exposures from cross-CCP transactions if they already collateralize their exposures with their respective local clients. However, it is the specific set-up of CCPs in power spot markets with

[11] Please refer to Chapter 10 for more information on congestion income collection and distribution under flow-based capacity allocation.

- Strong interconnectedness by means of shared order books and resulting inter-NEMO flows
- Deviating risk management approaches between NEMOs/CCPs without regulatory minimum risk management standards
- Each CCP remaining fully liable towards its own clients for the fulfilment of matched trades while it has no influence on its market coupling counterparties' risk management practices.

That requires CCPs to calculate margin requirements and collateralize their exposure towards each other. Where a third party (e.g. TSO) acts as shipping agent between two CCPs,[12] the shipping agent might need to fulfil margin requirements as a standard member of the respective CCPs. Special arrangements on collateral provision between some NEMOs/CCPs and TSOs in their role as shipping agent might exist, reflecting their "trading" profile (settlement of market coupling results) and economic strength as (often state-backed) critical infrastructure operators. For these reasons and as an exception to the rule, risk management between some shipping agents might be agreed in other forms than in the form of margin calculation and collateral requirements. A full picture of the applied practices cannot be drawn, since bilateral agreements between shipping agents and CCPs are not public. Usually, bilateral cross-CCP agreements include provisions on

1. The applicable margin methodology, which determines the margin needed to cover cross-CCP exposures
2. The type of acceptable collateral (often cash or bank guarantees) to cover calculated margin requirements
3. Operational timings and definition of default situations (e.g. under-collateralization) when collateral can be utilized for outstanding payments and exposures and Settlement Link Agreements can be terminated (exceptions might apply in force majeure and technical default events).

It must be considered a sector specific characteristic that NEMOs or their CCPs are competitors and collaborate on Market Coupling Operator functions, including post-coupling activities, at the same time. This has resulted in distinct cross-CCP risk management principles, but also revealed some adverse incentives on setting the right level of risk management. On the one hand, many CCPs apply the same risk management methods (margin model, acceptable type of collateral) to their own members and market coupling counterparties. This approach ensures non-discrimination between CCPs and prevents the market coupling counterparties

[12] Depending on the shipper configuration one shipping agent (e.g. an interconnector TSO) or two shipping agents (one TSO on each side of the bidding zone border) might operate between two NEMOs or their CCPs for the exchange of energy.

from asking too conservative margin requirements from each other. In other words, competition prevents a CCP from being asked excessive margins to collateralize the settlement of inter-NEMO flows. On the other hand, the CACM Guideline does not provide any minimum risk management standards for NEMOs/CCPs in power spot markets. With the aim to become more attractive to the market participants in NEMO competition, a "race to the bottom" between competing NEMOs/CCPs with regard to their security standards could occur and increase systemic risk in European market coupling.

Whilst ex-ante collateralization can be considered as a preventive risk management measure, timely and effective decoupling procedures as curative measures in case of a CCP or shipping agent default are similarly important. Decoupling contains and effectively isolates counterparty risk and thus prevents the interconnected NEMOs/CCPs from a spill-over of these risks and associated knock-on effects between them.

Depending on the time, cause and severity, different SDAC decoupling cases (partial vs. full decoupling) are procedurally defined and technically foreseen. For each of these decoupling cases applicable regional fallback methodologies apply. These might differ and range from shadow auctions to fallback regimes with maintained coupling between NEMOs and CCPs in the Nordic and Baltic Capacity Calculation Regions. Today, the design of the Nordic and Baltic fallback solutions prevents effective decoupling between NEMOs or their CCPs in these regions. This comes with increased risks for knock-on effects and the possibility of contagion between different NEMOs or CCPs. For SIDC auctions the SIDC continuous market shall be the fallback solution. For a decoupling of the SIDC continuous market no fallback solution is foreseen.

9.4 Outlook

The initiated review of the CACM Guideline has shown different proposals for further development of clearing and settlement in power spot markets. The discussion between relevant stakeholders included suggestions ranging from the evolution of existing decentralized clearing and settlement arrangements to a complete overhaul of the current architecture and full centralization of Market Coupling Operator activities, including shipping, in a single legal entity. Absent a detailed cost–benefit analysis the evaluation of such centralization remains unclear and the need for and prioritization of these far-reaching governance changes in relation to the completion and further development of European market coupling is strongly debated between different regulatory and industry bodies. Some elements seem to be of no regret to maintain or further improve the resilience and efficiency of European market coupling with regard to clearing and settlement:

1. Clear responsibilities and mandates for regulatory bodies to set a framework for cross-CCP clearing and settlement requirements are needed, including implementation monitoring, cost management and recovery for non-controllable cost

items. Some of the most contentious points in past market coupling projects have been related to cross-CCP clearing and settlement and resulted in repeated escalations and in some cases project delays. This was caused by regulatory uncertainties under the CACM Guideline with missing focus and accountability for the regulation of financial settlement and risk management aspects and missing or insufficient cost recovery. Post-coupling activities are particularly prone to high operational costs being incurred on the NEMOs or CCPs and shipping agents due to collateral and pre-financing costs, which are determined by several external and to a large extent non-controllable factors: market coupling and inter-NEMO flow calculation results, structural flow patterns between bidding zones or Capacity Calculation Regions, interest rate levels, third parties' margin and collateral requirements, the number of market coupling counterparties and non-energy related legal requirements (e.g. value added tax).

2. The narrowness of the existing framework for clearing and settlement of power spot products is in stark contrast with the close interconnectedness of the respective power exchanges and their clearing entities. Introducing minimum risk management standards to prevent a "race to the bottom" between CCPs regarding their security standards is needed. Competition between NEMOs and their CCPs is a core principle of the CACM Guideline and should be preserved, but an improved and coordinated regulatory framework for NEMO designation including clear criteria for sound clearing and settlement should be implemented. The magnitude of transactions, strong interlinkage of markets and importance of short-term power trading for system security calls for an improved focus on financial risk management and reduction of systemic risk. Improving the resilience of coupled power spot markets also requires further development of fallback solutions in some Capacity Calculation Regions to ensure timely and effective decoupling as curative risk management measures.
3. NEMOs, CCPs and market participants, who are operating as balance responsible parties in multiple markets or regions, are faced with a multitude of local TSO requirements. Harmonizing these requirements for CCPs, taking their legal status and rules into account, contributes to lower margin requirements for CCPs' members. The Europe-wide roll-out of priority nomination rights for CCPs and a close interaction between TSOs and CCPs regarding BRP suspension can reduce delivery risks and associated margins for clearing participants. Also, an exemption for CCPs from BRP collateral requirements due to their balanced position in market coupling would help to decrease costs. Moreover, further streamlining technical nomination formats, protocols and processes between TSOs and imbalance settlement operators would reduce operational complexity and costs for CCPs and would lower market entry barriers for NEMOs or CCPs.
4. The sharing of orderbooks is a necessary precondition for trading in coupled day-ahead and intraday markets and has the positive effect of increasing liquidity in these timeframes. However, opening national legal monopolies for trading services to competition between NEMOs is still pending. On the trading

layer, opening monopolistic markets could further incentivize product innovation and reduce trading and technology costs by providing a choice for market participants. For clearing they could benefit from a broader network of clearing partners, netting effects across different products and asset classes and additional services. Nevertheless, enabling innovation within a set market coupling and risk management framework also requires regulators to leave potential for differentiation between NEMOs and CCPs next to the shared orderbooks, where no differentiation is possible.

5. The CACM Guideline focuses on providing a regulatory and governance framework for market coupling in the power sector. This chapter has shown that additional aspects beyond the power sector need to be taken into consideration for a level playing field between NEMOs/CCPs and for efficient post-coupling arrangements. Hence, needed evolutions in the market coupling framework and reforms in other sectors need to go hand in hand. This holds particularly true for non-EU countries' reform processes as part of the EU accession process such as, but not limited to, the Value-Added-Tax (VAT) legislation in line with the EU international reverse charge mechanisms. This cross-sectoral legal harmonization of non-technical aspects is particularly important prior to the upcoming extension of market coupling to the Energy Community Contracting Parties as further described in Chapter 13.

9.5 Summary

Although several possibilities for trading in liberalized power markets exist, market participants need to access a NEMO with central clearing if they want to trade cross-border power spot markets in Europe. Hence, the importance of exchange-based trading as well as central clearing and settlement is set by the CACM Guideline as the regulatory framework. However, there are also clear benefits a NEMO (in its CCP-role) or its designated clearing house provides to its clients, such as a single access to a wide network of counterparties, anonymity, capital efficient collateralization and management for a multitude of risks and services. In addition to their own clients, CCPs also clear and settle with each other in the post-coupling phase of the SDAC and the SIDC. Either CCPs or TSOs in their role as shipping agents enable the exchange of energy between coupled markets in accordance with the market coupling results and collect and transfer congestion income where needed. The decentralized contractual and technical set-up of settlement links between them allows to take specificities into account, but joint principles on cross-CCP clearing and settlement have developed as a de facto standard. These contracts are embedded in and governed by a multitude of other market coupling contracts between NEMOs and TSOs.

CCPs, by the nature of their business, are a focal point for financial risk management in the market and operate many interfaces with other actors in European market coupling: market participants, who become members of the CCP; NEMOs for the reception of market coupling results; TSOs in order to enable physical settlement by means of nominations and where TSOs act as shipping agents; congestion income distribution agents for the transfer of congestion income. CCPs are also interacting with the financial sector to enable financial settlement and risk management through collateral. Hence, CCPs play an important role for and beyond power spot markets and often need to handle requirements from different sectors and regulatory frameworks.

Risk management in European market coupling is key and should play an even more important role with the aim to increase the resilience of short-term power spot markets. The price spikes in the recent past triggered by the gas supply crisis have underlined the importance of proper risk management in market coupling and beyond. In particular the number of state interventions and support needed to stabilize some market participants showed that more and continued emphasis on risk management is needed. With regard to the CACM Guideline this should cover clear NRA responsibilities for clearing and settlement, minimum requirements for CCP safety standards as well as curative risk management measures with effective decoupling as fallback solutions in all regions.

Bibliography

ACER. (2023). Decision No 10/2023 of 30 May 2023 on the amendment of the methodology for calculating scheduled exchanges resulting from single day-ahead coupling. Retrieved February 2024, from https://www.acer.europa.eu/sites/default/files/documents/Individual%20Decisions/ACER_Decision_10-2023_on_the_Day-Ahead_Scheduled_Exchanges_Methodology.pdf

All NEMO Committee. (2023). CACM Annual Report 2022, page 18. Retrieved January 2024, from https://nemo-committee.eu/assets/files/cacm-annual-report-2022.pdf

Gestore dei Mercati Energetici. (2024). Spot Electricity Market. Retrieved February 2024, from https://www.mercatoelettrico.org/En/Mercati/MercatoElettrico/MPE.aspx

Nord Pool. (2022). Nordic System Price Methodology for Calculation, page 8. https://www.nordpoolgroup.com/49b878/globalassets/download-center/day-ahead/methodology-for-calculating-nordic-system-price---may-2022-.pdf

Robert Wand has worked in the context of European market coupling for more than 10 years. As a member of ENTSO-E's network code drafting team he was involved in the development of the CACM Guidelines and subsequently contributed to its implementation in various positions. After working in academia and at a Transmission System Operator, Robert joined European Commodity Clearing AG, which operates a central counterparty infrastructure. As Expert Clearing Strategy he implements clearing and settlement solutions for spot and derivative commodity markets. His focus on post-coupling aspects in European Market Coupling led him to develop and introduce multiple-NEMO arrangements, inter-NEMO flow calculation requirements and fallback methodologies in cooperation with a multitude of partner exchanges, Transmission System Operators, European and national regulatory authorities across the EU and Energy Community Contracting Parties.

Congestion Income Distribution 10

Karel Sebesta and Martin Palkovsky

Abstract

This chapter introduces Congestion Income (CI) and the process of Congestion Income Distribution (CID), including the history and evolution of the process and how Congestion Income is regulated. An overview of the Congestion Income Distribution methodology for day-ahead allocation is provided, followed by the principles of the methodology. Negative Congestion Income linked to non-intuitive flows is discussed, as well as efforts by Transmission System Operators to deal with such occurrences. The important role of slack hubs is introduced, as well as the situations leading to remuneration inconsistencies and the impact of reserving some of the day-ahead capacity for balancing purposes. Finally, challenges associated with cross-Capacity Calculation Region Congestion Income transfers are presented.

10.1 Principles of Congestion Income Distribution

Congestion income arises in the European electricity market when the transmission capacity between bidding zones is not sufficient to fully couple supply and demand in connected market areas, resulting in different clearing prices in both markets.[1]

The following Congestion Income Distribution (CID) among TSOs is an integral part of the market coupling process. In this chapter we will discuss the concept of Congestion Income Distribution, its evolution, challenges, impact, and strategies for optimization during recent years.

[1] The functioning of market coupling is described in detail in Chaps. 1 and 2.

K. Sebesta (✉) · M. Palkovsky
Prague, Czechia

A. Estermann et al. (eds.), *European Electricity Market Coupling*, Future of Energy,
https://doi.org/10.1007/978-3-031-86315-8_10

CID is a system of rules and procedures to ensure that the congestion income generated from cross-zonal capacity allocation is distributed fairly and efficiently among the involved TSOs.

In Europe, electricity is bought and sold in different bidding zones (BZs)/ countries. The goal of market integration that has been under implementation for the last decade is to facilitate optimal trade across countries. Ideally, in the case of sufficient transmission capacity, cross-zonal capacities are made available for use free of charge. However, due to insufficient interconnection between zones, the congestion management had to be introduced. Therefore, the trades compete for the transmission capacity in a market-based manner and the capacity is "priced" as a result.

Congestion Income Distribution is a system which ensures that income generated from cross-zonal capacity allocation is distributed fairly among relevant TSOs (ACER, 2023).

We have different types of cross-zonal capacity pricing for different types of cross-zonal trading. The pricing in Day-ahead Market Coupling, which is directly linked to the pricing of transmission capacity, has already been explained in previous chapters. In case there is enough cross-zonal capacity, the coupled zones become a single price zone. In the opposite case, the offered cross-zonal capacity cannot fulfil all demand and prevents the markets from converging on their prices. The spread (price difference) between these two zones corresponds to the price of the cross-zonal capacity.

The situation is different for Intraday Market Coupling where bids are matched continuously, and capacity is allocated simultaneously while observing cross-border constraints as described in previous chapters. As a consequence of pay-as-bid system, cross-border capacity is not priced as we discover congestion only after matching is already done. As payment cannot be requested retroactively from market participants, no congestion income arises from continuous trading. To address the missing congestion income in single intraday coupling (SIDC), hybrid system with three intraday (implicit) auctions (IDAs) will be introduced. To conduct these auctions, continuous trading must be temporarily on hold to avoid double allocation of cross-zonal capacity and resulting congestion income will be generated and distributed as it is with single day-ahead couple (SDAC).

For congestion income itself, the same basic equation (Eq. 10.1) for its calculation can be always applied.

$$\text{Congestion Income}(e) = \text{Price}\left(\frac{e}{\text{MWh}}\right) * \text{Quantity}(\text{MWh})$$
$$\text{Congestion Income} \tag{10.1}$$

By applying this formula (in diverse processes of varying complexity) we get revenues stemming from cross-zonal capacity usage representing the congestion income to relevant TSOs. Since the capacity is allocated based on underlying trades and anticipated benefit from the inter-zonal trade, the Congestion Income indicates how much market participants value the possibility for cross-border trade. Congestion Income can be generated from different Capacity Allocation timeframes, e.g. forward, day-ahead, intraday or balancing and different capacity allocation mechanisms, i.e. explicit or implicit.

The concept of Congestion Income Distribution has been evolving over the years. With more and more complex capacity allocation moving from bilateral net transfer capacity (NTC) to regional flow-based principles, the congestion income distribution has become complex too. New rules and methodologies have been developed to ensure that the costs and benefits of modern congestion management are shared fairly.

10.2 History and Evolution

The CID has been a very simple process for a long time. With the NTC calculation and allocation, and generally only two TSOs owning the cross-zonal lines, the congestion income was derived directly from the allocation at that border and was shared equally for each side of the border. Sometimes, specific keys must be defined for situations where ownership stakes need to be considered, as well as participation of more than two TSOs at one bidding zone border (BZB), and others. These specific rules do not affect TSOs outside the relevant BZB and their definition therefore depends only on the relevant TSOs and NRAs. This system, where the BZB CI is distributed equally, has been in place for decades.

Implementation of capacity calculation and allocation developed under the CACM Guideline (European Union, 2021) brought new dimensions into the CID: the congestion that limits the exchange between the zones may not be on their interconnection, the flow on the border may not always follow the positive price spread (i.e. from lower to higher price) and there may be constraints other than network constraints (e.g. ramping, allocation constraints). Even if these factors have been addressed locally on individual borders, their relevance has been revealed only when the TSOs had to combine them into a single European methodology with the anticipated go-live of the Core Flow-based capacity calculation methodology.

In 2016, based on the CACM Guideline (Article 73), all TSOs started discussing and drafting a new methodology to reflect anticipated changes of capacity allocation in the day-ahead timeframe—flow-based market coupling in the Core region. As part of this, the CID Methodology should serve the objective of promoting effective competition in the trading and supply of electricity. It should aim to ensure non-discriminatory access to cross-zonal capacity. By laying down objective criteria and solutions for the distribution of CI to be applied by all involved TSOs, the Methodology should create a solid basis for CI distribution for the first time at European level. Simultaneously, it is also essential to ensure that the rules and regulations governing CID are clear and easy to understand, while keeping the necessary freedom for individual approaches, as each region is somewhat unique.

After several months all TSOs drafted a congestion income distribution methodology and submitted it to the regulatory authorities for approval in August 2016.

The fundamental principles of CID were fine, but the methodology was considered too extensive and not detailed in some aspects. This first draft was therefore not accepted by all regulatory authorities, and they filed a request for change in February 2017.

It did not take long for all TSOs to come up with a corrected 2nd draft of the methodology in April 2017. This draft triggered a formal approval process and many consultations and teleconferences between all regulators, TSOs and ACER. Since the NRAs were not able to reach a conclusion, the Agency became responsible for a decision concerning the submitted proposal.[2] In December 2017, the Agency issued a decision including an improved version of the CID methodology, generally as we know it until today. However, we did not yet know what new knowledge and controversial individual interpretation of the rules would be brought by the subsequent years of implementation of the target model of the electricity markets coupling.

During implementation TSOs realized in 2021 that a major revision of the methodology was needed. This amendment was initially triggered by the TSOs of the Core region who were exploring the positive implications of the Multi Slack hub approach when implementing FBMC. During the work on the amendment, additional needs for adjustments resulting from experience with regional implementations and legislative developments were identified. Other changes were associated with proper inclusion of the intraday timeframe into the scope with the Intraday Capacity Pricing Auctions, consistency with the FCA CID Methodology and to cover the possible occurrence of negative CI on the level of the whole CCR under specific circumstances. However, even after this revision, many questions remained open mainly due to insufficient time to fully decide on final solutions. Therefore, another amendment has been made at the end of 2023 introducing universal solutions for non-intuitive flows irrespective of their causes, changes due to balancing exchange cross-border capacity reservations, and switching to 15 min resolution in day-ahead timeframe.

Due to decisions on long-term flow-based capacity calculation and allocation in the Core and Nordic CCR in 2021, the TSOs started discussing the CID Methodology also based on the FCA regulation framework (European Union, 2013).

10.3 Congestion Income Distribution Methodology for Day-Ahead Allocation

The first step is always the collection of CI resulting from the relevant trading mechanism. Central Counter Parties or Shipping Agents collect the Congestion Income arising from the Single Day-Ahead Coupling and from the Single Intraday Coupling (intraday auctions only) and shall ensure that collected Congestion

[2] Based on the provisions of Article 9(12) of the CACM Guideline.

Incomes are transferred to the TSOs, or entities appointed by TSOs, no later than two weeks after the date of the settlement.[3]

The main task of CID is to determine amounts of CI for each BZB—where we apply the well-known equal distribution for each side of the border. This appears to be an easy task as we seem to know all information—i.e. prices in each BZ that can be translated into price spreads on relevant BZBs and resulting commercial flows on each BZB.

While the CID methodology is straightforward for CNTC allocation, more challenges arise when applying flow-based allocation.

In the Flow-based allocation we witnessed that many evolutions of the electricity market introduced very challenging questions into the CID methodology:

- How to deal with non-intuitiveness?
- Is there a need for slack hubs?
- When more regions meet—how to manage allocation constraints?
- When it comes to interactions over different timeframes, is balancing stealing money? How to manage the inconsistency of LT and DA methodologies and LTTRs remuneration?

10.4 Overview on the Underlying Formulae and Process

In this chapter, we describe the current CID principles in a simplified way as an overview that you can refer to in other chapters. Although the methodology is at the level of All TSOs this is mainly for harmonization purposes, as the actual implementation and calculation is then at the CCR level.

Within CID we operate with a pot of actually generated CI collected via Central Counter Parties or Shipping Agents. The main task is to distribute this pot between the individual BZBs of the CCR. CCR generated CI can be calculated using results from the SDAC i.e. Net Positions (NP) (netted import/export result) of each BZ times its price (P).

$$CI\ CCR = -\sum_{j=1}^{NBZ} NPj * Pj$$

CCR generated congestion income (10.2)

Calculation of CI in absolute value for each BZB comes from the resulting commercial flow on a BZB times relevant market spread (i.e. the difference between the BZ prices on both sides of the BZB). For the CNTC approach, commercial flow is equal to allocated cross-zonal capacities, which are directly resulting from the SDAC. For Flow-based, commercial flow needs to be calculated using contributions from each BZ NP multiplied with the PTDF factor transforming the resulting

[3] More details on clearing and settlement can be found in Chaps. 8 and 9.

trades into actual physical flows on BZBs, which we call additional aggregated flow (AAF).

$$AAFi = \sum_{j,k \in i} NPj * PTDFj, k$$
$$CI_{BZB_i,calculated} = commercial\,flow_i * market\,spread_i$$
Calculating Commercial Flow under Flow-based Approach (10.3)

The pot of calculated CI of a CCR is then the sum of all BZB CIs in the CCR. Next, we need to match this calculated CCR pot to the generated CCR pot using proportional adjustment. The ratio between these two pots will determine the correction factor by which calculated CI for each BZB is multiplied. This process is illustrated in Fig. 10.1. The core calculation can be seen in Eq. (10.4).

$$CI_{CCR\,calculated} = \sum_{i=1}^{NBZ} CI_{BZB_i,calculated}$$
$$CI_{BZB_i,final} = \frac{CI_{CCR\,generated}}{CI_{CCR\,calculated}} * CI_{BZB_i,calculated}$$
Congestion Income Core Calculation (10.4)

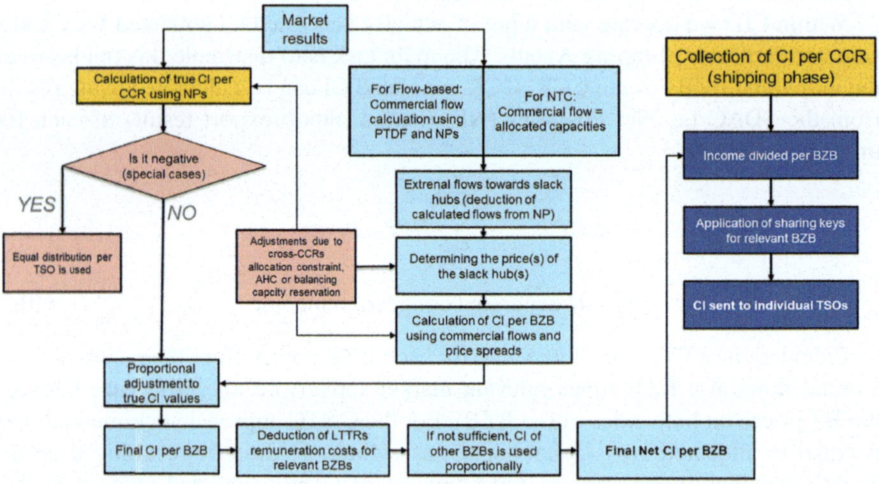

Fig. 10.1 Process overview congestion income distribution. *Source* Author's own illustration

10.5 Negative Income

The coupling algorithm aims to maximize the welfare of the coupled bidding zones. In 2019, we introduced so-called plain flow-based evaluation. Given the focus on the bidding zone borders (inherited from NTC world), we expect that the underlying commercial flow follows the price spread. However, we may experience that the flow goes against the spread. An intuitive flow is from a cheaper BZ to a more expensive one, creating a surplus of money on their border. A non-intuitive flow is the opposite—a flow from a more expensive BZ to a cheaper BZ.

A non-intuitive flow can occur through market optimization when there are multiple BZ affecting a network constraint—typically in the Flow-based approach. Accepting a non-intuitive flow can have a relieving effect on other elements in the grid that, thanks to netting of counter flows, form a greater economic welfare than the lost economic welfare due to the non-intuitive flow. Similar situations may happen in case there are allocation constraints, such as ramping limits of an interconnector or export/import limitations on a BZ. For the former, the cross-zonal flow against the spread yields a lower loss than the optimization of the individual BZ. For the latter, the formation of a price does not correspond to the network constraints. In case the BZ is used for transit, a non-intuitive flow may be seen on some borders that leads to a negative CI.

The best example of this is the case where the export/import allocation constraint is set to zero, and therefore, the BZ de facto does not participate in the SDAC. The price of such a BZ is then made up of purely domestic supply and demand and all cross-border flows are transits ignoring this price.

Albeit negative results appear on single BZBs, the global optimization is designed to increase the welfare of the coupled CCR and positive CI is present in the whole coupled area. But how should we interpret such signals? We see that the distribution of the total CI pot to the individual borders does not reflect the underlying optimization of the flow-based allocation. Therefore, some borders gain more benefit (CI) than others. TSOs have tried to find out a logic and rational economic theory to deal with negative CI on BZBs. A simple yet effective improvement was introduced in the CACM Guideline's CID Methodology to deal with non-intuitive flows which is an absolute value calculation of BZB CI. This leads to non-negative output on each BZB. This absolute calculation keeps information for each BZB reflecting its contribution to the whole economic welfare. However, the sum of individual CI is greater than the CI collected. Therefore, each BZB CI is adjusted proportionally in order to match the total CI generated by cross-border electricity exchanges within a CCR. This way each BZB has non-negative CI and the overall CI is shared among all BZBs.

The proportional adjustment approach with the absolute CI at BZB (BZB CI) as a basic factor became universal for other issues as well—it represents each TSOs' share on the total CI pot generated within the CCR. This approach is then used for all adjustments of the total CI pot and simplifies all relevant processes.

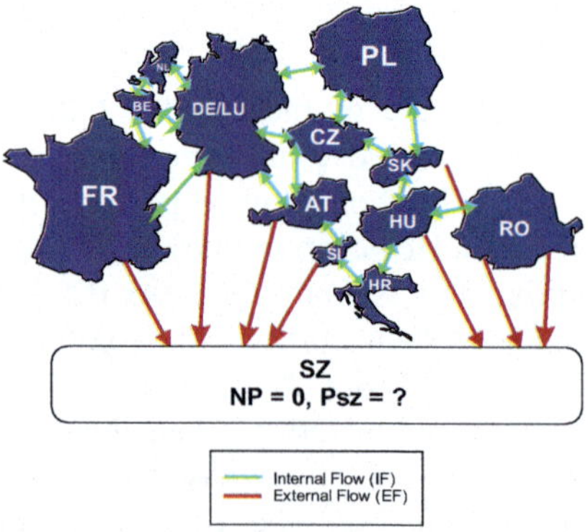

Fig. 10.2 Visualization of one or two slack hubs (SZ) for CCR Core with assigned net position (NP) and price (Psz). *Source* Adapted from ENTSO-E (2021)

10.6 Slack Hubs

In the Flow-based allocation, the sum of flows on intraregional BZBs can be lower than the net position of the relevant BZ. This is because of physical laws, when electricity also flows over BZBs that fall into another CCR. Such flows leave and re-enter the CCR using external borders. This flow has a value, but we miss a unique BZ price outside of the CCR to calculate the CI. Therefore, a virtual zone—a slack hub is created to resolve all external flows (as included in Fig. 10.1 and also illustrated in Fig. 10.2). Its net export–import position is always zero and we only need to calculate the price for it. This is calculated as a fitting price creating the minimum CI on the respective external borders. However, this calculation can also create a discrepancy between the CI calculated and that actually generated, so the principle of proportional adjustment works here as well.

A single slack hub approach can be unsuitable for some CCRs and the results can favour the external BZB and be to the detriment of the internal BZB. A multiple slack hubs approach should be always possible to achieve better or equal results as we split slack hub price optimization to more sub-tasks. A new determination of the number of slack hubs and their associated bidding zones was introduced in the CACM Guideline's CID Methodology which is unambiguous for each CCR from its specific topology. Currently two slack hubs are applied for Core Flow-based Market Coupling (FBMC).

10.7 Remuneration Inconsistency

The CID in the day-ahead timeframe is linked closely to other timeframes as well. The main link is to the long-term transmission rights (LTTR) with financial remuneration. An LTTR owner has a right to remuneration of the day-ahead BZB capacity price. This principle was well established in the NTC world for single

BZBs, where it was working in a straightforward way. If the day-ahead market "offered capacity" was the same or greater than the LTTRs, the resulting day-ahead CI should always be sufficient to cover remuneration. Individual BZBs did not influence each other and the BZB approach was working well. With the advent of shared constraints and, more recently, the Flow-based allocation, this delicate balance was disrupted, and we now witness the collision of two worlds (LT and DA) with different principles.

As described above, the TSOs strive to distribute the total CI pot to individual BZBs. The LTTRs are still determined and allocated for individual BZBs, so the remuneration is held by the respective TSOs from their day-ahead BZB CI. A problem emerges when we combine the border-only approach with shared constraints. This means that congestion on such shared constraints may cause a commercial flow on a BZB lower than the allocated LTTRs. An example is illustrated in Fig. 10.3 where from 200 MW allocated LTTRs from A to B we can end only with 120 MW flow in FBMC using the same "trade". This breaks the balance that the CI at a BZB is always the same or greater than the LTTR remuneration. As a result, the TSOs at the concerned BZB must pay more than they receive from the day-ahead allocation.

This already occurred in the NTC world because of the so-called technical profiles that connected multiple borders with one capacity limit (mainly PL<>CZ+DE+SK and PL<>DE(50 Hz)+CZ). On a much larger scale, this problem emerged in the Core Flow-based market coupling. As mentioned earlier, with Flow-based, all BZBs within the CCR are linked and influence each other. Thanks to so-called Long-Term Allocation (LTA) inclusion that ensures the day-ahead capacity domain is at least the same as the LTTRs, it is ensured that the total CI pot within the CCR covers the total LTTRs remunerations (in Fig. 10.3 ensured

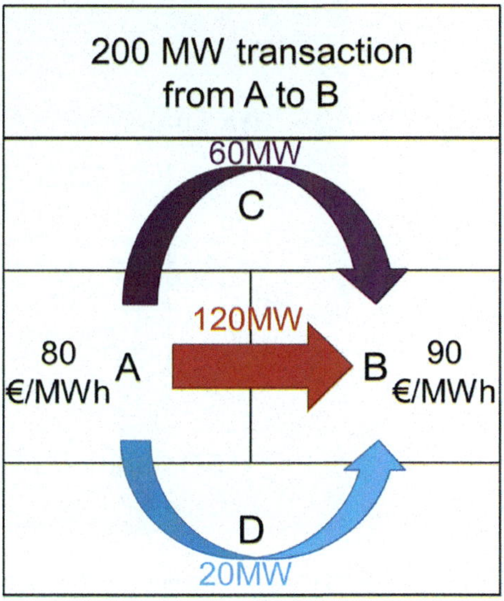

Fig. 10.3 Breakdown of market flow into physical flows in FBMC. *Source* Author's own illustration

by CI created by 60 MW and 20 MW flows). However, no border is guaranteed a collection of enough congestion income in day-ahead trading to cover its remuneration of LTTRs. An example with basic CI calculation for this challenge is illustrated in Fig. 10.4.

CID should ensure that the costs and benefits of market coupling are shared fairly, not only among market participants but also among TSOs. Since several TSOs had negative balance on some or even all BZBs, discussions among the TSOs about mutual compensation started. However, the question of fairness did not appear so clear and there were different opinions among TSOs around the scenarios and levels of compensation.

In fact, the topic of compensation of LTTRs was already in the scope of the methodology according to Article 61 of the FCA—the remuneration shall be shared based on the above-mentioned principle—proportionally to the BZB CIs. Nonetheless, the TSO struggled to find consensus on the topic of remuneration.

However, even in the case of agreement on compensation, there were multiple options. It was primarily a matter of deciding between full compensation or only

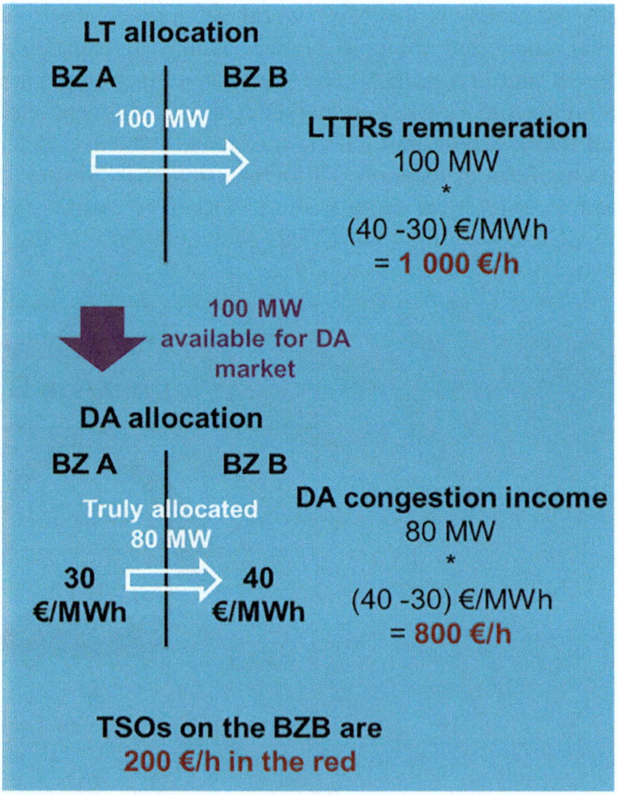

Fig. 10.4 Example of the creation of missing income for the remuneration of LTTRs. *Source* Author's own illustration

up to zero, to ensure that no one will have net negative CI. In addition, there were a few more complex proposals on the spectrum between these two options, but even with up-to-zero compensation there was a question of whether to include income from long-term auctions or not. The possibilities of tracing the reallocated capacity were also analyzed meaning the question of who should pay to whom. Should all TSOs contribute to compensation or only some? However, at least with FBMC, the system is so complex that any tracing seems impossible.

The question of the adequacy of the size of LTTRs was also raised. At the time when the issues first arose, the amount of long-term capacities offered was based on bilateral agreements on the respective BZBs because the Core LT Capacity Calculation methodology would not be implemented until 2024. This might lead to freeriding because of lacking coordinated calculation.

Therefore, several months of discussions and numerous analyses by TSOs followed in order to create a first draft of this methodology. The topic was strongly discussed both with individual regulators and with ACER, and in the end also with the European Commission. TSOs finally put together the methodology proposal in April 2020 which once again started an intensive consultation period. ACER published its decision and a revised version of the methodology in October 2020. However, this decision was appealed and the entire methodology was revised again. The final version approved by ACER in October 2021 included an up-to-zero compensation using a proportional division among all other BZBs in the region with a positive day-ahead CI. Albeit in force, this methodology is only applicable with a Core LTCC methodology in operation, which is anticipated as of 2025. Until then, the Core TSOs have not unanimously agreed on the form of mutual compensation. With the already functional FBMC in the Core region, there are several BZBs that do not earn enough CI in daily allocation to cover the LTTRs remuneration.

10.8 Reservation of Capacity for Balancing

The balancing market processes have always been considered rather independent of the power market. With the Electricity Balancing Guideline (EBGL), cross-border exchange of balancing capacity or sharing of reserves became optional for TSOs. TSOs can reserve part of the day-ahead capacity for balancing purposes and thus remove this capacity from the power market without removing obligations linked to this capacity. Both the market-based allocation process and the *Allocation process based on economic efficiency analysis* mean that part of the day-ahead offered capacity is allocated outside of the market coupling. If it is done efficiently and in line with the EBGL regulation it brings more welfare to the entire electricity market.

While drafting the corresponding EBGL methodologies, several CI experts noted that this can lead to a situation where the CI generated in the day-ahead market coupling is lower than the LTTR remunerations—available capacity for day-ahead market is lowered below LTA inclusion. This would represent a major problem especially in the Core region where only some TSOs would reserve capacity for balancing—the missing CI from the DA market would not be recovered equally in the balancing market. Therefore, both balancing and CI experts needed to update all concerned methodologies and processes and work on solutions to consider and distribute CI from balancing exchanges as CI from the day-ahead market.

10.9 Cross-CCRs Congestion Income Transfers

Other extensions around negative income issues are allocation constraints and Advanced hybrid coupling (AHC). We will discuss these two specific types in one section, as their implications are similar since both are ultimately about sharing of congestion income from common constraints / mutual influences between multiple borders. However, the main difference here is the aspect of crossing the boundary of a single CCR.

In the evaluation of day-ahead trading, we still try to reach a solution with the maximum economic welfare value. However, the optimization algorithm (Euphemia) optimizes the whole SDAC and not individual CCRs. Normally, we do not have to consider the impact of trading across regions on CI, but the situation may change with shared allocation constraints. For BZs applying allocation constraints located inside the CCR (e.g. Italy North) no impact on CI arises. The induced non-intuitive flows and allocation restrictions remain within a single CCR and these discrepancies always fall under the mechanism of proportional adjustment. However, for a BZ having allocation constraints affecting BZBs of more CCRs (e.g. limiting total export/import of the outer BZ), the CI distribution does not correspond to the CCR needs—one CCR gains more CI than the other.

AHC has a very similar impact, however it is not a constraint tied to an individual BZ, but directly selected BZBs. For adjacent CCRs where one applies FB and the other NTC we can observe strong interactions between BZBs even if they are not together in the same CCR. This mutual influence is then taken into account by the AHC application. The influence of realized NTC transactions on the physical margins of the critical branches in the FB model are taken into account during allocation and vice-versa.

This problem therefore requires a cross-CCR solution at the all TSOs level going beyond the current definitions and rules in the CACM Guideline's CID methodology. There is thus a clear need to amend the methodology and introduce a legislative obligation to coordinate CID at the level of at least all interdependent BZBs across CCRs. There are primarily two possible solutions to this problem.

The first solution is to extend proportional adjustment to more than one CCR to cover all interdependent BZBs. This is an effective and easy-to-understand solution. In implementation, all interdependent BZBs would have to be part of one CID calculation, which goes hand in hand with the contractual conditions with the relevant entities mediating the CID calculation. A theoretical disadvantage of this solution could be an unfair distribution of congestion income as we would combine FB and NTC allocation mechanisms. The second solution focuses more on identification and valuation of transit and cross-CCR flows and their effects on CID. Using robust mathematical interpretation backed up by extensive simulations, results with filtered-out effects of the cross-CCR allocation constraint can be calculated. The resulting additional pot is distributed between the borders and slack hub borders of the bidding zone in the same direction as the active cross-CCRs allocation constraint.

In the current CACM Guideline's CID Methodology the problem is known and solution based on the second approach described. The TSOs from CCRs mutually affected by allocation mechanisms with cross-CCR impact shall now jointly develop, test and validate the algorithms, tools and procedures for the cross-CCRs mechanisms defined in this methodology. Until then, the cross-CCR compensation is done on a voluntary basis based on a simplified formula close to the approved solution (with very few operational cases currently occurring).

10.10 How is CI Regulated?

Congestion income is not negligible and can reach tens to hundreds of millions of euros annually. Moreover, the size of these profits is closely linked to the size of the capacity offered for trading. It is therefore appropriate to monitor the CID processes and to keep an eye on the abuse of the offered capacity to increase profits. However, the main topic here is how TSOs should use this revenue as it should primarily eliminate the cause of its origin—to increase the capacity limiting commercial transactions.

Rules for the use of congestion income are established by Article 19 of Regulation (EU) 2019/943 of the European Parliament and of the Council of 5 June 2019 on the internal market for electricity. In accordance with Article 19(4) transmission system operators drafted a methodology, which was approved by ACER on 23 December 2020. The methodology was applied to congestion income collected from 1 January 2021.

The methodology for the use of congestion income describes cost categories contributing to priority objectives either with Article 19(2) (a) and/or Article 19(2) (b) of Regulation (EU) 2019/943 for which congestion income can be allocated to. Allocation of the congestion income to individual cost categories is subjected to ex-ante communication between the TSO and their National Regulatory Authority (NRA). If priority objectives are deemed fulfilled by the relevant NRA, remaining congestion income can be used for calculating network tariffs or fixing network tariffs, or both. The residual revenues shall be placed on a separate internal account line until such a time as it can be spent on the priority objectives.

The legally defined priorities mentioned above are aimed at:

- guaranteeing the actual availability of the allocated capacity including firmness compensation; or
- maintaining or increasing cross-zonal capacities through optimization of the usage of existing interconnectors by means of coordinated remedial actions, where applicable, or covering costs resulting from network investments that are relevant to reduce interconnector congestion.

Generally speaking, congestion income should be used to eliminate the cause of its origin, to increase the capacity limiting commercial transactions i.e. to maintain or to increase cross-zonal capacities considering both CAPEX (investments) and OPEX (redispatches and compensation) oriented solutions.

10.11 Outlook

The concept of CID is important for ensuring that electricity markets remain efficient and competitive. By sharing the costs and benefits of market coupling fairly, CID helps to ensure that electricity is bought and sold at the most economical prices. In addition, CID helps to reduce the overall cost of electricity as it provides incentives to reduce bottlenecks in the grid. However, it can also have a negative impact on market competition if it is not implemented properly. For the future of CID, it is important to keep the processes and legislation up-to-date and monitor and evaluate market results on a regular basis.

Bibliography

ACER. (2023, December). *ACER amends congestion income distribution methodology.* https://www.acer.europa.eu/news-and-events/news/acer-amends-congestion-income-distribution-methodology

ENTSOE-E. (2021). *CACM CIDM Amendment Explanatory note 9 July 2021.* Retrieved October 26, 2024, from https://eepublicdownloads.entsoe.eu/clean-documents/nc-tasks/210709_CACM%20CIDM%20Explanatory%20note_ext.pdf

European Union. (2013). *Commission Regulation (EU) 2017/459 of 16 March 2017 establishing a network code on capacity allocation mechanisms in gas transmission systems and repealing Regulation (EU) No 984/2013.* https://eur-lex.europa.eu/legal-content/EN/TXT/?uri=CELEX%3A32017R0459

European Union. (2021). *Commission Regulation (EU) 2015/1222 of 24 July 2015 establishing a guideline on capacity allocation and congestion management.* https://eur-lex.europa.eu/legal-content/EN/TXT/?uri=CELEX%3A02015R1222-20210315

Karel Šebesta graduated in 2018 from the Czech Technical University with a focus on energy engineering. During his studies, he started working for the Czech TSO ČEPS a.s., and within this cooperation, he also wrote his diploma thesis. This thesis was, of course, on the topic of Congestion income distribution. When subsequently the update of the methodologies within ENTSO-E and Core CCR started to be discussed, Karel was able to apply his already achieved conclusions and soon became a convener of the respective working groups. Among other things, he was actively involved in the implementation and launch of both SIDC and SDAC, focusing on both national and European sides, from financial flows and contracting, to technical aspects of allocation, to systems testing. Long-term timeframe has not been neglected either and Karel has taken on the role of Secretary and co-chair of the Single Allocation Platform (SAP) Board for long-term cross-border capacity trading. Linked to this position, he was also the FCA SPOC of ENTSO-E and FCA SPOC of Core CCR. Lately, he has been focusing mainly on market analyses and the market design reform.

Martin Palkovský is a seasoned professional in the energy sector with over 14 years of experience. With a knack for the technical and regulatory aspects of energy, he's played a key role in shaping European electricity markets through his work on the target electricity market model, CACM network code, market coupling projects like CZ-SK-HU MC and 4MMC and the formation of JAO, S.A., a joint allocation office backed by 25 TSOs. At CEPS, the Czech TSO, Martin has been at the forefront of managing transmission services, focusing on everything from capacity calculation to cross-border trade. He chaired the Single Allocation Platform Council and JAO Supervisory Board. Not just confined to electricity, he has also ventured into the gas sector, contributing to projects like Capacity4Gas and now sitting on the supervisory board of NET4GAS, the Czech gas TSO. With a degree in Economics and Management of Power from the Czech Technical University in Prague, Martin combines academic knowledge with real-world experience, making him a pragmatic and influential figure in the energy landscape.

Part IV

News from the Regions

Day-Ahead Flow-Based Capacity Calculation and Market Coupling in Core CCR

11

Ferenc Nagy, Melinda Nagy, Luca Toth, Ágnes Takacsne Esze, Ákos Arnold, Dániel Diveny, and Gábor Szathmari

Abstract

This chapter starts with a brief overview of flow-based capacity calculation and market coupling. A comparison of the results from the Net Transfer Capacity approach compared to flow-based market coupling highlights the differences regarding social welfare, net positions, and price convergence. The chapter also analyzes the market coupling results from the perspective of topics such as active constraints, shadow prices and price spreads. The chapter concludes with an assessment of the social welfare gains, as well as consumer and producer surplus, generated following the application of the flow-based capacity calculation and market coupling.

F. Nagy (✉) · M. Nagy · L. Toth · Á. T. Esze · Á. Arnold · D. Diveny · G. Szathmari
Hungarian Transmission System Operator MAVIR Ltd, Budapest, Hungary
e-mail: nagyf1@mavir.hu

M. Nagy
e-mail: nagym@mavir.hu

L. Toth
e-mail: tothl1@mavir.hu

Á. T. Esze
e-mail: esze@mavir.hu

Á. Arnold
e-mail: arnolda@mavir.hu

D. Diveny
e-mail: divenyid@mavir.hu

G. Szathmari
e-mail: szathmarig@mavir.hu

A. Estermann et al. (eds.), *European Electricity Market Coupling*, Future of Energy,
https://doi.org/10.1007/978-3-031-86315-8_11

11.1 Introduction

11.1.1 Flow-Based Capacity Calculation and Market Coupling

The flow-based (FB) method introduces a completely new approach with the coordinated capacity calculation, in accordance with Article 20(2) of the CACM Guideline, compared to the well-known bilateral calculation method—i.e. Net Transfer Capacity (NTC) (ENTSO-E, 2000). It is the first concept to have introduced the modelling of the superposition effect of the power flows over network elements resulting from the simultaneous commercial energy trades. By doing so, it is possible to calculate the available margin (MW) for each modelled individual critical network element and contingency (CNEC) which will act as potential limiting constraints to the market. The main aim of the flow-based method is to maximize the available cross-zonal capacity for the electricity market while respecting operational security limits. Thus, the social welfare maximization by means of energy exchanges can be achieved. In addition, the flow-based coordinated capacity calculation has paved the way for the introduction of transparent and harmonized operational processes which will help to bring the physical and commercial aspects of the energy grid closer. For a general introduction to capacity calculation see Chap. 3.

The implementation of the Day-Ahead Capacity Calculation Methodology (ENTSO-E, 2019) has proven to be an extremely complex task. The method introduces certain controversial aspects, such as the minimum capacity requirement, according to Article 17 of the Day-Ahead Capacity Calculation Methodology (DA CCM), which has given one of the most difficult tasks to TSO experts as to what risk it can pose for the grid security. In the project's life, one of the major achievements was to be able to compute and generate the first flow-based domain back in early 2020. The flow-based domain can be characterized as a n-dimensional convex polyhedron shape while an NTC domain, given its nature of bilateral coordination with independent limit per border/direction, is usually modelled as several independent two-dimensional rectangles.

With the successful go-live, TSOs, market participants and various expert groups have had the long-awaited opportunity to gain experience and knowledge of the operational results. Unlike the test (external parallel run) phases, in the operational process after the go-live, the market coupling results have been based on real supply and demand bids from the energy traders. This has offered the unique opportunity to study the results of the capacity calculation in combination with the market coupling while the global energy supply and markets have been going through extreme circumstances especially in Q3-Q4, 2022.

In order to give a good understanding of the capacity calculation, hereafter referred to as the pre-coupling and the market coupling, it is important to first introduce the basics of the flow-based capacity calculation method. This includes definitions, modelling of the grid, the mathematical formulas and the methodological steps. The main flow-based parameters for the market coupling are described in the form of domain limiting linear constraints with their capacity (Remaining

Available Margin—RAM) and sensitivity (Power Transfer Distribution Factors—PTDF). Based on the CNECs as domain limiting constraints, the cross-zonal capacity can be calculated for allocation during the market coupling. While the pre-coupling step produces the domain limiting constraints, the market coupling outcome lists the active constraints which are effectively limiting the market direction—i.e. market clearing point (MCP). The MCP in the balanced n-dimensional domain can be described as a vector of net positions of the bidding zones. Net position in the FB context should be read as the net position of a market given the market exchanges via the meshed network. The active constraints, a subset of the domain limiting constraints, play a key role in the price formation of the bidding zones. These identified constraints are associated with their shadow prices (EUR/MW) which show how sensitive the market direction to a certain constraint is. The higher the associated shadow price is, the greater the impact of the active constraint on the bidding zone price results is. As a result of the allocated cross-zonal capacity, key indicators, such as bidding zone prices, net positions, active constraints with associated shadow prices are generated. These indicators are used for analysis and simulations in order to better understand the linkage between the domain limiting constraints and the outcome of the market coupling optimization algorithm.

In sum, the chapter aims to provide a good insight into the basic definitions of flow-based methodology in Sect. 11.1.1. In Sect. 11.2.1, a comparison of the pre-go-live results from the NTC era and the market coupling results is provided, with an overview of social welfare, the net positions, and the price convergence in the Core region (Fig. 11.1). Furthermore, the chapter focuses on the demonstration of the analyzed market coupling results, e.g. active constraints, shadow prices, price spread from the operational process in order to have a good understanding of what has been learned and concluded by the various experts since the go-live of the first operational business day (Sects. 11.2.2, 11.2.3 and 11.2.5).

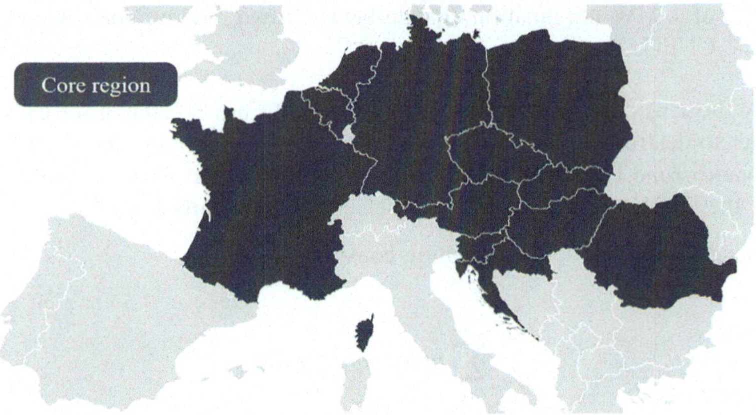

Fig. 11.1 Map of the core CCR. *Source* Author's own illustration

11.1.1.1 Pre-coupling

The pre-coupling stage consists of the steps necessary for preparation and calculation of the final flow-based (FB) domain that is sent to the market coupling.

The Core DA FB CC business process can be summarized with the following steps:

1. **Delivery of input for capacity calculation:**
   a. Common Grid Models (CGMs): Individual grid models provided by each TSO are merged into CGMs. They include a forecast of both the grid topology and power generation and consumption within the TSO's control area based on the reference day. The CGM encompasses the synchronous continental ENTSO-E grid. It is needed to model the physical influences of these grids with two days ahead congestion forecast—i.e. D2CF CGM.
   b. Critical Network Elements (CNEs) with Contingency (N-k) (CNECs): It is the list of network elements with associated contingency. These network elements can be various elements, e.g. AC power lines, transformers defined with their characteristics.
   c. Generation and Load Shift Key (GLSK): GLSKs are a list of generation and load factors that define how a change in net position is mapped to the generating units and loads in a bidding zone.
2. **Capacity Calculation with CNEC selection and Remedial Action Optimization:**
   a. Flow-based domain generation: The flow-based domain can be described by the FB parameters. Of the FB parameters, the Remaining Available Margin (RAM) and Power Transfer Distribution Factors (PTDF) factors determine the domain limiting constraints.
      i. RAM: specifies the available margin for the market exchange in MW.
      ii. PTDF: PTDFs are sensitivities of CNECs to the changes of net positions of bidding zones. PTDF factors are calculated using GLSKs and a sensitivity calculation based on the CGM. It represents an incremental flow of 1 MW net position change between two bidding zone A and B on a CNEC.

During the sensitivity calculation, the zonal PTDFs are calculated for each CNEC in order to determine which CNEC can be a domain limiting constraint based on the zone-to-zone PTDF threshold criterion. According to Article 15 of Core DA CCM, the PTDF threshold for CNEC selection is currently 5%.

The general form for domain limiting constraints:

$$\begin{aligned} \sum_{i=1}^{N} ptdf_i \cdot np_i &\leq RAM \\ \sum_{i=1}^{N} np_i &= 0 \end{aligned} \tag{11.1}$$

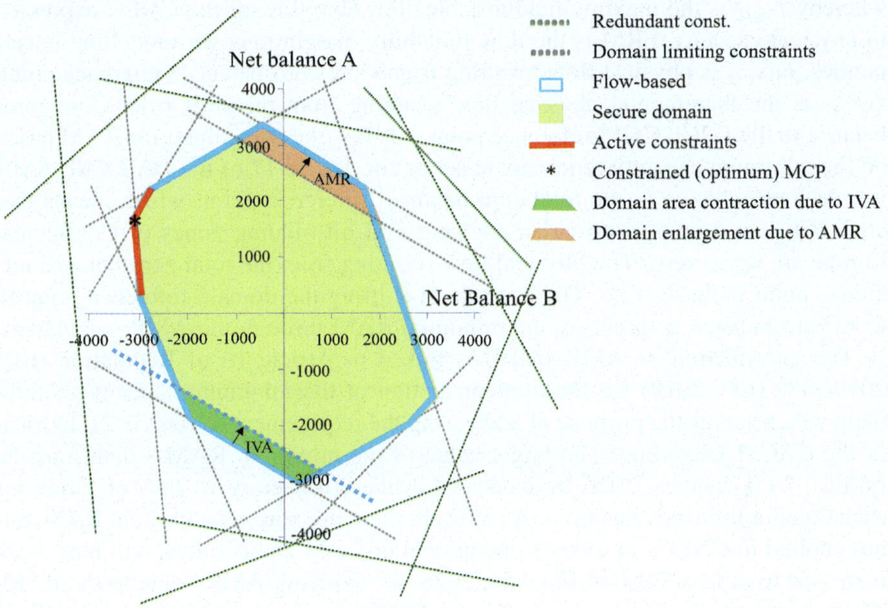

Fig. 11.2 Two-dimensional slice of FB domain. *Source* Based on (NEMO, 2019)

whereby N is the number of considered zones (hubs), np_i are variables in the balance search space, $ptdf_i$ and RAM are the corresponding FB parameters of the considered constraints.

The calculated RAM is relative to the reference point in the balance search space. The reference point in the balance space is defined by a vector of net positions. In Fig. 11.2 the reference point is the origin point of the domain area.

Figure 11.2 illustrates a two-dimensional slice of the balanced domain. The secure domain is defined by limiting constraints.

The flow-based constraints define a search space for the net positions to be determined within the market coupling. In the Core region, there are 14 bidding zones. This means that the search space for the net positions is defined by a fourteen-dimensional balance space.[1]

The RAM before capacity validation can be described by the following formula:

$$RAM = F_{max} - FRM - (F_{0,core} + \Delta F_{ref}) + AMR \tag{11.2}$$

[1] Because the sum of the net positions equals zero, one net position can be written as a combination of the remaining ones. Due to this balance constraint, when considering n zones, the space of the independent variables has actually a dimension of n-1.

whereby F_{max} is the maximum admissible allowable flow of the CNEC expressed in physical capacity, FRM is the flow reliability margin to cover modelling uncertainties, $F_{0,core}$ is physical flow resulting from Core zero-balance[2] reference point, ΔF_{ref} is the incremental physical flow resulting from the shift from Core zero-balance to the D2CF CGM reference point, AMR is the adjustment for RAM based on the minimum capacity requirement set out in Article 17 of the DA CCM. AMR is calculated relative to the total zero-balance[3] reference point, which means that all D2CF based reference market exchanges of all bidding zones of Continental Europe are set to zero. The physical flow resulting from the total zero-balance reference point is the $F_{0,total}$. The purpose of shifting the domain reference point to Core zero-balance is to ensure the minimum RAM through the AMR calculation.

The introduction of AMR (MW) is linked to Article 16 of Regulation (EU) 2019/943[4] (EU, 2019) for the implementation of the minimum capacity requirement which serves the purpose of addressing the requirement of Article 21(1)(b)(ii) of the CACM Guideline. The target value of the minimum RAM which must be fulfilled by 1 January 2026 by means of a linear trajectory is 70% of Fmax for each domain limiting constraint. As a result, there are various minimum RAM values applied to CNECs in today's operational process. These values will have been increased to at least 70% of Fmax by 2026. By applying AMR value to the RAM, TSOs commit to ensuring capacity on the CNEC that is physically not available in the grid. This is why it is referred to as virtual capacity which undoubtedly carries an increased risk of compromised grid security. Figure 11.2 illustrates the enlarged area of the domain due to AMR.

The remedial action optimization expands the domain area around the forecasted market clearing point by increasing the RAM over the CNECs. The type of remedial actions can be preventive (pre-fault) and curative (post-fault) by means of phase-shifting transformers and/or topological measures. The availability of the remedial actions is provided in the input data by the TSOs.

3. **Capacity Validation (TSO)**: During the capacity validation step, TSOs assess if the calculated RAM per CNEC can be provided to the market coupling while the grid security is ensured. TSOs individually perform the validation by assessing multiple likely market clearing points. In case all available remedial actions are not able to secure the likely market clearing point, TSOs might have to resort to reducing the RAM by applying the individual validation adjustment (IVA). This leads to a loss of domain area as illustrated by Fig. 11.2.

Given the minimum capacity requirement, there has been an increased emphasis on the grid security during the capacity validation phase.

---

[2] Core zero-balance means the situation when all market exchanges among the Core bidding zone borders inside Core CCR are set to zero.

[3] Total zero-balance means the situation when market exchanges among all bidding zones of Continental Europe are set to zero.

[4] Also known as Europe's Clean Energy Package (CEP).

4. **Final Capacity Calculation**: After the input from the capacity validation step is received by the central capacity calculation (tool) process, the final FB domain is calculated. The reference point of the final domain is shifted to Core zero-balance. Thus, the RAM after validation for the market coupling is calculated as follows:

$$RAM = F_{max} - FRM - F_{0,core} + AMR - IVA \tag{11.3}$$

whereby IVA is the individual validation adjustment according to Article 20 (5) of DA CCM for the implementation of which can be calculated and applied by each Core TSO to the RAM of their own CNECs in order to ensure grid security, thus making these CNECs limiting constraints. Limiting constraints with IVA have a higher probability of becoming active constraints.

The final FB parameters determine the domain limiting constraints, while the redundant constraints are identified as illustrated by Fig. 11.2.

There is an important distinction that needs to be made between the domain limiting constraints and the active constraints as pointed out earlier in the chapter. The limiting constraints are a set of constraints that give the bounded domain. However, only a subset of these limiting constraints in fact limit the market clearing point inside the final FB domain. These constraints are called the active constraints which are focused on with regard to shadow prices and price spreads in the remainder of the chapter.

11.1.1.2 Market Coupling

The Market Coupling (MC) process involving a group of sellers and buyers from different areas is an initiation of the European Power Exchanges organizations in order to integrate different energy markets into one coupled market for the purpose of liquidity (see Chaps. 1 and 2 for a more detailed introduction). Geographically, the electricity market in the European Union (EU) is based on the "zonal model". The zonal model implies that there is one single price per bidding zone (and market time unit). In SDAC, the market time unit (MTU) is 60 min. The iterative clearing algorithm defined in (Singh Gill, et al., 2022), developed for solving the problem associated with the coupling in SDAC, is known as EUPHEMIA (Nordpool Group, 2020) (see also Chap. 6). The main objective of EUPHEMIA is to maximize the social welfare that is calculated as the sum of the consumer surplus, the producer surplus and the congestion income across the region—i.e. total market value. In this chapter, the social welfare gain attributed to flow-based market coupling in comparison to the NTC regime is demonstrated in Sect. 11.2.1.

The market coupling process applying the flow-based model receives the final FB constraints from the central capacity calculation process. The final FB constraints are given by means of two components, namely RAM and PTDFs.

The constraint that is imposed is the following:

$$PTDF \cdot np \leq RAM \tag{11.4}$$

whereby *np* is the vector of net positions which are subject to the flow-based constraints. Each constraint corresponds to a single row in the *PTDF* matrix and has one corresponding *RAM* value.

The sum of the bidding zone net positions in Core CCR must equal zero. A net position with a positive sign means exports and a negative sign means imports.

11.1.1.3 Market Coupling Results

EUPHEMIA returns, among other results, the BZ net positions, BZ prices (market clearing prices), scheduled exchanges and active constraints with their associated shadow prices. The balanced market coupling net positions are used to adjust the RAM of CNECs. In doing so, the new RAM represents the RAM on the CNEC after the net positions of the FB MC are applied.

The identification of the active constraints with their positive (*p*) zone-to-zone $PTDF_{y \rightarrow z,k}$ (Eq. 11.5) (Dufour, 2007) reveals the limit to the market exchange direction inside the FB domain. The contribution to the price spread depends on the shadow price and zone-to-zone $PTDF_{y \rightarrow z,k}$. The trend in price spread among the bidding zones is analyzed in detail in Sect. 11.2.5.2.

$$pPTDF_{y \rightarrow z,k} = max(PTDF_{y,k} - PTDF_{z,k}, 0) \tag{11.5}$$

where:

$PTDF_y$: zonal PTDF of zone y of constraint k.

$PTDF_z$: zonal PTDF of zone z of constraint k.

An active constraint in combination with its shadow price has the maximum contribution to the price spread of the two corresponding BZ borders having the maximum zone-to-zone PTDFs. The maximum positive (*p*) zone-to-zone PTDF for the constraint *k* with its shadow price can be calculated by any combination of the bidding zone borders from the Core CCR.[5]

$$pPTDF_{z1 \rightarrow z2,k} = (PTDF_{z1,k}) - (PTDF_{z2,k}) \tag{11.6}$$

If EUPHEMIA finds a welfare maximizing solution inside the FB domain, this implies that none of the FB constraints are active. As a result, all clearing prices should fully converge. When EUPHEMIA finds a solution on one of the edges of the domain, one or more constraints will be active. As a result, clearing prices will differ. Figure 11.2 illustrates the example of active constraints. Active constraints show the network congestion in the form of RAM = 0 MW relative to the market clearing point. The quantity used for the representation of the market sensitivity to a constraint is the shadow price. Shadow prices in the FB model represent the increase in social welfare caused by making 1 MW more capacity (RAM) available to the market (Kristiansen, 2020). In a FB model, price differences (spread) among

[5] Any combination of bidding zone borders does not necessarily mean it is a physical border.

bidding zones are the result of shadow prices on all congested CNECs. The general formula that gives the relation between the overall price spread and the shadow prices, including all congested active constraints, can be expressed in the following way:

$$P_z - P_y = \sum_k (PTDF_{y,k} - PTDF_{z,k}) shadow\,price_k \geq 0 \qquad (11.7)$$

whereby zones z and y are a subset of all Core bidding zones, P_z is price in zone z, $PTDF_{y,k}$ is influence from zone z on constraint k. The price spread is proportionate to the sensitivity (i.e. shadow price) and the difference in PTDFs (Kristiansen, 2020).

In general, it can be concluded that if there is no active constraint, the full price convergence (i.e. all bidding zone prices in the Core CCR) can be obtained.

Although the objective of this chapter is to focus on the flow-based domain constraints exclusively, it is important to note that EUPHEMIA in the SDAC market coupling for the Core CCR models the Long-Term Allocation (LTA) inclusion in combination with the Flow-Based model. LTA inclusion is modelled with the LTA domain which includes the long-term capacities allocated explicitly for the Core borders. Given the convex combination of the FB domain and the LTA domain, the market clearing point is always found inside the computed convex hull (N-Side, 2022). Thus, there can be active constraints from both the FB domain and the LTA domain that can limit the market. If there are active constraints purely from the FB domain, it means that the market clearing point is constructed by the constraints of the FB domain. A variable (i.e. alpha = [0..1]) is introduced for the level of usage of the FB domain. If the alpha variable is 1, then the bidding zone net positions from the market coupling are given by the FB domain. Distribution of alpha factor is shown in Fig. 11.2. Figure 11.3 illustrates the convex combination of the FB domain and the LTA domain. It is important to point out that there are active FB constraints with RAM > 0 MW relative to the MCP when the market clearing point in between the FB domain and the LTA domain. The FB active constraint (1) has a non-zero shadow price even though the constrained MCP is not positioned on the boundary line of constraint (1) because shifting the constraint (1) to (1') would enlarge the convex hull, thus allowing the MCP to move further inside the convex hull. The FB domain limiting constraint (2) with RAM increased by the AMR during the capacity calculation is an active constraint (2′) with negative RAM at the MCP due to the enlarged convex hull by the LTA domain inclusion.

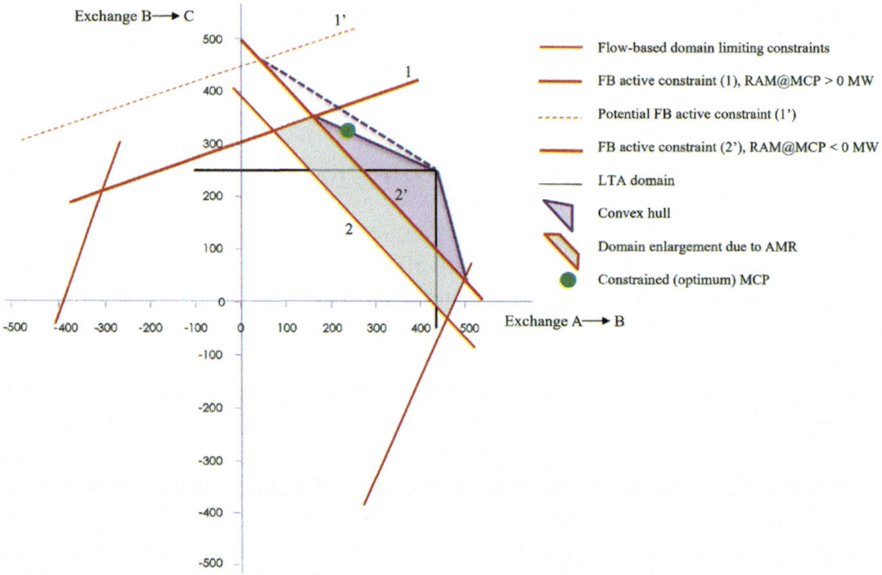

Fig. 11.3 Illustration of the convex combination of the FB domain and the LTA domain. *Source* Author's own illustration

11.2 Analysis and Results

11.2.1 Comparison of NTC and FB MC (EXT//RUN) Results

Prior to the implementation of FB MC in the Core CCR, the period of external parallel run (ext//run) was a demonstration and test phase, which ensured the smooth change of the basis of the single day-ahead market coupling and extended the complex logic of FB MC to 14 BZs. During this period the key input parameters and results were published on (JAO, 2021). The external parallel run began in October 2021 and lasted until the go-live on June 8th, 2022. The analysis included in the Appendix shows the results of the period of 01.10.2021–17.04.2022, with a total of 139 business days, where there were available and sufficient results both for the parallel run and the historical (or production) run with NTC.

The advantage of the comparison of these two scenarios is that we can have the opportunity to get a deep insight into the impact of the FB MC compared to the NTC based capacity calculation. In that sense, the TSOs used this referred period to analyze the results and implement further changes in the procedures before the go-live date. Due to this experimental aspect of the parallel run, we cannot conclude that the results we present below show the most accurate impact of the FB MC for the present days as well. However, the key messages are clear and visible.

As the Polish TSO (PSE) regularly applies allocation constraint,[6] we excluded the PL region from the analyses in order not to distort the results (therefore for this chapter Core CCR refers to 12 BZs without Poland and the HVDC zone between BE and DE). For this chapter we introduce the abbreviation of Prod which refers to the historical NTC based results. Historical data for the comparison was produced by the Simulation Facility environment. Simulation Facility is a test environment of the EUPHEMIA, therefore the results are very comparable to the historical day-ahead market results, but still minor differences can be observed.

11.2.1.1 Social Welfare

From the perspective of social welfare, the external parallel run results confirmed the expectations as the sum of the consumer and producer surplus increased with application introduction of FB MC during the external parallel phase.

Social welfare is the sum of the consumer surplus, the producer surplus and the congestion income of TSOs based on EUPHEMIA Public Description (NEMO, 2019), however under this chapter we focused only on the consumer and the producer surplus. The reason for this is twofold; first, Simulation Facility does not calculate congestion income for bidding zones and TSOs, as this is a separate post-coupling process, second, the volume of the regional congestion income is negligible (around 0.05% of total social welfare) compared to the volume of the consumer and the producer surplus.

With the application of the FB MC, the consumer surplus increased by 0.89% and the producer surplus did by 1.08%. Overall, a total gain of 0.92% was generated in social welfare during the external parallel run period. In general, the majority of the social welfare is gained from the consumer surplus, which was about five times greater than the producer surplus. This rate has not changed significantly after the FB MC go-live.

With regard to a certain accepted (demand) bid, the consumer surplus is equal to the bid price minus the market clearing price for 1 MWh. Otherwise, the price difference should be multiplied by the accepted volume of the certain bid in order to obtain the whole surplus. If there is a price-taker demand (buy) bid, which means that the bid price is the technical maximum, then the consumer surplus is the difference between the technical maximum and the market clearing price for an accepted price-taker bid of 1 MWh. During the time of the external parallel run, the technical maximum was 3000 EUR/MWh. On the other edge of the curve, the technical minimum price level is −500 EUR/MWh. If there is a price-taker offer, the producer surplus is the market clearing price minus the −500 EUR/MWh for an accepted bid of 1 MWh. Due to the level of the average prices, the consumer surplus is typically much higher than the producer surplus. As a result, the absolute sum of the surpluses is fundamentally influenced by the volume of the

[6] Polish allocation constraint: Regulation (EU) 2015/1222 establishing a guideline on capacity allocation and congestion management (EU, 2015) introduces the term of allocation constraints. Allocation constraints are regularly used by the Polish TSO to limit the import and export volumes via cross-border capacity limitations.

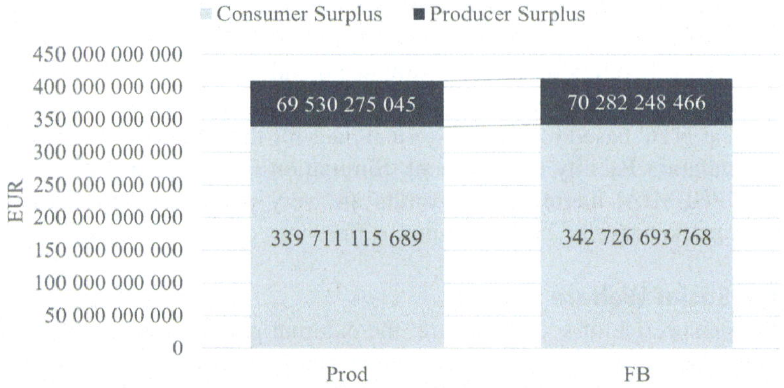

Fig. 11.4 Social welfare gain due to the FB (FB-Prod) in EUR from 01.10.2021 to 17.04.2022. *Source* Author's own illustration

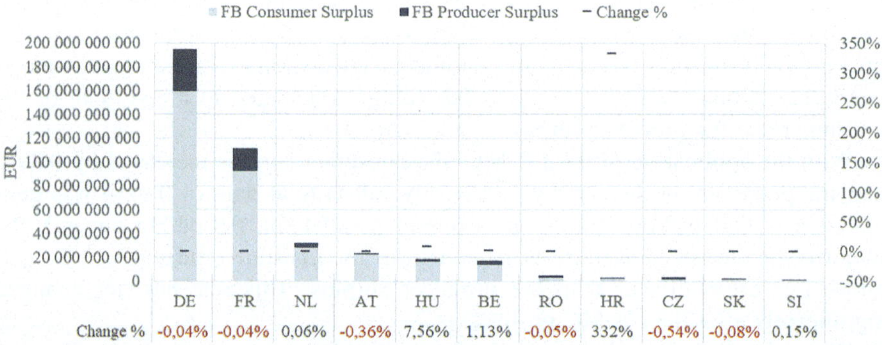

Fig. 11.5 Sum of the social welfare in EUR based on FB external // run results from 01.10.2021 to 17.04.2022. *Source* Authors own illustration

price-taker bids. Thus, showing the zonal distribution with the absolute values can be misinterpreted. However, for the purpose of the whole context, it is necessary to show the absolute values because in the case of Croatia, the extreme increase in percentage, in itself, with the application of the FB MC would be misleading. See Figs. 11.4 and 11.5.

11.2.1.2 Net Positions

During the external parallel run, the average net positions changed greatly in the DE bidding zone. The average German export position jumped from 3.8 GW to 5.2 GW due to the FB MC, while the second largest change occurred in the FR bidding zone as the import position of France increased by 1.3 GW to 3.5 GW. With respect to the other bidding zones, the average change (independent from the direction) was 0.23 GW and the lowest impact occurred in Austria (where NP decreased by only 12 MW) (Fig. 11.6).

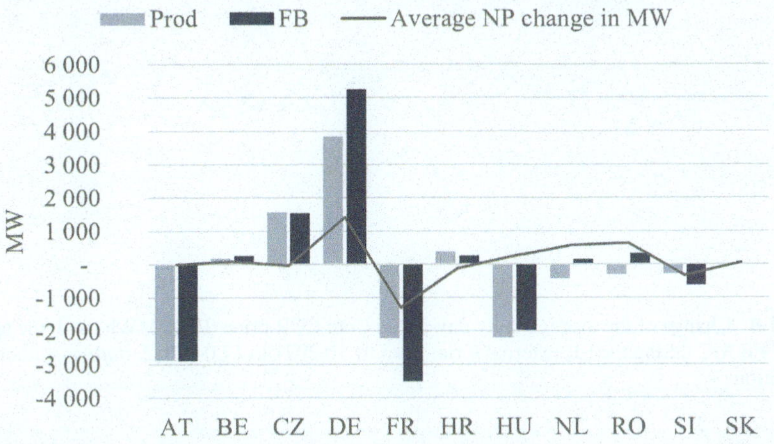

Fig. 11.6 Average SDAC net position changes (FB-Prod) in MW from 01.10.2021 to 17.04.2022. *Source* Author's own illustration

11.2.1.3 Prices and Spreads

Regarding prices, the external parallel run showed that the average daily prices with the FB MC scenario increased slightly compared to the Prod scenario (see Fig. 11.7). As a result, the price convergence weakened. The frequency of the total price convergence, when all the bidding zones in the Core CCR have the same price for a certain hour, dropped to 4.02% from 5.10% with the application of the FB MC, while the frequency of the high spreads (above 200 EUR/MWh) increased significantly. The spread of the Core region is the difference between the maximum and the minimum price for a certain hour.

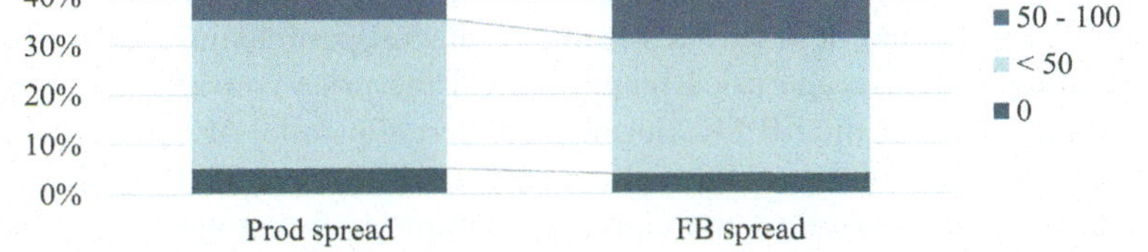

Fig. 11.7 Frequency changes of the Core CCR spread (EUR/MWh) based on FB external // run results from 01.10.2021 to 17.04.2022. *Source* Author's own illustration

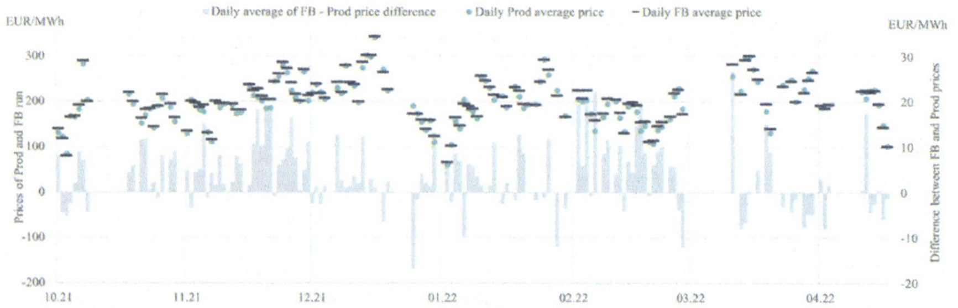

Fig. 11.8 Change of the average daily day-ahead Core CCR price (EUR/MWh) with the application of FB MC during the FB external // run from 01.10.2021 to 17.04.2022. *Source* Author's own illustration

The average daily (or base-load) prices were most often at roughly the same level with the two types of methodology. Nevertheless, the FB MC results frequently showed slightly higher prices than the Prod scenario during the parallel run period (Fig. 11.8).

Figure 11.9 shows statistical changes in the daily minimum, maximum and average prices (with and without outstanding data points). The widest range occurred at daily maximum prices, where the difference between the two scenarios reached +306.90 EUR/MWh and the average change was +36.02 EUR/MWh with the specific application of FB MC. The changes in the minimum prices remained in a narrower range of −78.19 and 126.23 EUR/MWh including the outstanding differences, while the most moderate changes occurred at the average prices (due to the leverage effect of averaging).

Despite the results showing that the FB MC caused increasing prices and worsening price convergence, we cannot conclude that this impact is purely due to the FB MC. During the external parallel run period, TSOs had the opportunity to learn more and gain first-hand experience with the FB MC methodology in operation. Statistical analysis showed a period of an intensive application of IVA by TSOs during the external parallel run which is understood to have influenced the price spreads of FB MC negatively. Although it is not possible to have a one-to-one comparison of the two methodologies (NTC and FB) on the basis of capacity results due to their fundamentally different approaches, this comparison still provides a good insight into understanding the results of the two scenarios with the same set of supply and demand bids from the market participants.

Even though the average price increased and the price convergence weakened with the application of the FB MC during the external parallel run, in Sect. 11.2.1.1 it is demonstrated that the social welfare increased with the FB MC. It was also demonstrated that the sum of the consumer surplus is five times higher than the producer surplus, which could induce lower social welfare in combination with higher market prices in general. However, the reason behind the increase of social welfare is that more demand and supply bids were accepted in DE BZ leading

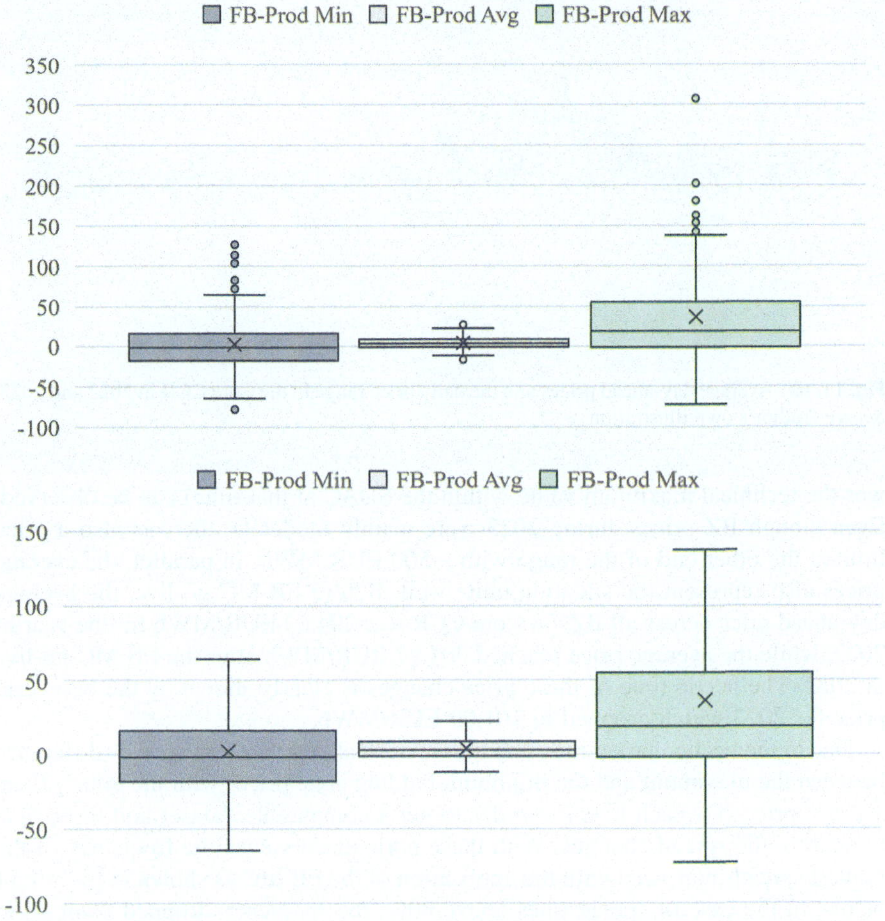

Fig. 11.9 Changes of the daily Core CCR minimum, average and maximum prices in EUR/MWh with and without outstanding points with the application of the FB MC during the FB external // run from 01.10.2021 to 17.04.2022. *Source* Author's own illustration

to double the consumer and producer surpluses. This explains the sum of the increased social welfare for all the Core BZs.

11.2.2 Price Convergence from the Go-Live

Examining the price spreads during the year of the go-live with the right approach can be difficult due to the energy price spikes and the energy crisis over the course of the year 2022 in combination with the sharp change at the beginning of 2023, when prices started to subside. The high volatility of the market is visible in Fig. 11.10, where a maximum price is close to 3 000 EUR/MWh (which latter

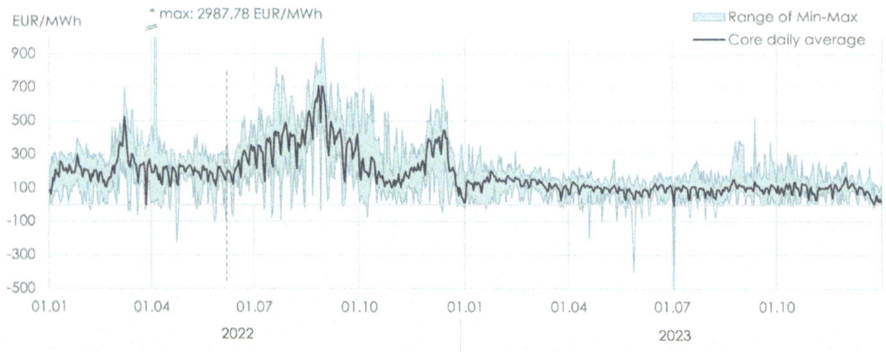

Fig. 11.10 Average day-ahead prices and the daily price range in the Core CCR in 2022 and 2023. *Source* Author's own illustration

was the technical maximum value within the SDAC at that time) can be observed. Even though BZ prices during 2023 were mainly moderate, there is also a price limit at the other end of the range with −500 EUR/MWh. In parallel, the average prices also represent the situation quite well. Before FB MC go-live, the average day-ahead price across all BZs in Core CCR was 204.63 EUR/MWh for the year of 2022, while the average price reached 301.97 EUR/MWh after the FB MC go-live in 2022. The hectic time of these price changes is clearly shown by the base-load price for 2023 which dropped to 101.09 EUR/MWh.

Due to the hectic market movements, we normalized the Core spread (difference between the maximum and the minimum bidding zone price) with the hourly Core average price (for each hour) in order to get a comparable dataset and to be able to analyze the spread changes. With these normalized values the frequency of the highest spreads narrowed with the application of the FB MC as shown in Fig. 11.11 below. In the case of spread rates above 80%, the frequency dropped from 16 to 11% during the external parallel run and 13% after the go-live while the frequency rate of spreads above 60% (including the category of 80% <) decreased from 25 to 17%. Regarding the full price convergence of the Core bidding zones, it appeared in 20% of the hours before the go-live in 2022. This rate jumped to 28% in 2022 after the go-live and 27% in 2023. The cases in the "low spread" category (0–20% in the legend) did not change with the go-live in 2022 (it was 25% for both halves of the year). However, the frequency of the "low spread" category increased to 31% in 2023. As a result, the frequency of the hours in which the spread range was maximum 20% (including convergence) increased from 45% (before go-live in 2022) to 58% (2023).

On the whole, even despite the extreme situations on the day-ahead markets, the analysis results are in line with the expectations that the positive contribution of the FB MC have been confirmed through the frequency of the full price convergence and the "low spread" category occurrence.

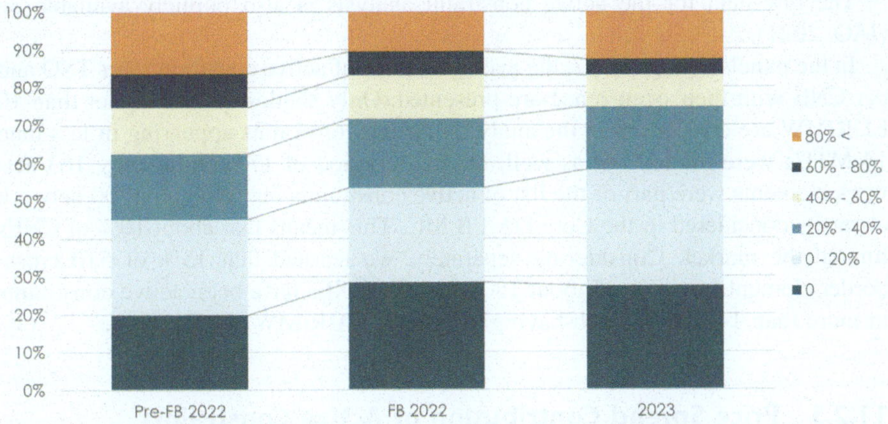

Fig. 11.11 Frequencies of day-ahead spreads normalized by hourly average prices in the Core CCR in 2022 and 2023. *Source* Author's own illustration

11.2.3 Analysis of Parameters Affecting the Market

The figures mentioned in this section are included in the Appendices and focus on demonstrating the flow-based parameters which have the most direct impact on the key indicators of the market coupling results. The IVA analysis aims to highlight the CNECs which are the most frequently subject to capacity reduction (Fig. 11.25). The RAM reduction makes the domain limiting constraints well positioned to become active constraints for a certain market direction. Therefore, the statistical analysis on the IVA associated with the simultaneous shadow price provides an insight into the impact of the capacity reductions on the price convergence (Fig. 11.28). In addition to the IVA and shadow price indicators, the usage of virtual capacity (AMR, LTA) at the market clearing point is analyzed in order to see to what extent the market makes use of the capacity that is beyond the physical F_{max} of the CNEC (Fig. 11.29). The virtual capacity definition in the extended LTA context constitutes the AMR from the minimum RAM requirement and the margin related to the enlarged convex hull by the LTA domain inclusion.

11.2.4 Statistics of Active Constraints

Active constraints, CNECs, with flow-based domain shadow price are the subset of domain limiting elements which in fact limit market exchanges, illustrated in Fig. 11.2.

In what follows, a general overview of active constraints is presented through the statistics of limiting elements (see in Fig. 11.27), appearing in at least 14 MTUs (0.1% of time) from the go-live until December 31st 2023.

The raw data for the active constraint analysis is also publicly available on (JAO, 2021).

In the panels of Fig. 11.27, the average values of active constraints per TSO and per CNE with their occurrence are presented. Only shadow prices higher than 10 EUR/MW are considered in the analysis. Active constraints appearing in less than 14 MTUs were filtered out as well. During a period of 17 months, only 163 network elements were part of the list of active constraints out of about 1700 network elements considered in the Core DA FB MC. This means that about 10% of CNEs limited the market. Considering separately, we learned that 15% of 510 cross-border elements and 7% of about 1200 internal CNEs have been active constraints in more than 14 MTUs, each having at least 10 EUR/MW shadow price.

11.2.5 Price Spread Contribution of Active Constraints

The most significant limitation appears between two bidding zone borders to which the active constraint has the maximum positive zone-to-zone PTDF. Nonetheless, active constraints with any zone-to-zone PTDF contribute to the market limitation shown in the price spread.

In this section we present case studies that show the effect of flow-based domain active constraints on price spreads. First, we focus on the price spread contribution of one particular active constraint, and second, we show examples of the calculation of price spreads of bidding zones taking into account each active constraint with their associated shadow price.

Shadow price data higher than 10 EUR/MW were analyzed from the go-live until December 31st, 2023 and were linked to DA market clearing prices.

11.2.5.1 Cross-Border Network Element

The active constraint selected for this study was the cross-border element [CZ-D8] Hradec–Rohrsdorf 445 [OPP] [D8] with N-1 Hradec–Rohrsdorf 2 contingency, limiting trades from DE to CZ bidding zones as its maximum zone-to-zone PTDF shows. According to the analysis shown in Fig. 11.27, this element is one of the top 15 most frequent active cross-border constraints.

The timeline showing below in Fig. 11.12 shows a short period of time in which the price spread appeared between bidding zones DE and CZ due to the active constraint. The market clearing price (MCP) in the CZ BZ increased compared to the price in the DE BZ.

The price differences of the two analyzed bidding zones to the Core average prices are plotted in correlation with the shadow prices. We can see that the spread increases with the shadow price (Fig. 11.13). However, the price difference is not symmetrical with regard to the two bidding zones. In this particular case, the slope of the price decrease in the DE bidding zone is higher than the increase rate of the CZ prices.

Although the maximum positive zone-to-zone PTDF represents the maximal impact of an active constraint, the very same constraint with its other positive

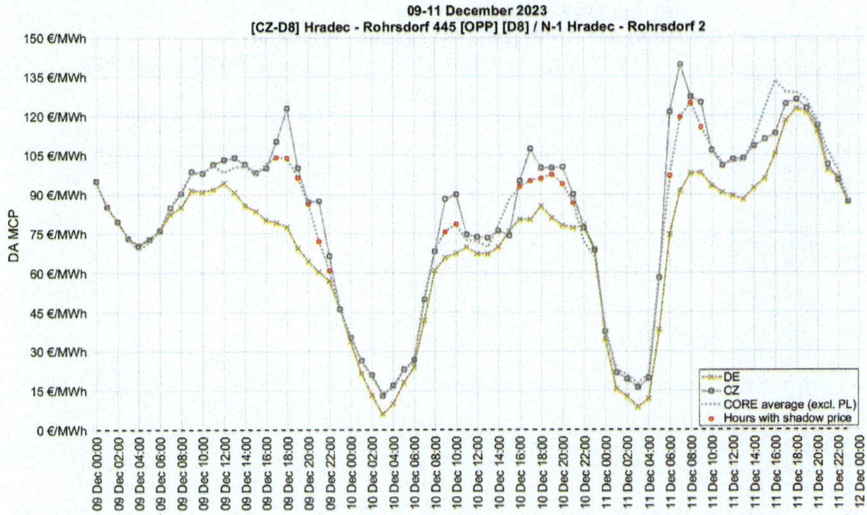

Fig. 11.12 DA market clearing prices (DA MCP) of CZ and DE bidding zones with Core average prices and epochs of shadow prices higher than 10 EUR/MW for BDs 20231209-11 in the case of [CZ-D8] Hradec–Rohrsdorf 445 [OPP] [D8]/N-1 Hradec–Rohrsdorf 2 CNEC. *Source* Author's own illustration

zone-to-zone PTDFs can have significant limiting impacts on other trades. This is presented below (Fig. 11.14) as the distribution of the zone-to-zone PTDFs of the active constraint. It is clearly shown that in addition to the DE-CZ border, the CNE limits market trades more significantly from DE to AT, from AT to CZ among many other market directions. It should be noted that the zone-to-zone PTDF with a negative sign to the market direction, e.g. CZ-AT, has the positive sign to the opposite market direction, e.g. AT-CZ.

Further price spread analysis shows that this single active constraint can contribute to the price decoupling of the entire Central and Eastern part of the Core region, including CZ, SK, HU, SI, HR, RO bidding zones.

11.2.5.2 Internal Network Element: Example 1

The active constraint selected for this study was the internal element [SK-SK] V.Dur - Levice 1 [DIR] with N-1 V.Dur - Levice 2 contingency from SEPS TSO grid, limiting trades mostly on CZ-HU market direction, given its maximum zone-to-zone PTDF. According to the analysis shown in Fig. 11.27, this element is one of the top 15 most frequent active internal constraints.

As our long-term analysis shows in Fig. 11.15, this network element can lead to a noticeable 280 EUR/MWh price spread in CZ-HU direction in extreme shadow price cases. With regard to the average Core price compared to HU BZ price, this active constraint can even cause a price difference of 190 EUR/MWh. It can also be observed that this active constraint most of the time is subject to IVA

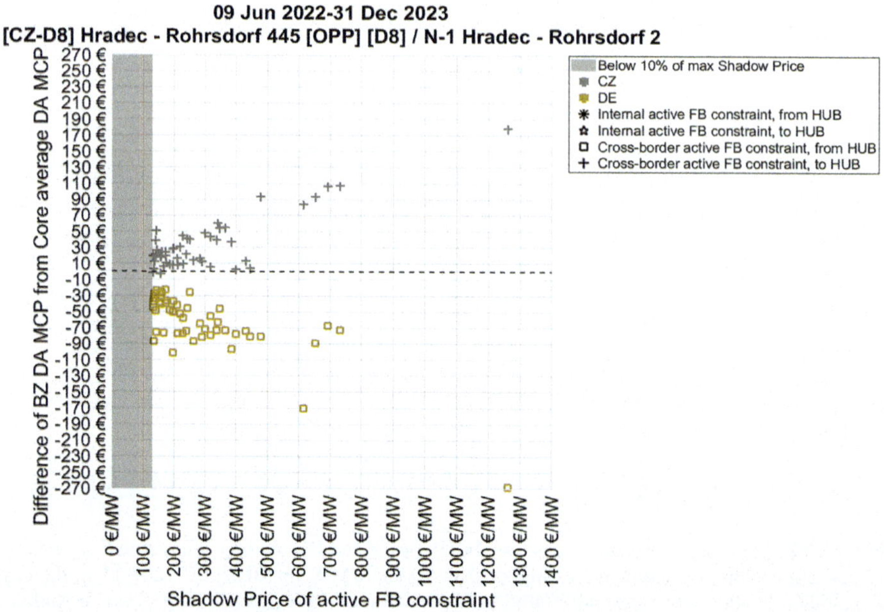

Fig. 11.13 Correlation of shadow prices and price decrease/increase of from hub DE/to hub CZ prices from Core average prices represented by x-axis based on max zone-to-zone PTDF of active constraint in the case of [CZ-D8] Hradec–Rohrsdorf 445 [OPP] [D8]/N-1 Hradec–Rohrsdorf 2 CNEC. Statistic is based on 215 MTUs when max zone-to-zone PTDF was from DE-CZ direction. *Source* Author's own illustration

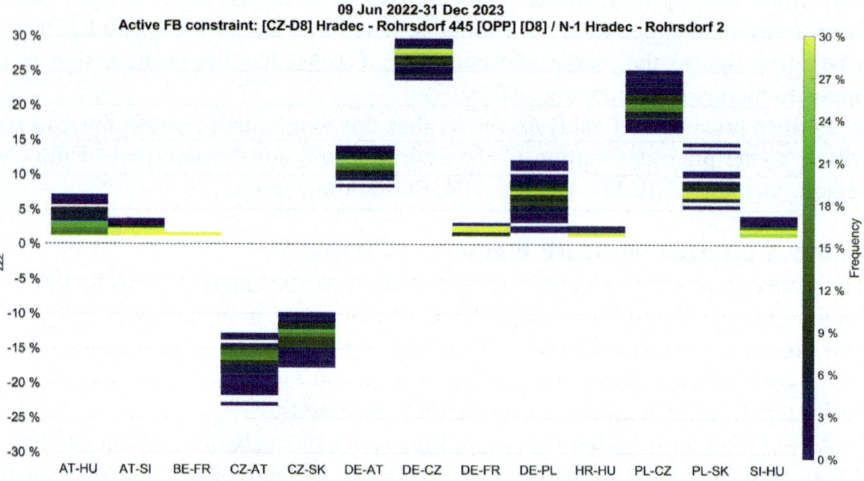

Fig. 11.14 Higher than 1.5% zone-to-zone PTDF distributions of active constraint [CZ-D8] Hradec–Rohrsdorf 445 [OPP] [D8]/N-1 Hradec–Rohrsdorf 2, analyzed from go-live until 31st Dec 2023. Statistic is based on 215 MTUs when max zone-to-zone PTDF was from DE-CZ direction. *Source* Author's own illustration

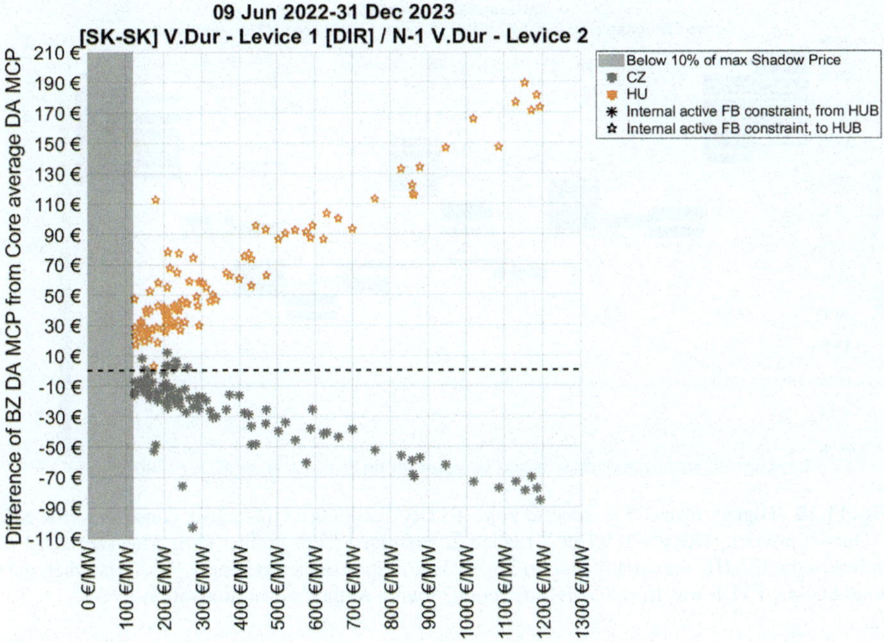

Fig. 11.15 Correlation of the shadow prices and price decrease/increase of from hub CZ/to hub HU prices from Core average prices represented by x-axis based on max. zone-to-zone PTDF of active constraint, for the entire analyzed period in the case of [SK-SK] V.Dur - Levice 1 [DIR]/ N-1 V.Dur - Levice 2 CNEC. Statistic is based on 274 MTUs when max zone-to-zone PTDF was from CZ-HU direction. *Source* Author's own illustration

application, which even increases the probability of becoming an active constraint with a higher shadow price.

The distributions of the positive zone-to-zone PTDFs demonstrate that this active constraint has a significant limiting impact over most of the HU import from all market directions. As a result, statistical analysis shows that the HU BZ can experience price decoupling from the rest of the Core BZs due to the electrical characteristics of a single active constraint (Fig. 11.16).

11.2.5.3 Internal Network Element: Example 2

The selected internal element [RO-RO] TR Rosiori 400/220 1 [DIR] with N-1 Rosiori - Gadalin contingency from Transelectrica TSO grid, is an interesting active constraint from its geographical location point of view. Given its highest positive zone-to-zone PTDF mainly from SK to RO and HU to RO direction shown in Fig. 11.18, this network element can contribute to the price decoupling of the RO BZ from the rest of the bidding zones in the Core region. According to the analysis shown in Fig. 11.27, this element is among the top 3 of the most frequent active internal constraints.

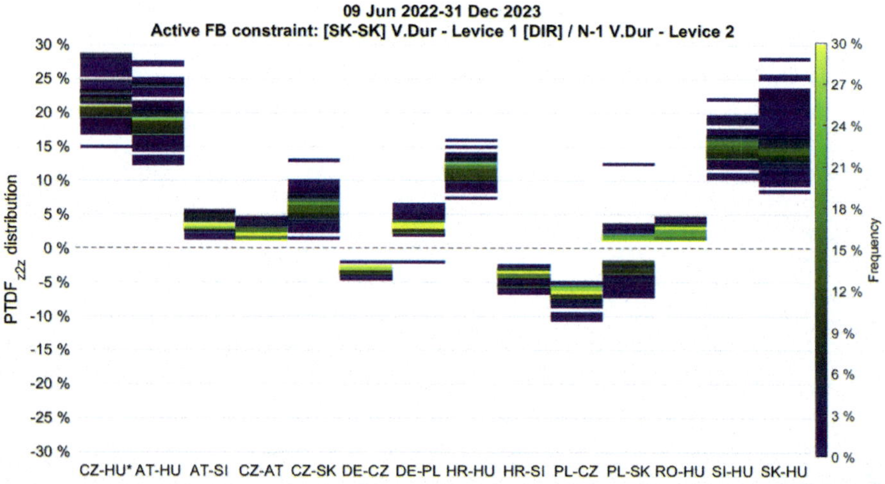

Fig. 11.16 Higher than 1.5% zone-to-zone PTDF distributions of active constraint [SK-SK] V.Dur - Levice 1 [DIR]/N-1 V.Dur - Levice 2, examined from go-live until 31st Dec 2023 (* indicates the CZ-HU maximum zone-to-zone PTDF). Statistic is based on 274 MTUs when max zone-to-zone PTDF was from CZ-HU direction. *Source* Author's own illustration

As shown in Fig. 11.17 this network element can lead to as high a price spread as 430 EUR/MWh in between RO and SK bidding zones direction. With regard to the average Core price compared to SK BZ price, this active constraint can be attributed to a price decoupling of 380 EUR/MWh for the RO BZ. It can also be observed that in addition to this active constraint, there can be other active constraints that cause a price increase in the SK bidding zone compared to the average DA MCP. These other active constraints in the Core CCR in combination with the RO internal element even further aggravate the price decoupling of the RO BZ from the rest of the Core BZs. Figures 11.27 and 11.28 also show that this RO internal network element is among the top internal network elements with high frequency of IVA application which can be attributed to being among top internal active constraints in combination with the simultaneous reduction of RAM by IVA.

11.2.6 Calculation of Price Spreads Based on Shadow Prices of FB Active Constraints

Based on Eq. (11.7) from Sect. 11.1.1.3, the contribution of any price spread made by a certain active constraint can be calculated. The following examples show this calculation process with active constraints of the FB domain limiting the market from further exchanges in the selected hours. It is worth noting that for the calculation of the spreads, zone-to-zone PTDFs are needed, which can be obtained using zonal PTDFs. Figure 11.19 via the table columns show the breakdown of the overall contributions to the selected price spread at active constraint level. In

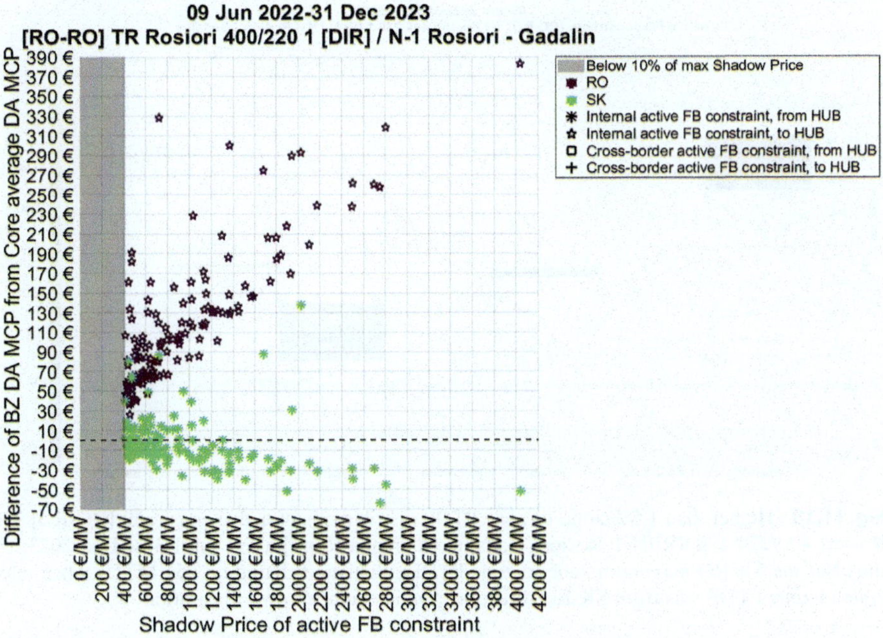

Fig. 11.17 Correlation of the shadow prices and price decrease/increase of from hub SK/to hub RO prices from Core average prices represented by x-axis based on max zone-to-zone PTDF of active constraint in the case of [RO-RO] TR Rosiori 400/220 1 [DIR]/N-1 Rosiori-Gadalin CNEC. Statistic is based on 651 MTUs when max zone-to-zone PTDF was from SK-RO direction. *Source* Author's own illustration

general, it can be noted that a high positive zone-to-zone PTDF with a low shadow price does not necessarily have higher contribution to the price spread than a lower positive zone-to-zone PTDF with a high shadow price.

The necessary inputs (active constraints with their associated shadow price and zonal PTDFs, are available on JAO Publication Tool, Core CCR (JAO, 2024), "ShadowPrices" page, while the DA BZ prices are published on ENTSO-E Transparency Platform (ENTSO-E, 2024) on page "Day-ahead Prices". It should be noted that when the alpha variable is not equal to 1, then the spread contribution is the combination of active constraints from the FB domain and the LTA domain. The sum of the increasing and decreasing contributions gives the overall price spread between two given bidding zones. Therefore, the contribution rate of a certain constraint to the overall price spread can exceed 100% when the overall price spread is based on a combination of increasing and decreasing contributions. The breakdown of the contributions shows which one of the active constraints has either an increasing or a decreasing contribution to the overall price spread. A certain active constraint with a decreasing contribution has a zone-to-zone PTDF with a negative sign to the corresponding bidding zone border (Figs. 11.20 and 11.21).

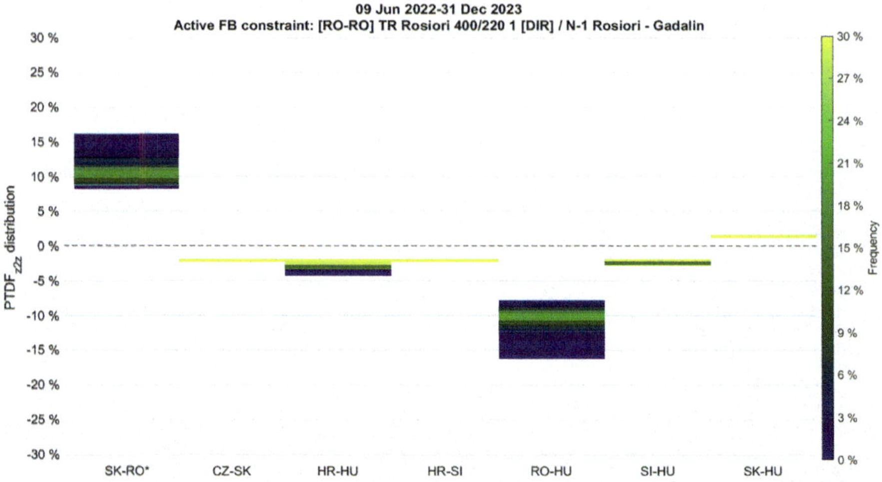

Fig. 11.18 Higher than 1.5% zone-to-zone PTDF distributions of active constraint [RO-RO] TR Rosiori 400/220 1 [DIR]/N-1 Rosiori–Gadalin, examined from go-live until 31st Dec 2023. (* indicates the SK-RO maximum zone-to-zone PTDF). Statistic is based on 651 MTUs when max zone-to-zone PTDF was from SK-RO direction. *Source* Author's own illustration

The examples show that several active constraints from different TSOs can make significant contributions to the price spreads. The overall impact of each active constraint to the price spread from DE to CZ market exchange is shown by Fig. 11.22 for a period of 17 months from the go-live. Cases with lower than 10 EUR/MWh spreads and lower than 1% contribution to the total price spread were excluded from the statistical analysis. Finally, active constraints appearing in at least 50 MTUs (0.36% of time) were plotted in the heat maps below. The sign of contribution indicates whether the constraint increases or decreases the spread. For example, a CNE with a positive contribution rate increases the spread between two certain bidding zones.

As Fig. 11.22 shows, one of the most significant positive rate contributions in terms of the average value is represented by [CZ-D8] Hradec–Rohrsdorf [OPP] [D8] with respect to the DE-CZ border. Its positive contribution rate to the overall price spread is over 50% around 15% of the time while the average positive contribution rate is around 60%. The results are in line with expectations as this network is a 400 kV double circuit cross-border element between DE-CZ bidding zones. Its parallel line as contingency makes this network element limiting most of the time.

In Fig. 11.23 the number of active constraints limiting contributing to the CZ and HU direction is shown. The most significant positive contribution rate can be assigned to the 400 kV [SK-SK] V.Dur – Levice 1 [DIR] internal CNEC. Its positive contribution rate to the overall price spread is around 90% around 15% of the time while the average positive contribution rate is around 60%. There are

H19 23.11.2023; alpha = 1

Network element name	Contingency name	FB Shadow price (EUR/MW)	DA price (EUR/MWh)		PTDF			Contribution to spread
			DE	CZ	DE	CZ	DE-CZ	
		(1)	(2)	(3)	(4)	(5)	(6) = (4) - (5)	(7) = (1) * (6)
[CZ-D8] Hradec - Rohrsdorf 445 [OPP] [D8]	N-1 Hradec - Rohrsdorf 2	133,07	70,63	114,63	0,126	-0,151	0,276	36,77
[SK-HU] Levice - God [DIR] [HU]	N-1 Felsozsolca - Sajoivanka	191,81	70,63	114,63	0,109	0,122	-0,013	-2,42
[NL-D2] Meeden-Diele 380 Z [OPP] [NL]	N-1 Diele - Meeden WEISS/W	745,19	70,63	114,63	0,010	0,003	0,007	5,19
[NL-NL] Krimpen a/d IJssel-Geertruidenberg 380 W [DIR]	N-1 Krimpen a/d IJssel-Geertruidenberg 380 Z	203,53	70,63	114,63	0,022	0,006	0,017	3,40
[D8-PL] Mikulowa PST1 [OPP] [PL]	N-1 Hagenwerder - Mikulowa 1	16,58	70,63	114,63	0,146	0,083	0,064	1,05
			Sum of contributions = CZ - DE spread = (3) - (2) :					44,00

H16 20.12.2023; alpha = 1

Network element name	Contingency name	FB Shadow price (EUR/MW)	DA price (EUR/MWh)		PTDF			Contribution to spread
			DE	CZ	DE	CZ	DE-CZ	
		(1)	(2)	(3)	(4)	(5)	(6) = (4) - (5)	(7) = (1) * (6)
[CZ-D8] Hradec - Rohrsdorf 445 [OPP] [D8]	N-1 Hradec - Rohrsdorf 2	60,71	73,49	88,00	-0,013	-0,293	0,280	16,99
[NL-D2] Meeden-Diele 380 Z [OPP] [NL]	N-1 Diele - Meeden WEISS/W	153,18	73,49	88,00	-0,049	-0,058	0,009	1,37
[RO-RO] TR Rosiori 400/220 1 [DIR]	N-1 Rosiori - Gadalin	247,31	73,49	88,00	0,000	0,001	-0,001	-0,24
[SK-SK] V.Dur - Levice 1 [DIR]	N-1 V.Dur - Levice 2	160,43	73,49	88,00	0,001	0,023	-0,023	-3,61
			Sum of contributions = CZ - DE spread = (3) - (2) :					14,51

Fig. 11.19 DE-CZ Price spread calculation examples (with alpha = 1). *Source* Author's own illustration

several other active constraints mainly cross-border network elements, e.g. 400 kV [SK-HU] Levice – God 1 [HU] closely following with comparable mean value.

In Fig. 11.24 the price spread between RO and SK bidding zones is shown. The geographical location of the RO BZ inside the Core CCR well reflects the positive contribution rate with respect to the RO-SK price spread. The top 9 active constraints can have over a 90% increasing contribution to the RO-SK price spread over 30% of the time. This shows that much of the price formulation in the RO BZ is influenced by a single active constraint. In other words, the contribution to the overall price spread with respect to the RO BZ can be determined by a single active constraint. This observation is confirmed by the figure below with respect to [RO-RO] TR Rosiori 400/220 1 [DIR].

H16 29.12.2023; alpha = 1

Network element name	Contingency name	FB Shadow price (EUR/MW)	DA price (EUR/MWh)		PTDF			Contribution to spread
			CZ	HU	CZ	HU	CZ-HU	
		(1)	(2)	(3)	(4)	(5)	(6) = (4) - (5)	(7) = (1) * (6)
[CZ-D8] Hradec - Rohrsdorf 445 [OPP] [D8]	N-1 Hradec - Rohrsdorf 2	86,87	31,00	75,06	-0,12	-0,01	-0,118	-10,24
[NL-D2] Meeden-Diele 380 Z [OPP] [NL]	N-1 Diele - Meeden WEISS/W	358,36	31,00	75,06	0,00	0,00	0,003	0,90
[D8-PL] Mikulowa PST1 [OPP] [PL]	N-1 Hagenwerder - Mikulowa 1	14,66	31,00	75,06	0,07	0,00	0,074	1,09
[SK-SK] V.Dur - Levice 1 [DIR]	N-1 V.Dur - Levice 2	255,22	31,00	75,06	0,17	-0,04	0,205	52,31
			Sum of contributions = HU - CZ spread = (3) - (2) :					44,06

H17 29.12.2023; alpha = 1

Network element name	Contingency name	FB Shadow price (EUR/MW)	DA price (EUR/MWh)		PTDF			Contribution to spread
			CZ	HU	CZ	HU	CZ-HU	
		(1)	(2)	(3)	(4)	(5)	(6) = (4) - (5)	(7) = (1) * (6)
[NL-D2] Meeden-Diele 380 Z [OPP] [NL]	N-1 Diele - Meeden WEISS/W	283,29	21,68	72,28	0,003	0,001	0,003	0,72
[SK-SK] V.Dur - Levice 1 [DIR]	N-1 V.Dur - Levice 2	243,62	21,68	72,28	0,167	-0,038	0,205	49,88
			Sum of contributions = HU - CZ spread = (3) - (2) :					50,60

H18 29.12.2023; alpha = 1

Network element name	Contingency name	FB Shadow price (EUR/MW)	DA price (EUR/MWh)		PTDF			Contribution to spread
			CZ	HU	CZ	HU	CZ-HU	
		(1)	(2)	(3)	(4)	(5)	(6) = (4) - (5)	(7) = (1) * (6)
[NL-D2] Meeden-Diele 380 Z [OPP] [NL]	N-1 Diele - Meeden WEISS/W	256,23	24,09	62,35	0,003	0,001	0,003	0,66
[RO-RO] TR Rosiori 400/220 1 [DIR]	N-1 Rosiori - Gadalin	73,27	24,09	62,35	0,063	0,062	0,001	0,06
[SK-SK] V.Dur - Levice 1 [DIR]	N-1 V.Dur - Levice 2	183,61	24,09	62,35	0,167	-0,038	0,205	37,55
			Sum of contributions = HU - CZ spread = (3) - (2) :					38,27

Fig. 11.20 HU-CZ Price spread calculation examples (with alpha = 1). *Source* Author's own illustration

H18 18.10.2023; alpha = 1

Network element name	Contingency name	FB Shadow price (EUR/MW)	DA price (EUR/MWh)		PTDF			Contribution to spread
			SK	RO	SK	RO	SK-RO	
		(1)	(2)	(3)	(4)	(5)	(6) = (4) - (5)	(7) = (1) * (6)
[FR-D7] Vigy - Ensdorf VIGY2 S [DIR] [D7]	N-1 Ensdorf - Vigy VIGY1 N	154,11	211,20	300,06	-0,008	-0,001	-0,007	-1,15
[PL-PL] Krosno Iskrzynia - Tarnow [OPP]	N-1 Nosovice - Varin	249,06	211,20	300,06	-0,021	-0,006	-0,015	-3,77
[D8-PL] Mikulowa PST1 [OPP] [PL]	N-1 Hagenwerder - Mikulowa 1	356,98	211,20	300,06	-0,029	-0,003	-0,026	-9,22
[RO-RO] TR Rosiori 400/220 1 [DIR]	N-1 Rosiori - Gadalin	945,27	211,20	300,06	0,077	-0,032	0,109	103,01
			Sum of contributions = RO - SK spread = (3) - (2) :					88,86

H06 19.12.2023; alpha = 1

Network element name	Contingency name	FB Shadow price (EUR/MW)	DA price (EUR/MWh)		PTDF			Contribution to spread
			SK	RO	SK	RO	SK-RO	
		(1)	(2)	(3)	(4)	(5)	(6) = (4) - (5)	(7) = (1) * (6)
[CZ-D8] Hradec - Rohrsdorf 445 [OPP] [D8]	N-1 Hradec - Rohrsdorf 2	57,92	82,67	112,34	-0,010	-0,001	-0,009	-0,54
[SK-HU] Levice - God [DIR] [HU]	N-1 R.Sobota - Sajoivanka	169,83	82,67	112,34	0,175	-0,009	0,183	31,14
[CZ-PL] Wielopole - Nosovice [DIR] [PL]	N-1 Albrechtice - Dobrzen	22,06	82,67	112,34	-0,043	-0,001	-0,042	-0,92
			Sum of contributions = RO - SK spread = (3) - (2) :					29,68

Fig. 11.21 SK-RO Price spread calculation examples (with alpha = 1). *Source* Author's own illustration

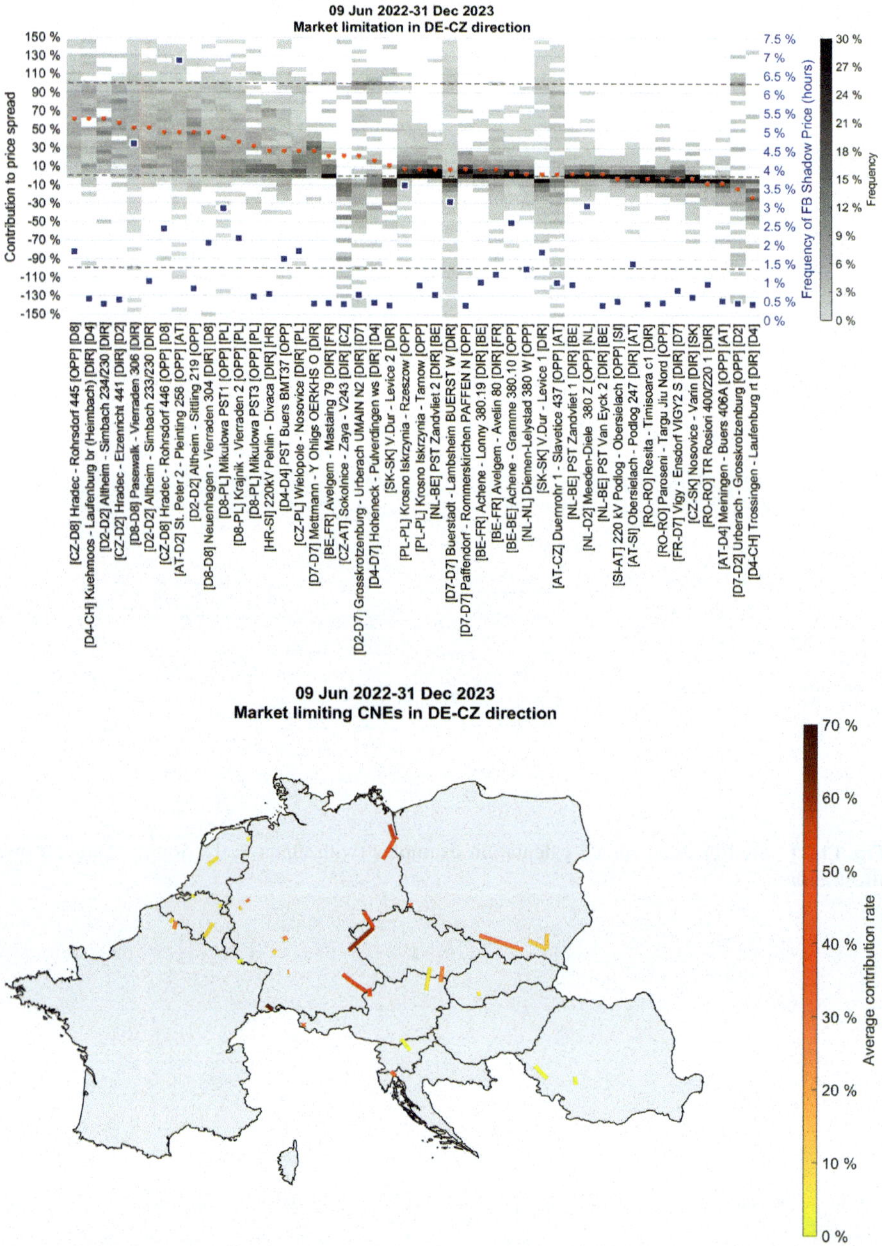

Fig. 11.22 Top panel: Contribution to the DE-CZ price spread of network elements being active in at least 50 MTUs in the order of the average contribution. Only cases with higher than 10 EUR/MWh spreads and higher than 1% contribution rate were plotted. Red dots indicate the average contribution rates. On the right y-axis we plot time frequency of cases to accomplish these criteria. The bottom panel shows the geographical locations of CNEs with the positive average contribution rates, coloured according to their mean contribution rates shown with red dots in the top. *Source* Author's own illustration

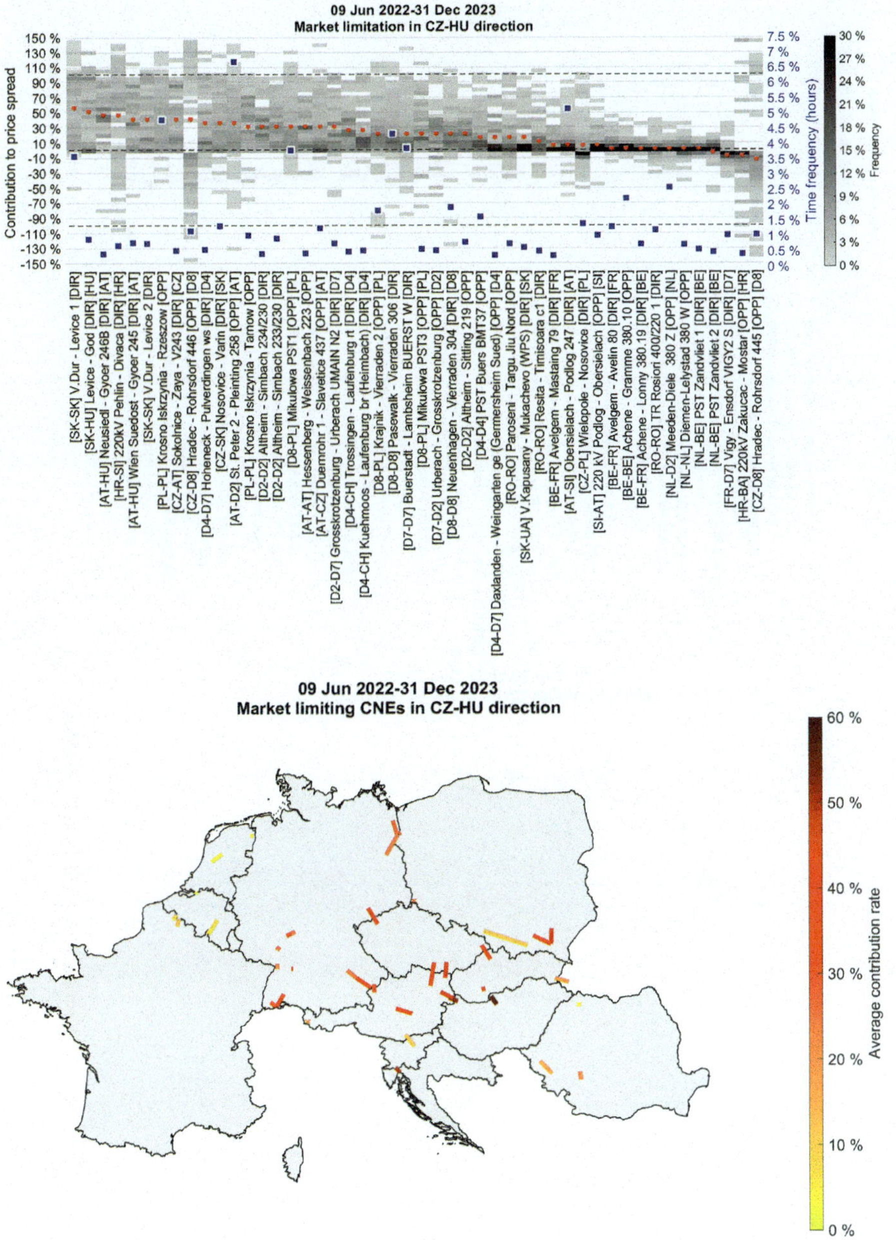

Fig. 11.23 Top panel: Contribution to CZ-HU price spread of network elements being active in at least 50 MTUs in the order of the average contribution. Only cases with higher than 10 EUR/MWh spreads and higher than 1% contribution rate were plotted. Red dots indicate the average contribution rates. On the right y-axis we plot time frequency of cases accomplish these criteria. The bottom panel shows the geographical locations of CNEs with the positive average contribution rates, coloured according to their mean contribution rates shown with red dots in the top. *Source* Author's own illustration

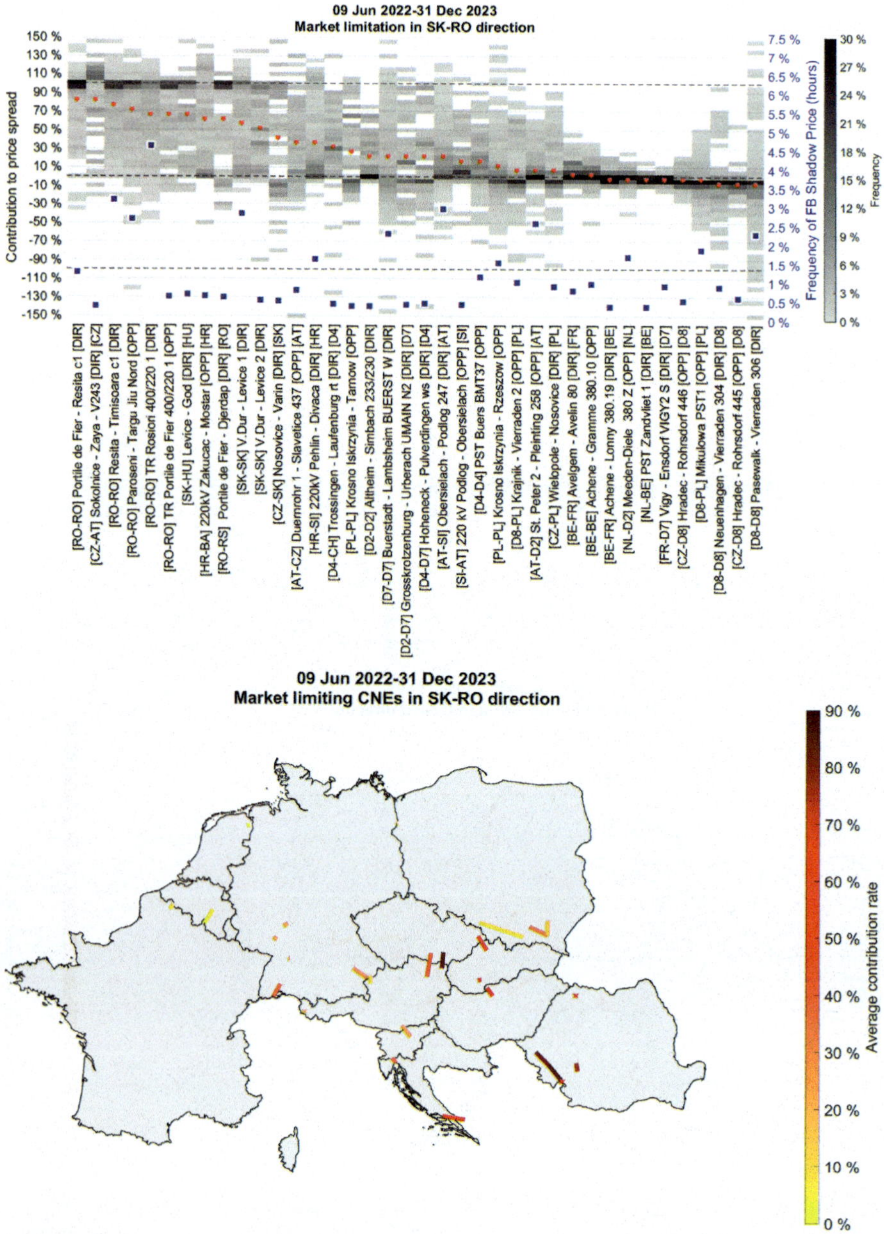

Fig. 11.24 Top panel: Contribution to SK-RO price spread of network elements being active in at least 50 MTUs in the order of average contribution. Only cases with higher than 10 EUR/MWh spreads and higher than 1% contribution rate were plotted. Red dots indicate the average contribution rates. On the right y-axis we plot time frequency of cases accomplish these criteria. The bottom panel shows the geographical locations of CNEs with positive average contribution rates, coloured according to their mean contribution rates shown with red dots in the top. *Source* Author's own illustration

11.3 Conclusion

With respect to the NTC and FB comparison in Sect. 11.2.1, the consumer surplus increased by 0.89% and the producer surplus did as well by 1.08% with the application of the FB MC during the period of the external parallel run. Overall, a total gain of 0.92% was generated in social welfare, which is a significant positive outcome. Thus, the substantial benefit of the new FB MC methodology was realized during the external parallel run.

Section 11.2.2 highlights that from the FB MC go-live, the frequency of the "low spread" category occurrence and the full price convergence has increased in line with the preliminary expectations even despite the extreme situations and high volatilities experienced with the day-ahead markets. Furthermore, it can be concluded from the statistical analysis of Sect. 11.2.4 in line with Sect. 11.2.5 that the contribution of a certain active constraint to the corresponding price spread is proportionate to its positive zone-to-zone PTDF value. It can also be observed that one single active constraint can contribute to the price spreads of several bidding zone borders. As a result, one or even more bidding zones, mainly depending on the geographical location inside the Core CCR, can be subject to price decoupling from the rest of the bidding zone prices. This phenomenon is particularly linked to those bidding zones, e.g. SK, HU, RO that have external borders along the Core CCR.

Acknowledgements We would like to express our immense gratitude to Wan-Ying Huang (wan-ying.huang@ge.com) as Analytics Engineer, Matthew Ligtenberg (mattlig59@gmail.com) as Language Specalist and the Logarithmo team for the insightful review and the valuable contribution to the completion of this chapter. This Chapter is based on the revision of the CIGRE Paper Introduction of the Operational Core Day-Ahead Flow-Based Capacity Calculation and Market Coupling through Active Constraints and Price Spread, CIGRE ID 11358 – Ferenc NAGY, Melinda NAGY, Luca TÓTH, Ágnes TAKÁCSNÉ ESZE, Ákos ARNOLD – 2024

Appendices

See Table 11.1, Figs. 11.24, 11.25, 11.26, 11.27, 11.28 and 11.29.

Table 11.1 List of business days for the external parallel run (experimental phase) analysis

Business days of the comparison of NTC and FB MC (external parallel run) results						
01.10.2021	03.11.2021	25.11.2021	18.12.2021	15.01.2022	12.02.2022	17.03.2022
02.10.2021	04.11.2021	26.11.2021	19.12.2021	16.01.2022	13.02.2022	19.03.2022
03.10.2021	05.11.2021	27.11.2021	25.12.2021	18.01.2022	14.02.2022	20.03.2022
04.10.2021	06.11.2021	28.11.2021	26.12.2021	19.01.2022	15.02.2022	23.03.2022
05.10.2021	07.11.2021	29.11.2021	27.12.2021	20.01.2022	16.02.2022	25.03.2022
06.10.2021	08.11.2021	30.11.2021	28.12.2021	21.01.2022	17.02.2022	26.03.2022
07.10.2021	09.11.2021	01.12.2021	29.12.2021	23.01.2022	18.02.2022	28.03.2022
08.10.2021	10.11.2021	02.12.2021	30.12.2021	24.01.2022	19.02.2022	29.03.2022
18.10.2021	12.11.2021	03.12.2021	02.01.2022	25.01.2022	20.02.2022	30.03.2022
19.10.2021	13.11.2021	04.12.2021	03.01.2022	26.01.2022	21.02.2022	01.04.2022
21.10.2021	14.11.2021	07.12.2021	04.01.2022	28.01.2022	22.02.2022	02.04.2022
22.10.2021	16.11.2021	08.12.2021	05.01.2022	30.01.2022	24.02.2022	03.04.2022
23.10.2021	17.11.2021	09.12.2021	06.01.2022	02.02.2022	25.02.2022	11.04.2022
24.10.2021	18.11.2021	10.12.2021	07.01.2022	03.02.2022	26.02.2022	12.04.2022
25.10.2021	19.11.2021	11.12.2021	08.01.2022	04.02.2022	27.02.2022	13.04.2022
26.10.2021	20.11.2021	12.12.2021	09.01.2022	05.02.2022	11.03.2022	14.04.2022
28.10.2021	21.11.2021	13.12.2021	10.01.2022	06.02.2022	13.03.2022	15.04.2022
29.10.2021	22.11.2021	14.12.2021	11.01.2022	08.02.2022	14.03.2022	16.04.2022
01.11.2021	23.11.2021	15.12.2021	12.01.2022	09.02.2022	15.03.2022	17.04.2022
02.11.2021	24.11.2021	16.12.2021	13.01.2022	11.02.2022	16.03.2022	

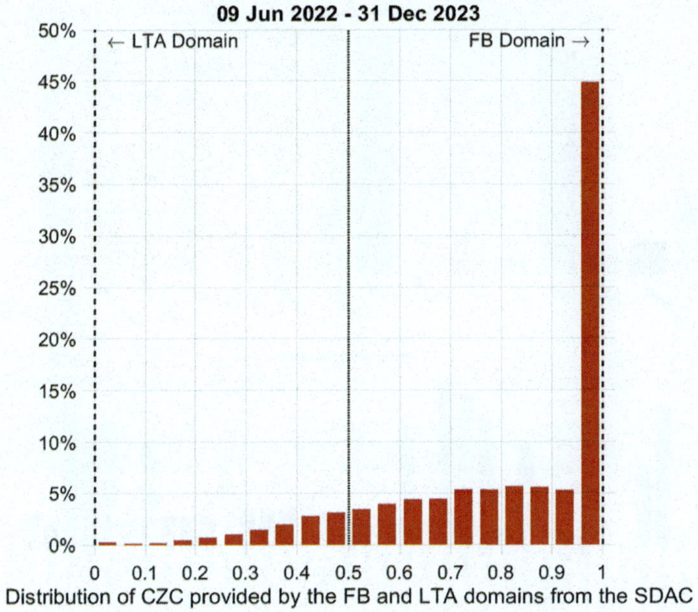

Fig. 11.25 Distribution of alpha factor. If the alpha variable is 1, then the bidding zone net positions from the market coupling are given by the FB domain, while alpha = 0, it is done by the LTA domain. *Source* Author's own illustration

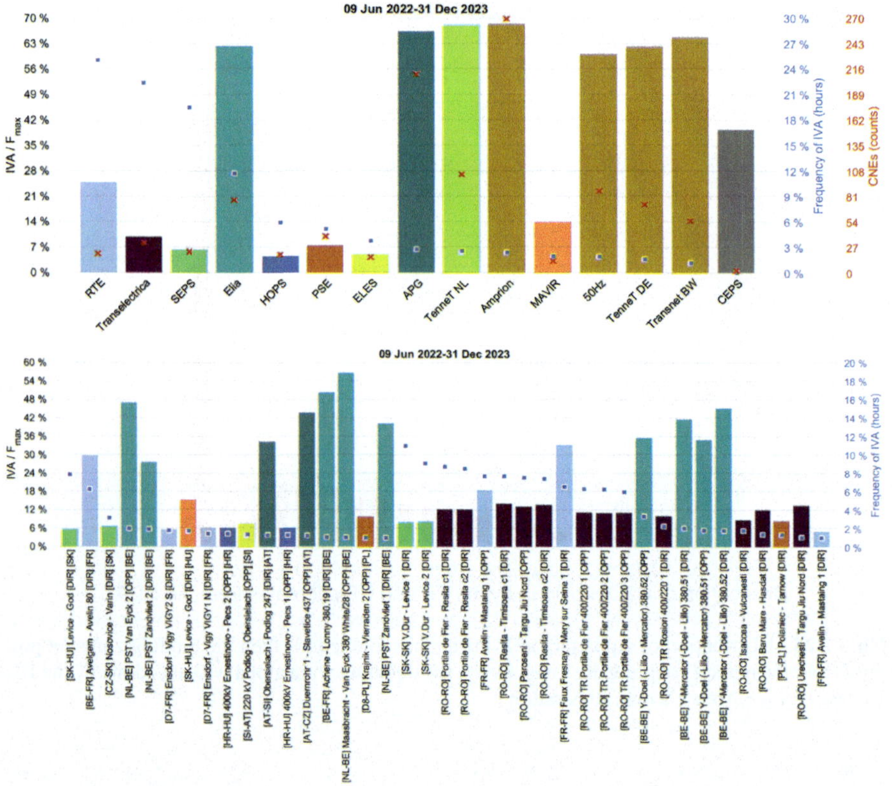

Fig. 11.26 Top panel: average IVA values (left y-axes) frequency (right y-axes, blue) and CNE count (right y-axis, red) per TSO in decreasing order of frequency. Bottom panel shows the statistics for CNEs appearing in more than 1% of time with higher than 10 MW IVA. Colour codes indicate TSO according to the top panel. Data from BD20230719 are excluded from the analysis due to large quantity of fallback IVA. Ranking is based on the frequency of IVA application in descending order and two groups are separated; first group is for cross-border network elements and the second is for internal network elements. *Source* Author's own illustration

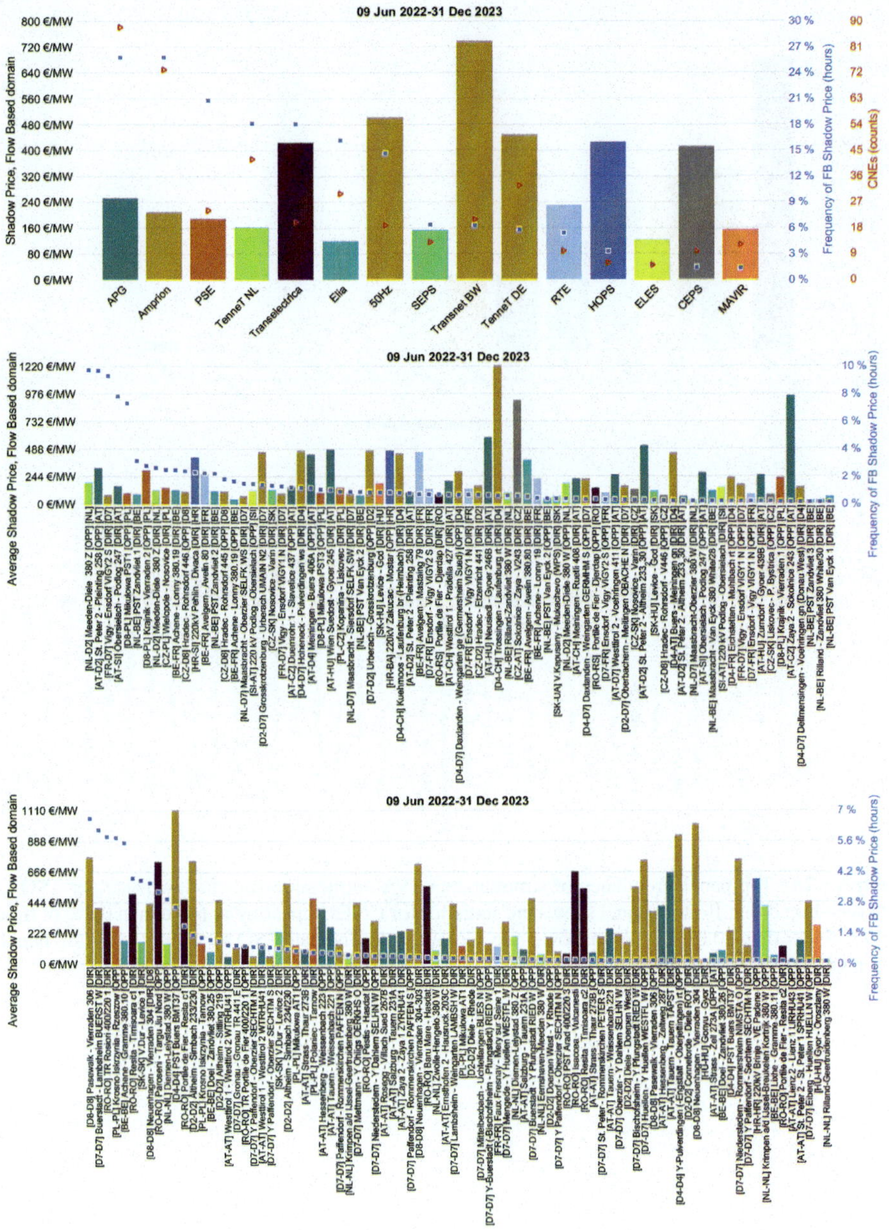

Fig. 11.27 Top panel: average flow-based shadow prices (left y-axes) frequency (right y-axes, blue) and CNE count (right y-axis, red) per TSO in decreasing order of frequency. Only shadow prices higher than 10 EUR/MW were considered. The middle panel show the statistics for cross-border network elements, while bottom panel for the case of internal network elements. Only CNEs limiting in at least 14 MTUs (0.1% of time) with shadow prices higher than 10 EUR/MW were considered. Colour codes indicate TSO according to the top panel. Data from BD20230702 are excluded from the analysis due to extreme, −500EUR/MWh prices. *Source* Author's own illustration

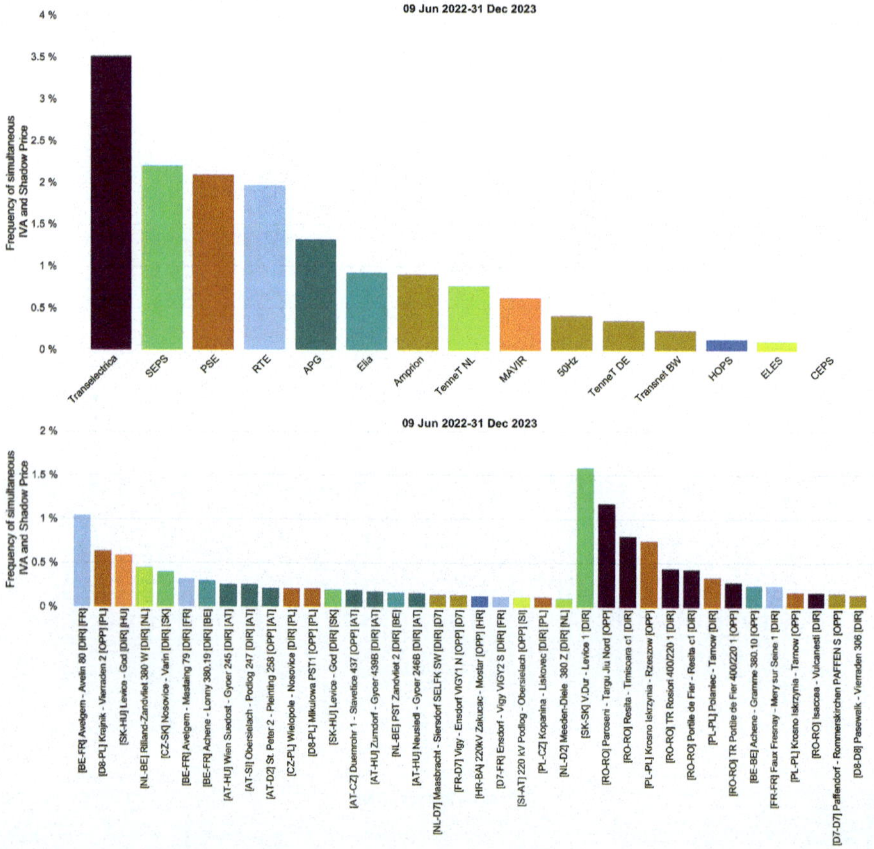

Fig. 11.28 Top panel: frequency of simultaneous IVA application and shadow price per TSO in descending order. Bottom panel shows the statistics for CNEs appearing in more than 0.1% of time with higher than 10 EUR/MW shadow price. Colour codes indicate TSO according to the top panel. *Source* Author's own illustration

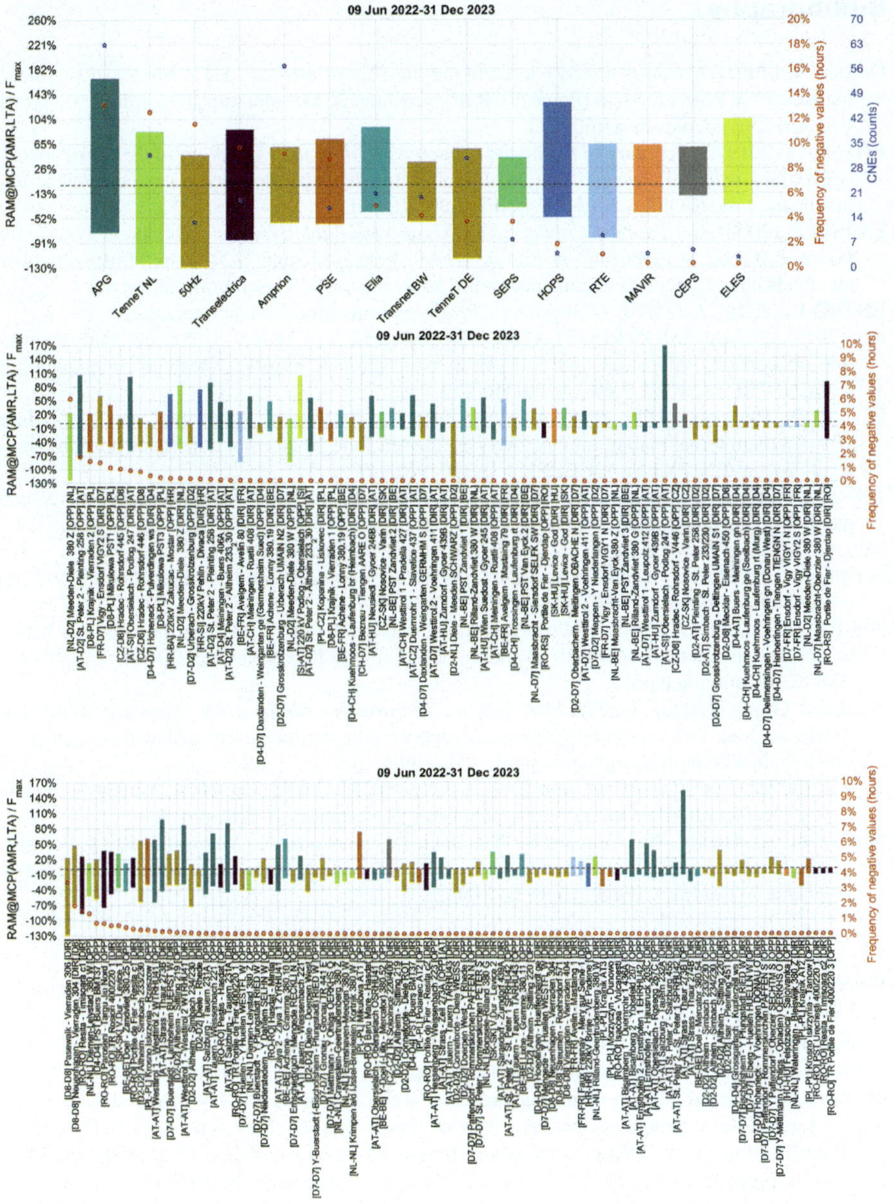

Fig. 11.29 Maximum and minimum available RAM at market clearing point of CNECs with shadow price. Top panel shows per TSO statistics ordered by frequency of negative or zero RAM. On the right axis we plot the frequency of negative RAM (red) and the number of CNEs with negative RAM (blue). The middle and bottom panel show statistics for cross-border and internal network elements with negative RAM higher than 10% of F_{max}. *Source* Author's Own Illustration

Bibliography

Dufour, A. (2007). *Capacity allocation using the flow-based method* . KTH Electrical Engineering Electrical Power System [Preprint]. Retrieved June 2024, from http://www.diva-portal.org/smash/record.jsf?pid=diva2:609137

ENTSO-E. (2019, 21 February). *Day-ahead capacity calculation methodology of the Core capacity calculation region.* Retrieved May 2024, from https://eepublicdownloads.entsoe.eu/clean-documents/nc-tasks/CORE%20-%20ANNEX%20I_III.pdf

ENTSO-E. (2000). *Net Transfer Capacities (NTC) and Available Transfer Capacities (ATC) in the Internal Market of Electricity in Europe (IEM).* Retrieved June 2024, from https://eepublicdownloads.entsoe.eu/clean-documents/pre2015/ntc/entsoe_NTCusersInformation.pdf

ENTSO-E. (2024). *ENTSO-E Transparency Platform.* https://transparency.entsoe.eu/

EU. (2015). COMMISSION REGULATION (EU) 2015/1222 of 24 July 2015 establishing a guideline on capacity allocation and congestion management. https://eur-lex.europa.eu/legal-content/EN/TXT/?uri=CELEX%3A32015R1222

EU. (2019). Regulation (EU) 2019/943 of the European Parliament and of the Council of 5 June 2019 on the internal market for electricity. Retrieved June 2024, from https://eur-lex.europa.eu/legal-content/EN/TXT/?uri=CELEX%3A32019R0943

JAO. (2021). *Core parallel Run - Publication Tool.* Retrieved June 2024 from JAO.EU: https://core-parallelrun-publicationtool.jao.eu/

JAO. (2024). *JAO Publication Tool - Core CCR.* https://publicationtool.jao.eu/core/

Kristiansen, T. (2020). *The flow based market coupling arrangement in Europe: Implications for traders.* https://doi.org/10.1016/j.esr.2019.100444

NEMO. (2019). *EUPHEMIA public description.* Retrieved June 2024 from https://www.nemo-committee.eu/assets/files/190410_Euphemia%20Public%20Description%20version%20NEMO%20Committee.pdf

Nordpool Group. (2020). *EUPHEMIA public description, single price coupling Algorithm.* Retrieved June 2024 from https://www.nordpoolgroup.com/globalassets/download-center/single-day-ahead-coupling/euphemia-public-description.pdf

N-Side. (2022). *Extended formulation for LTA inclusion.* Retrieved June 2024, from https://www.jao.eu/sites/default/files/2022-03/LTA_Inclusion_Description_202202.pdf

Singh Gill, A., Huang, W., Ammari, S., Yiannaki, I., Kanelakis, N., Mentesidi, K., & Perrakis, K. (2022). *Implementation of a wholesale electricity market based on bilevel programming algorithm in Cyprus.* CIGRE. Cigré ID 10932.

Ferenc Nagy is a Senior Expert in Flow-Based Capacity Calculation at the Hungarian Transmission System Operator MAVIR Ltd. Ferenc has been involved in ENTSO-E and European regional projects in the last 8 years. Throughout his years with active engagement in the Transmission System Operator (TSO) community, Ferenc has gained in-depth experience and knowledge in the Core Flow-Based Market Coupling project. Ferenc has been actively involved in the Core project for the development of the capacity calculation methodologies in both the Day-Ahead and Intraday projects. As a flow-based expert, Ferenc has been supporting IT development activities in the product owner role of the central Capacity Calculation Tool in the last 5 years in the Core Day-Ahead and Intraday projects. His contribution to representing the interest of the project participants has led to several new releases of the central capacity calculation tools in operation. In addition to his involvement in the international projects, Ferenc focuses on analyzing the flow-based results daily in order to improve and support the TSO operator's understanding of the flow-based results generated in the real-time operational process. Ferenc graduated as an electrical engineer majoring in Electric Power Systems at Óbuda University, Kandó Kálmán Faculty of Electrical Engineering in Budapest, Hungary. Ferenc is one of the authors of the CIGRE Paper

"Introduction of the Operational Core Day-Ahead Flow-Based Capacity Calculation and Market Coupling through Active Constraints and Price Spread, CIGRE ID 11358".

Dr. Melinda Nagy is a Data Analyst at MAVIR Ltd. She holds a Ph.D. degree in Physics from Eötvös Loránd University, Budapest. She gained extensive knowledge of data analysis in visualization of satellite measurements, solar cycle prediction and dynamo models. At MAVIR she focuses on the monitoring and research of day-ahead and intraday flow-based capacity calculation (FBCC) processes. She is also involved in the development and analysis of real-time and forecasted field measurements to determine the dynamic line ratings (DLR) for network elements at MAVIR. She, as a researcher, is dedicated to find new methods to analyze and visualize the models and measurements of the grid from network calculations in order to understand more of the operation of the system. Melinda is one of the authors of the CIGRE Paper "Introduction of the Operational Core Day-Ahead Flow-Based Capacity Calculation and Market Coupling through Active Constraints and Price Spread, CIGRE ID 11358".

Luca Tóth is a Market Analyst at MAVIR Ltd. She holds a master's degree in Economic Analysis from Budapest University of Technology and Economics. At MAVIR she focuses on day-ahead and intraday market integration and development projects from an economical perspective. Previously, she was involved in establishing processes of cross-zonal capacity methodologies such as co-optimisation or market-based methods and an R&D project via FARCROSS. Before joining MAVIR, she had gained experience at the trading company, MVM Partner Ltd also as a Market Analyst, focusing on the day-ahead and forward power and commodity markets. She is dedicated to understanding and sharing knowledge on the mechanism of the energy markets. Luca is one of the authors of the CIGRE Paper "Introduction of the Operational Core Day-Ahead Flow-Based Capacity Calculation and Market Coupling through Active Constraints and Price Spread, CIGRE ID 11358".

Ágnes Takácsné Esze is a Data Analyst at MAVIR Ltd. She graduated in Budapest University of Technology and Economics as a Mathematician. At MAVIR she focuses on the monitoring and research of day-ahead and intraday flow-based capacity calculation processes. As a flow based analyst she has wide knowledge of the mathematical model for both the flow based methodology in day-ahead and the capacity calculation methodology in intraday, and supports the Core project with defining and calculating KPIs. Ágnes is one of the authors of the CIGRE Paper "Introduction of the Operational Core Day-Ahead Flow-Based Capacity Calculation and Market Coupling through Active Constraints and Price Spread, CIGRE ID 11358".

Ákos Arnold is Head of Operational Security Development Department at the Hungarian Transmission System Operator MAVIR Ltd., and has been part of various projects in the European energy sector in the last 15 years. With his team he is focusing on the development of concepts, processes and tooling for the network operational planning. Since 2017 he has been involved in the cooperation within the Core Capacity Calculation Region including the day-ahead capacity calculation implementation project.

Dániel P. Divényi was born in Budapest, Hungary, in 1985. He received the M.Sc. degree in electrical engineering and the Ph.D. degree from the Budapest University of Technology and Economics, in 2009 and 2015, respectively. He is currently an Associate Professor with the Department of Electric Power Engineering, Budapest University of Technology and Economics. Since 2021, he has been with MAVIR Hungarian Transmission System Operator Company Ltd. His research interests include electric energy market, power system modeling and control, and power converters.

Gábor Szathmári is Congestion Management Development Team Lead at the Hungarian Transmission System Operator MAVIR Ltd., and has been actively involved in both domestic and international developments in the European energy sector. He has contributed to creating technical concepts and internal tools for various projects, including all timeframes of Flow-Based Capacity Calculation. Gábor has been deeply committed to improving and refining load-flow calculation methods, with a focus on developing automated processes. He also actively mentors colleagues, enhancing the team's skills and knowledge. Prior to joining MAVIR, Gábor participated in numerous research projects in electrical power engineering, including work in laboratory environments.

Day-Ahead Flow-Based Capacity Calculation in CCR Nordic

12

Pieter Schavemaker

Abstract

This chapter describes the reasoning behind the proposed introduction of flow-based market coupling to the Nordic power system. The chapter then discusses flow-based market coupling specificities in the Nordic Capacity Calculation Region, including the use of Advanced Hybrid Coupling and the impact of power system dynamics. Moreover, this chapter shows how the combination of flow-based market coupling and the creation of the Nordic Regional Coordination Centre will result in advanced collaboration between the countries including a more coordinated and data-centred capacity calculation process. The External Parallel Run of the flow-based capacity calculation methodology, currently being performed by the Nordic Transmission System Operators, is described and the initial results are discussed. The chapter closes with a brief overview of stakeholder engagement and the next steps for implementation of flow-based market coupling.

P. Schavemaker—Managing Director of E-Bridge Consulting B.V., written in his role as project manager of the Nordic Capacity Calculation Methodology project, on behalf of the Nordic TSOs Energinet, Fingrid, Statnett, and Svenska kraftnät.

P. Schavemaker (✉)
E-Bridge Consulting, Oosterbeek, The Netherlands
e-mail: pschavemaker@e-bridge.com

A. Estermann et al. (eds.), *European Electricity Market Coupling*, Future of Energy,
https://doi.org/10.1007/978-3-031-86315-8_12

12.1 Introduction

In the electricity markets, the transmission grid constrains how much electricity can be transferred between any two points in the grid. Even if these limitations can be removed by new investments, investments in transmission capacity are capital intensive and have a diminishing marginal value. Thus unlimited expansion of the transmission grid is unrealistic due to economics. This limiting nature of the transmission grid creates a need to have a methodology to optimize the utilization of the transmission grid according to the demand for electric power, while respecting the complex physical limits of the grid. This is also part of the Nordic Solutions Report 2022 (Nordic TSOs, 2022), where it mentions:

"The vision of the Nordic TSO strategy is to have clean and competitive electricity that enables a climate-neutral, secure and integrated energy system. In selecting the means necessary to reach these goals, three particular areas of action will need to be addressed:

- *Adequate infrastructure is needed.*
  (a) *Build adequate infrastructure including the Baltic Sea and North Sea regions*
  (b) *Speed up connection to grid*
  (c) *Optimal utilization and performance of the existing system*
  (d) *Use the full transmission technology mix for further grid expansion*
- *A secure system and integrated markets.*
- *Optimized energy infrastructure*"

Optimization will be about developing improved data-based IT solutions for operational planning and control, such as the implementation of FB market coupling. By introducing FB market coupling, an optimized balance between market facilitation and TSO system operations can be established: maximum trading opportunities for market actors, without sacrificing the operational security of the Nordic power system.

The power system is a non-linear system with endless complexities. However, the algorithms used to calculate the electricity prices and volumes are simplified in order to meet operational requirements. One of the simplifications is the representation of transmission grid capacities. In the price calculation algorithm, transmission capacities are represented as linear constraints where all constraints are modelled as fixed numbers. This gives the TSOs the task of supplying accurate information to the algorithm while respecting the constraints on linearity. Another simplification is the representation of bidding zones. In reality, a power system consists of nodes that are geographically located. In the simplification, a large set of nodes are clustered together in a bidding zone and the transmission grid is represented by bidding zone borders, thus congestions occur on these borders in the electricity market, but in reality these congestions could be caused by any internal node and/or line and not only at the bidding zone borders. The better the representation of the transmission grid is in the electricity market, the more accurate the TSO can feed physical constraints into the price calculation algorithm.

The Nordic Capacity Calculation Region (CCR) is responsible for the calculation of cross-zonal transmission capacity in the Nordic region. The CCR is a regional organization consisting of Transmission System Operators (TSOs) from Denmark, Finland, Norway, and Sweden. The CCR is responsible for managing the coordinated cross-zonal transmission capacity calculation in the day-ahead and intraday timeframes. The Nordic CCR is depicted in Fig. 12.1.

The CCR implements a coordinated capacity calculation methodology to determine the available cross-border transmission capacity between different bidding zones. This flow-based (FB) methodology is a better representation of the transmission grid in the electricity market, than the NTC system used today. The methodology is designed to ensure that transmission capacity is allocated in a fair and transparent manner, and that market participants have equal access to the available cross-zonal capacity.

The Nordic development of a FB approach was initiated in 2012 due to the increasing complexity of the Nordic power system. This complexity makes it more

Fig. 12.1 The Nordic CCR (dark green). *Source* Adapted from ENTSO-E (2020)

difficult to provide cross-zonal capacities by using the NTC approach, while ensuring efficient grid utilization and operational security. This is, among others, driven by the rapidly growing amount of wind, solar, and new large-scale electrification, as well as a steady increase of interconnection with the European continent.

To manage the congestions in the grid, the FB approach will provide tools to consider the grid elements and their contribution and limitations to host power flows. This is acknowledged in the legislation as well, as the FB approach is the default method in the European legislation. The Nordic capacity calculation methodology (CCM) ('Nordic Capacity Calculation Region capacity calculation methodology in accordance with Article 20(2) of Commission Regulation (EU) 2015/1222 of 24 July 2015 establishing a guideline on capacity allocation and congestion management') has been approved by the regulatory authorities (Nordic RCC, 2020).

12.2 Principles and Specificities

This section will zoom into the CCR Nordic FB specificities, including the use of the so-called advanced hybrid coupling and the consideration of power system dynamics.

12.2.1 Advanced Hybrid Coupling in CCR Nordic

The Nordic region is strongly interconnected with other synchronous areas through HVDC links: 10 HVDC links interconnecting the Nordic synchronous area to other regions, and 5 HVDC links within the Nordic CCR. Therefore, the proper modelling of the HVDC links within the FB capacity calculation and allocation is key. This modelling approach should ensure that the commercial exchange over a HVDC link competes for the scarce capacity both on the HVDC link, as well as in the AC grid where the converter stations are located.

Power flows on HVDC interconnections are by nature fully manageable, and a radial AC transmission grid has no meshed structure for the power to fan out. Thus, in a pure HVDC network, or in a radial AC transmission grid, both the NTC and FB perception of the power flows corresponds fully to the real physics of the power system. However, in a meshed AC network, the FB approach is the only one of the two which can manage real physical power flows.

In the Nordic countries, all interconnections to adjacent CCRs are either HVDC or radial interconnections. These parts of the Nordic transmission grid are by definition a physical embodiment of NTC, and it doesn't make sense to implement a FB approach on these parts of the transmission grid. With this realization in mind, the Nordic CCR must apply a hybrid coupling, i.e. both NTC and FB constraints to be considered in the allocation mechanism, to integrate with the other CCRs.

The hybrid coupling might be either the standard hybrid coupling or the advanced hybrid coupling as described also in Chap. 3. Before discussing standard hybrid coupling and advanced hybrid coupling in the Nordic CCR, it is important to bear in mind that when the power flow from another CCR (originating from a HVDC or a radial AC interconnection) enters the meshed AC transmission grid, the power flow will fan out in the AC transmission grid and will use the scarce transmission capacity like all other power flows in the AC transmission grid.

The distinction between standard hybrid coupling and advanced hybrid coupling is the difference in how power flows from another CCR (originating from a radial AC or HVDC interconnection) are managed by the market coupling algorithm in the meshed AC transmission grid. At a high level, the standard hybrid coupling is granting priority access in the meshed AC transmission grid for power flows coming from a radial AC or a HVDC interconnection, while in the advanced hybrid coupling, these power flows are subject to competition for transmission capacity with all other power flows in the transmission system:

- Standard hybrid coupling: Reserves MWs (margin) in the AC grid for the HVDC cable exchanges and Available Transfer Capacity (ATC) exchanges with the other CCRs: i.e. ATC exchanges as well as HVDC cable exchanges receive a priority access to the grid;
- Advanced hybrid coupling: The influence of HVDC cable exchanges and ATC exchanges on the MWs (margins) in the AC grid in the FB model are taken into account during the allocation stage: i.e. Power Transfer Distribution Factors (PTDFs) need to be computed that reflect the impact of the ATC exchanges and HVDC cable exchanges on the margins of the FB constraints.

These two approaches are illustrated in Fig. 12.2.

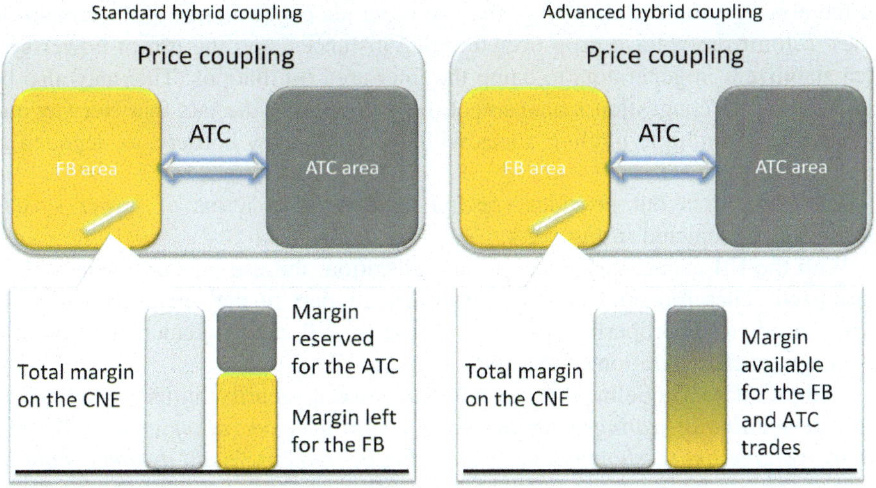

Fig. 12.2 Standard and advanced hybrid coupling. *Source* Author's Own Illustration

Under the standard hybrid coupling, the TSO needs to apply a capacity split: capacity is reserved ex-ante for the flows induced in the AC grid by the HVDC cable exchanges. If it turns out, during the allocation process, that not all this reserved capacity is used by the energy exchange over the HVDC link, it cannot be used anymore to allow for more exchanges in the AC grid. This implies an efficiency loss, as scarce capacity cannot always be fully used in this timeframe. This issue is mitigated in the advanced hybrid coupling approach.

The advanced hybrid coupling fits perfectly well with the FB capacity calculation and allocation and establishes true competition between all relevant exchanges, including the ones on the HVDC links, for the scarce capacity. The Nordic CCR applies the so-called advanced hybrid coupling on the HVDCs and radial AC connections to neighbouring CCRs that are subject to the single day-ahead coupling.

In the advanced hybrid coupling approach, a virtual bidding zone—a bidding zone without any demand or supply—is introduced at the converter station on the Nordic side of the HVDC link. Like for any other bidding zone, the impact of an import or export from the virtual bidding zone on the AC network elements is assessed and captured in the form of PTDFs. In this way, any commercial exchange of power over the HVDC link is competing to make use of the scarce capacity in the Nordic AC grid. This implies that the HVDC link becomes subject to FB properties. Or in other words: a price differential can occur over a HVDC link, while the transmission capacity of the link is not fully utilized.

12.2.2 Power System Dynamics

The Nordic power system spans a long North–South distance. This is reflected in the power system itself by e.g. the use of series capacitors in the transmission lines to limit the voltage drop over the long distances, and the use of power system stabilizers in generators to damp the inter-area oscillations. This may also be reflected in the congestion management, for example by the fact that Norway and Sweden have multiple bidding zones within their country borders, as depicted in Fig. 12.3.

This system layout provides specific challenges in terms of power system dynamics, as depicted in Fig. 12.4.

With the FB capacity calculation and allocation, the use of the system will be optimized. Therefore, power system stability, as part of the operational security limits, needs to be properly considered during the FB capacity calculation process to safeguard the operation of the grid.

In the CACM Guideline Article 2(7), operational security limits are defined as *"the acceptable operating boundaries for secure grid operation such as thermal limits, voltage limits, short-circuit current limits, frequency, and dynamic stability limits"*.

The list of operational security limits, obtained from offline computations, that is applied in the Nordic coordinated capacity calculation, includes:

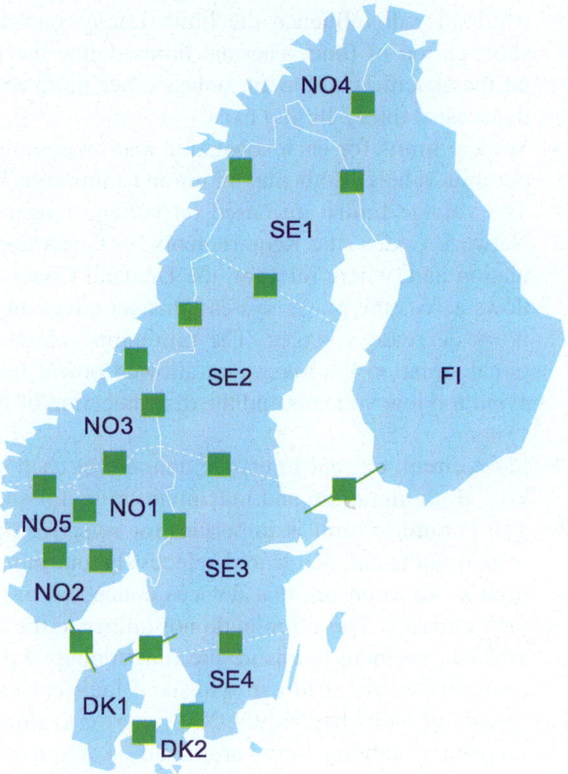

Fig. 12.3 The bidding zones in the Nordic CCR. *Source* Adapted from ENTSO-E (2023)

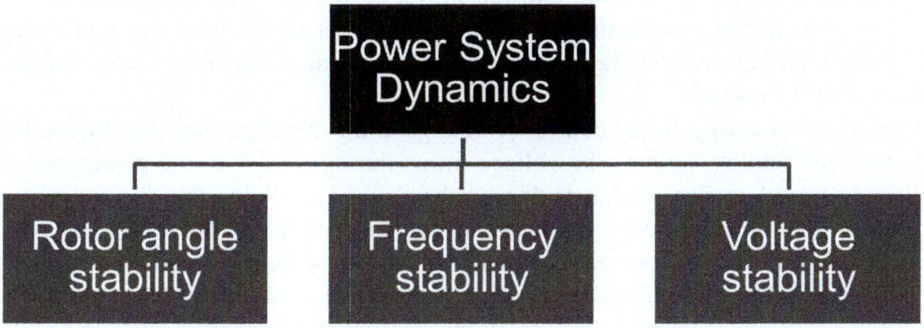

Fig. 12.4 Categorization of power system dynamics. *Source* Author's Own Illustration, adapted from (Kumar, 2014)

- Thermal limits are limits on the maximum power carried by transmission equipment due to the heating effect of electricity current flowing through the equipment. They depend on the physical structure of the equipment and the voltage level. Ambient conditions like temperature, wind and the duration of overload will influence the limit. Larger power flows may be allowed for a short period of time. Thermal limits define the maximum allowed power flow on the specific equipment, unless other more restricting limits (e.g. voltage or dynamic stability limits) exist.

- Voltage limits for each substation and its equipment are defined in kVs and/or per-unit values. Both maximum and minimum limits for voltages are defined. The voltage limits are based on voltage ranges as defined in the Connection Network Codes (the Requirements for Generators Regulation, the HVDC Regulation and, where relevant, the Demand Connection Code Regulation). Power flows across the power system have an effect on the voltages; increasing power flows decrease voltages. The minimum voltage limit defines for each operational situation the maximum allowed power flows in the transmission grid to avoid too low voltages and the disconnection of the equipment by the protection systems.
- Short-circuit current limits are defined for each substation and its equipment in kAs. Both minimum and maximum limits for short-circuit currents are defined. The minimum limit is important for selective operation of protection devices, so that faults can be cleared selectively and in a timely manner. The maximum limit is set to ensure that devices connected to the grid can withstand induced fault currents. These limits do not influence the allowed power flows in the AC grid but are there to ensure the functioning of protection systems, that devices connected to the grid can withstand fault currents, and that the probability of cascading faults beyond the N-1 criterion is minimized.
- Frequency stability limits are based on frequency ranges set in the Connection Network Codes and in the System Operation Regulation. Frequency stability limits are taken into account during dynamic stability studies to see if the limits would have affected the allowed power flows on the transmission grid. It is foreseen that these limits will have more effect in the future system operation, due to changes in the generation mix including more energy from variable renewable sources.
- Dynamic stability limits consist of voltage and rotor angle stability limits. For voltage stability studies, the voltage limits during a fault in the power system and after clearance of that fault shall be studied to define the allowed power flows within the power system, respecting the voltage limits. For rotor angle stability studies, the power flow and generator rotor angle oscillations are studied for each operational situation to define the allowed power flows within the power system with predefined damping coefficients for power and rotor angle oscillations. The magnitude of oscillations and their damping depends on the structure of the power system and the power flows across the power system.

The acceptable operating boundary for secure grid operation is defined by the maximum flow on a Critical Network Element and Contingency (CNEC). It can be written as CNEC (Fu_{max}, $u \in \{$thermal limits, static voltage limits, dynamic voltage limits, transient stability limits, and damping limits$\}$). It is monitored in the operational security analyses and in real-time operation. It is defined as a MW limit for maintaining the voltage and short-circuit current level, frequency and dynamic stability within its limits.

In case operational security limits cannot be transformed efficiently into maximum flow on specific CNECs, the TSOs may use the combined dynamic constraint, which limits the sum of power flows on a set of network elements, for the purpose of respecting the dynamic stability limits.

Example 12.1 The need to use combined dynamic constraints In the south-west part of Norway, there are several production units in remote areas. These are connected with the eastern part (Oslo) where the consumption is high, especially during winter. The transmission lines between the production and the consumption area can be very long. When a contingency (N-1 situation) occurs, i.e. the forced outage of one of the transmission lines, this can cause oscillations on the remaining lines. In this example, there are more than 2 lines in parallel serving the Oslo region. In some dynamical limitations, there are up to 4–5 lines in which oscillations can occur. It is not possible to single out these stability problems to one or two network elements, due to the production being so distributed and spread over a large area. The dynamic limitations are therefore best described as a sum of flows through several lines. It also depends on which generators are in operation at a given time, which is impacted by many factors such as price in the neighbouring areas and legal limitations regarding hydrological considerations in the water ways. Another example, this time in Sweden, which is closely monitored, is related to voltages in the network. Large amounts of electricity transferred through long transmission lines consume a significant amount of reactive power, which is mostly compensated by generators. This is the case in Sweden where most of the production is located in the north, and consumption in the south. Large transmission lines from SE1 and SE2 transfer power to SE3 and SE4. Eleven lines are connecting SE3 (Stockholm area) and SE2 (see also Fig. 12.3). The voltage limit on these lines can vary depending on the dispatch situation (i.e. which generators are in operation) and planned outages. Generators operate in different operation modes and with varying reactive power producing capabilities and they contribute to voltage control differently. In case of a contingency where reactive power consumption increases or reactive power producing capabilities are reduced, the electricity system may end up in a situation where reactive power sources are depleted. This could further reduce the voltage and result in a voltage collapse. Voltage collapses can lead to large-scale blackouts, such as those which occurred in Sweden in 1983 and 2003. To prevent this, the power flow through the corridor (between SE2 and SE3) and possible contingencies need to be studied together. The variables used to monitor the voltage limits are the active power flow through the north–south corridors. From these, the maximum possible flow that can be transferred on all the lines together without leading to a voltage collapse after a contingency is calculated and used as an operational security limit in the dispatch centre. It is not possible to study the voltage problems by only looking at a subset of the 11 lines connecting SE2 and SE3. They must be monitored together.

In the Nordic CCR, the work on the development of a dynamic Common Grid Model (CGM) to be used in the FB capacity calculation process has started, as well as the work on a common dynamic analysis.

12.2.3 Intraday Capacity Calculation

The Nordic TSOs developed a transitional solution for the calculation of cross-zonal capacities for the intraday time frame until the single intraday coupling can support FB parameters. The transitional solution is a so-called ATC Extraction (ATCE) methodology. In general, the ATCE methodology determines the left-over capacity after the Day-ahead (DA) stage, and it translates the FB capacity domain into an ATC domain.

The translation from a FB domain to an ATC domain is not straightforward. A simple example is shown in Fig. 12.5. Indeed, in this figure, only one possible ATC domain—that fits within the FB domain—is depicted; many other ATC domains can be extracted without violating the FB domain. Therefore, the Nordic TSOs developed an optimization-based approach to extract a single set of optimal ATC values.

As stated before, the translation from a FB domain to an ATC domain is not straightforward. Indeed, the ATC concept describes the grid limitations on a higher and less-detailed level than FB does. As an example: the ATC capacity is an option—it can be used, but is not guaranteed to be used. As such, only the loading effect of the use of the ATC capacities can be accounted for on the FB CNECs; the relieving effect cannot be considered. This consideration, together with the DA being optimized by Euphemia, may lead to a situation where the Intraday (ID) gate opening capacity may be very limited. Therefore, the Nordic TSOs have designed the ID ATCE methodology such that relaxations can be applied in the

Fig. 12.5 ATC extraction from a FB domain, for the bidding zone borders A-C and B-D. *Source* Author's Own Illustration

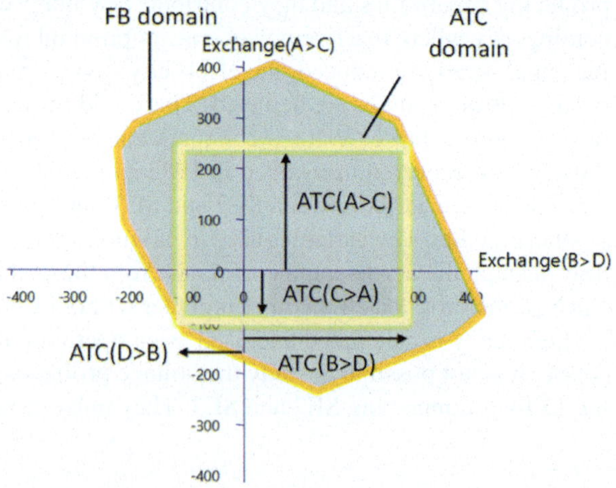

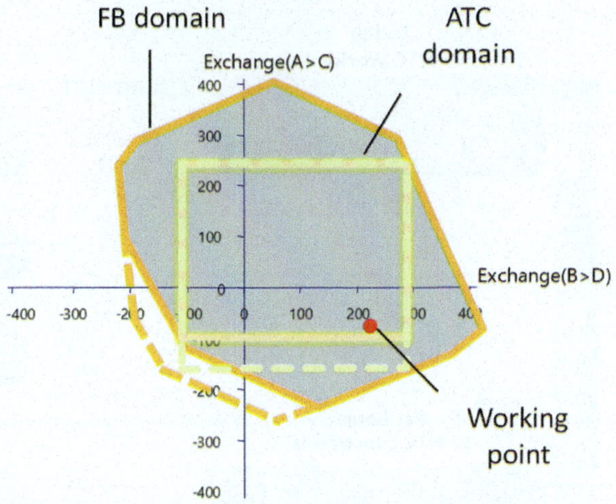

Fig. 12.6 Relaxation of the FB domain around the DA market clearing point (working point). *Source* Own illustration

ATCE process in order to provide as much ID capacity to the market participants as possible—from an operational security point of view. This is schematically depicted in Fig. 12.6.

In Fig. 12.6, the amount of MWs that can be used by the market, i.e. the Remaining Available Margin (RAM), on some CNECs has been increased, as indicated by the dashed lines. Thereby the FB domain is increased, and the ATCE can determine a larger ATC value from the bidding zone C to A. This can be beneficial for the ID market, as the market can now move more power in the preferred market direction (i.e. from bidding zone C to A): i.e. the DA market clearing point (the "working point" as indicated by the red dot in Fig. 12.6) can move further down without violating operational security limits.

Thanks to this approach, the resulting ID ATC values for the ID gate opening trades are always larger than or equal to zero, and capacity may be released on CNECs that was not available at the DA stage. Like in any coordinated capacity calculation process, a TSO domain validation is applied where capacities may have to be reduced—in case potential overloads resulting from the ID ATCE cannot be coped with.

12.2.4 Nordic RCC: More Cooperation and Digitalization

The Nordic Regional Coordination Centre (RCC) is an independent company owned by the four electricity Transmission System Operators (TSOs) in the Nordic region (see also Chap. 5 for a more general introduction to RCCs and their tasks). The RCC supports the national TSOs in maintaining the operational security of the power systems across Finland, Norway, Sweden and Denmark. An RCC does many of the same advanced calculations that happen in the TSO control rooms

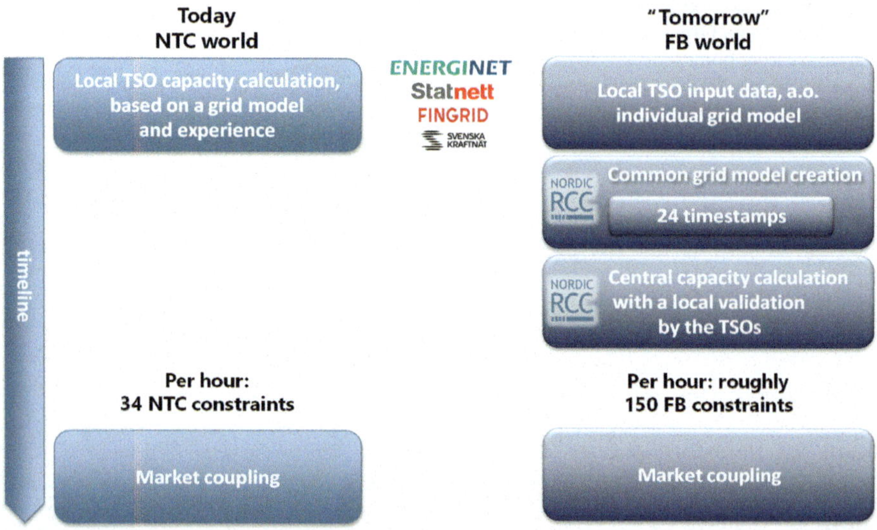

Fig. 12.7 Capacity calculation today (NTC world) and when Nordic FB is live. *Source* Author's Own Illustration

like capacity calculation and security analysis but on a regional level. The regional analyses ensure increased coordination and effectiveness.

With the introduction of the RCC and the FB capacity calculation, a more coordinated and data-centred capacity calculation process has been set up, as can be observed from Fig. 12.7.

12.3 Benefits of Flow-Based Market Coupling

As of early 2024, the Nordic TSOs are performing an external parallel run (EPR) of the new FB capacity calculation methodology. In the next sections an explanation is provided of what this entails, and what the results of the EPR are highlighting.

12.3.1 External Parallel Run

In the CACM Guideline, Article 20(8), it states: "To enable market participants to adapt to any change in the capacity calculation approach, the TSOs concerned shall test the new approach alongside the existing approach and involve market participants for at least six months before implementing a proposal for changing their capacity calculation approach". This is what is referred to as the "external parallel run" (EPR) in this section.

Operation	NTC Capacity calculation	NTC Market Coupling	Market results	**Compare**
Simulation	FB Capacity Calculation	FB Market Coupling	Market results	

Fig. 12.8 Overview of parallel run. Source Author's Own Illustration

In the EPR, the FB capacity calculation is performed by the TSOs and RCC, alongside the operational Net Transfer Capacity (NTC) capacity calculation. While the operational NTC values are used in the Single Day-Ahead Coupling (SDAC), the FB parameters are used in a FB market coupling simulation, by using the order books submitted in the SDAC. This is schematically depicted in Fig. 12.8. Therefore, the only difference between the two, is the parameters applied in the DA market; thus, it is possible to compute the isolated (static) impact of applying FB parameters instead of NTC.

As such, the FB market results are indicative only, they do not represent a FB market forecast.

The EPR is a learning-by-doing period for all parties involved. A period that enables the identification of mistakes, errors, or hiccups, so that they can be mitigated before the actual go-live. The Nordic capacity calculation methodology decision requires at least 12 months of EPR, of which 6 months are required by the CACM Guideline. The Nordic National Regulatory Authorities (NRAs) required an evaluation phase of the operational functioning of the Nordic FB during the external parallel run, to take place before the last 6 months of external parallel run can be conducted and finally the FB capacity calculation can go-live. During this consecutive three-month evaluation period, the TSOs had to demonstrate that the FB capacity calculation and allocation performed well enough—in terms of the quality of the process and the quality of the FB parameters—by measuring and reporting on criteria defined by the NRAs. The findings of the three-month evaluation period have been captured in a report (Nordic TSOs, 2023) that was subject to a public consultation, before submitting it to the NRAs for their assessment. The Nordic TSOs started the remaining 6 months of EPR in July 2023. The Nordic DA FB go-live went live in October 2024.

In the following sections, some of the results from the Nordic three-month EPR evaluation report are presented to demonstrate the functioning of the Nordic FB, and to highlight some of the changes it will bring. In short, during the three-month evaluation period (12.12.2022–12.3.2023 (week 52—week 10)), the Nordic FB approach has demonstrated the following features:

- The FB capacity calculation process is stable and functions as planned.
- The utilization of the Nordic power system has been improved by the FB approach.
- The total socio-economic welfare has increased in the Nordic CCR.

12.3.2 Results of the Socio-Economic Welfare Computation

This section presents the DA welfare comparison between the operational NTC system and the simulated FB system. The DA results contain simulations from the three-months reporting period 12.12.2022–12.3.2023 (week 52–week 10). The total change in socio-economic welfare (FB–NTC) and the distribution of its components is shown in Fig. 12.9. For the three-month EPR period, the total socio-economic welfare gain with FB is 73.4 M€ in the Nordic CCR. This consists of increases in congestion income and producer surplus (PS) and a decrease in consumer surplus (CS). The positive result for FB originates from positive results on a day-by-day basis, and not from individual days with extreme results.

When a Bidding Zone (BZ) price increases/decreases due to the application of FB, it means that power is imported or exported differently than with NTC and that there is a redistribution of welfare between consumers and producers. This is also shown in Fig. 12.10 where the changes in CS and PS are split per BZ. The BZs obtaining an increase in CS and a decrease in PS are mainly areas that in NTC had high prices, such as southern Norway (NO1, NO2, and NO5), as these areas had a decrease in prices with FB. The opposite occurs for northern Norway (NO3, NO4) and SE3 that obtained an increase in PS and a decrease in CS as these areas saw an increase in prices with FB. The impacts for Denmark (DK1, DK2), Finland, but also Northern Sweden (SE1, SE2) and the southernmost Swedish bidding zone SE4 are smaller in magnitude.

During the three-months reporting period, there is an overall positive socio-economic welfare impact for the Nordic CCR, but a decrease in CS. However, as shown in Fig. 12.10, some BZs have an increase in the CS, while the others have a decrease. The BZs with the initially highest prices typically show a decrease in prices, and thereby an increase in CS, with FB. However, these increases have not

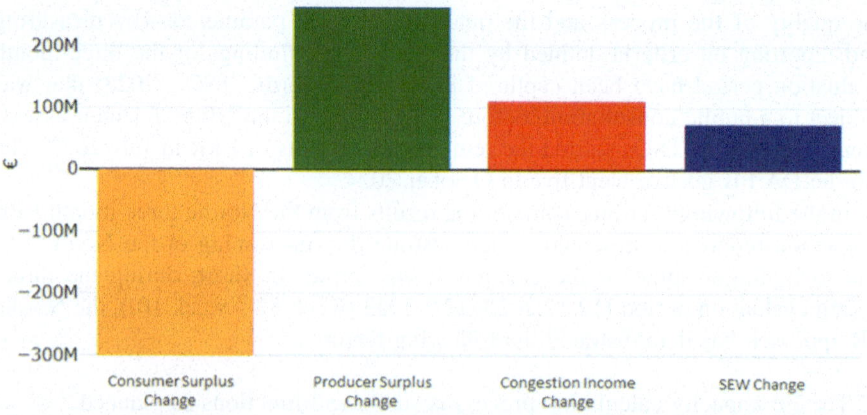

Fig. 12.9 Welfare change for the Nordic CCR resulting from shift from NTC to FB. *Source* Nordic TSOs (2023)

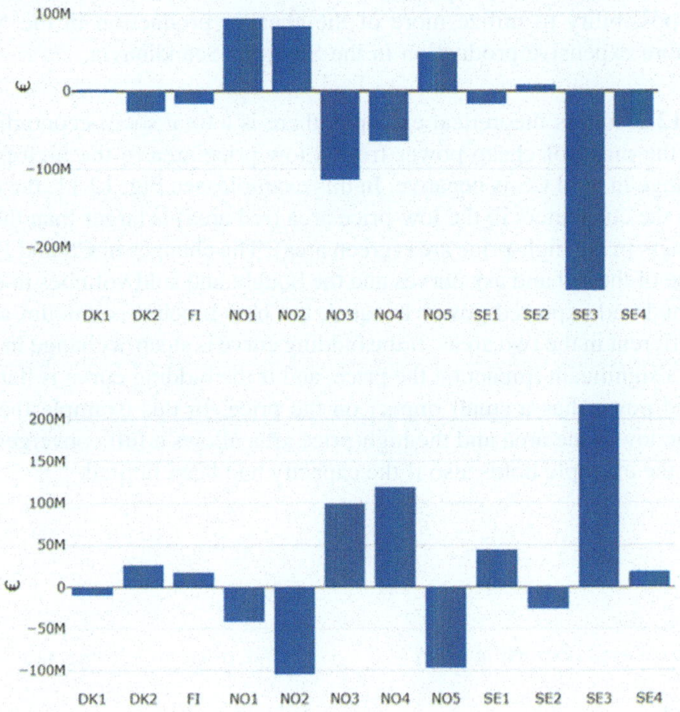

Fig. 12.10 Change in consumer (top) and producer (bottom) surplus in the Nordic BZs (FB-NTC). *Source* Nordic TSOs (2023)

been enough to cover the decreases of CS in other BZs. In the Example 12.2 it is illustrated, for two bidding zones, how the slopes of the demand and supply curves can impact the CS. The slope of the curves is of significant importance for the impact assessment result when more power is exchanged between, initially, low-price and high-price BZs.

SE3 is an area with relatively steep supply order curves, while areas in the south of Norway have a lot of flexible hydro production, resulting in flatter bid curves on the supply side. When FB increases the available capacity between these areas, and the south of Norway is more expensive than SE3, a consequence can be that the increase in CS in the south of Norway is not enough to cover the consumer loss in SE3. Yet, as Example 12.2 shows, this will still increase the overall socio-economic welfare for the system. The conclusions do not change if more bidding zones were included in the example above, because the key outcome from the analysis does not depend on the number of BZs, but the impact of increasing exchange possibilities (= capacity) in the power system.

One of the reasons for the positive socio-economic welfare when introducing FB is the possibility to utilize more of the cheaper production in the North and save the more expensive production in the South of Scandinavia.

Example 12.2 In this theoretical example, there is a total socio-economic welfare gain from the move of cheap power from a low-price area to the high-price area, but the change in total CS is negative. In this example, see Fig. 12.11, the change in surplus for the consumers in the low-price area (red area) is larger than the gain for the consumers in the high-price area (green area). The changes in CS and PS depend on the slope of the bid and ask curves and the bought and sold volumes in each area. The imported and exported power is equal, but how much this amount affects the prices is different in the two areas. If the bidding curve is steep, a change in produced power has a significant impact on the price, and if the bidding curve is flat a change in produced power has a small impact on the price. In this example the capacity between the low-price area and the high-price area allows a full convergence of the prices, but the example holds also if the capacity had been limited.

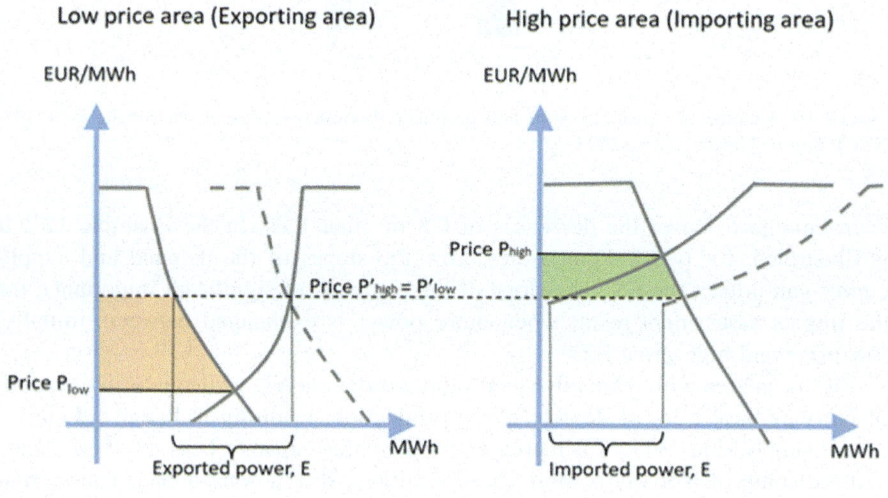

Fig. 12.11 Example of exports from low-price area to high-price area. *Source* Nordic TSOs (2023)

12.4 Outlook

12.4.1 Implementation Status and Foreseen Go-Live

As mentioned above, the Nordic capacity calculation methodology decision requires at least 12 months of external parallel run, of which 6 months are required by the CACM Guideline. The Nordic TSOs started the remaining 6 months of EPR in July 2023; the EPR continued until the go-live which happened in October 2024.

The FB capacity calculation process is stable and functions as planned. The challenge before the actual go-live is mainly linked to the integration of all the IT systems involved and the testing. Indeed, all the systems involved—those from TSOs, RCC, NEMOs, JAO, and ENTSO-E—need to be able to exchange the data with consistently good quality and performance, to guarantee a continued and reliable SDAC and Single Intraday Coupling (SIDC) operation after the Nordic FB DA go-live. Some of the next steps on the implementation timeline are the functional integration tests (FIT), system integration tests (SIT), SDAC tests, and NEMO member tests. With many projects queuing up to go-live on the SDAC and SIDC timelines while considering the summer freeze period, it is needless to state that finding a go-live date is not a trivial exercise.

Twelve months after the DA go-live, the Nordic FB long-term capacity calculation will go-live. This is followed by the application of the FB Intraday (ID) allocations, and the application of the FB approach in the balancing timeframe.

12.4.2 Stakeholder Involvement

Stakeholder involvement is key when implementing a fundamental change to the market design. The introduction of the FB market coupling is considered to be such a fundamental change.

The Nordic market started as a joint Norwegian/Swedish market in 1996 and was fully integrated in 2000. It was based on the NTC system. The NTC system brought the market to where it stands today, and the NTC values are the references in the mind of all stakeholders acting on the power market. The introduction of FB may be viewed with different eyes. “A necessary step to manage the power grid while facilitating the market during the energy transition”, from a TSO perspective. “An unnecessary change, that is only implemented to follow EU legislation, and that makes it more difficult for stakeholders to forecast the market prices”, from potential market participants’ perspectives. Indeed, this underlines a clear need to listen to one another, hear each other’s arguments and concerns, to have the external parallel run as the learning period for all stakeholders involved, and to work together in a constructive way towards a successful go-live of the Nordic FB.

The Nordic stakeholder involvement for the preparation of the Nordic FB market coupling is depicted in Fig. 12.12.

As indicated in Fig. 12.12:

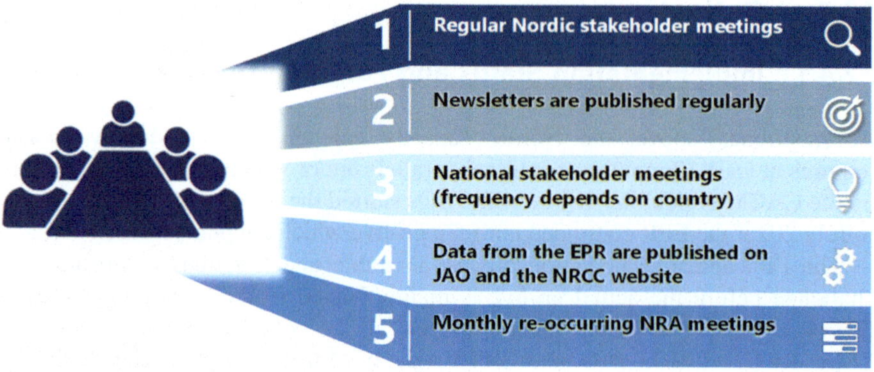

Fig. 12.12 Stakeholder involvement in the CCR Nordic. *Source* Author's Own Illustration

- All FB parameters of the Nordic FB external parallel run can be found on the JAO platform: (JAO, 2024)
- The flow-based section on the Nordic RCC website provides access to the methodology, relevant documentation, simulation results, news and updates: (NORDIC RCC, 2024)

Bibliography

ENTSO-E, 2023. ENTSO-E Capacity Calculation and Allocation Report. (2023). Available at: https://eepublicdownloads.blob.core.windows.net/public-cdn-container/clean-documents/nc-tasks/ENTSO-E%20CC%26A%20Report%202023.pdf. Accessed 26 June 2024.

ENTSO-E, 2020. *Network Codes.* Available at: https://www.entsoe.eu/network_codes/eb/nordic-afrr-capacity-markets/. Accessed 26 June 2024.

JAO. (2024). *JAO Publication Tool - Nordic CCR.* Retrieved October 29, 2024, from https://test-publicationtool.jao.eu/nordic/. Accessed 29 October 2024.

Kumar, A., 2014. *Engineering Articles.* Available at: https://top10electrical.blogspot.com/2014/04/classification-of-power-system-stability.html. Accessed 29 June 2024.

Nordic RCC. (2020). *Nordic Capacity Calculation Region capacity calculation methodology in accordance with Article 20(2) of Commission Regulation (EU) 2015/1222 of 24 July 2015 establishing a guideline on capacity allocation and congestion management.* Available at: https://nordic-rcc.net/wp-content/uploads/2024/01/CACM-Nordic-CCR-DA-and-ID-CCM-2020-1.pdf. Accessed 26 June 2024.

NORDIC RCC. (2024). *Nordic capacity calculation methodology project.* Retrieved October 29 2024, from https://nordic-rcc.net/flow-based/. Accessed 29 October 2024.

Nordic TSOs. (2022). *Solutions for a green Nordic energy system - Strategies to meet the climate challenge.* Available at: https://en.energinet.dk/media/iwahnjlz/solutions_report_2022.pdf. Accessed 26 June 2024.

Nordic TSOs. (2023). *External parallel run evaluation report - For assessment by the NRAs of the Nordic CCR, as required by the Nordic DA/ID CCM.* Available at: https://nordic-rcc.net/wp-content/uploads/2023/06/Parallel-run-report_final_public.pdf. Accessed 26 June 2024.

Dr. Pieter Schavemaker Managing Director of E-Bridge Consulting B.V., has worked in the energy sector for more than 20 years. Pieter has a varied experience as assistant professor at the Delft University of Technology, with a large manufacturer of power system equipment (ABB), with a Transmission System Operator (Tennet TSO), and as a consultant. During the past years, Pieter was heavily involved in the international work on the boundary between system operations and the electricity market. Pieter is one of the founding fathers of the Flow-based Market Coupling that is currently in operation. Pieter holds a Ph.D. in Electrical Power Systems from Delft University of Technology, received his Didactic Qualification for University Education from Delft University of Technology, and is a certified PRINCE2® Practitioner. Pieter Schavemaker is author of the book 'Electrical Power System Essentials', published by Wiley.

The Energy Community 13

Lisa-Marie Mohr

Abstract

This chapter describes the latest developments which have expanded the European Union's legal framework to the Energy Community Contracting Parties and enabled their integration into the internal market for electricity, in particular their adherence to the SDAC and the SIDC. This is not only crucial for the Energy Community but also for the European Union and its European Green Deal, the success of which is linked to its neighbours and partners. The envisaged extension is substantial in size and its implementation requires all involved parties to proceed fully committed and coordinated. Hence, this chapter introduces the Energy Community, its institutional set-up and legal framework, including the Electricity Integration Package. Besides getting to know the electricity markets of the Contracting Parties, the reader is familiarized with important implementation steps and preconditions to successfully enlarge the SDAC and the SIDC across Europe.

The Author is a Managing Consultant at Magnus Energy. The Author's Contribution Was Provided in Personal Capacity, Neither Reflecting the Position of Magnus Energy Nor of the Energy Community Secretariat, Where the Author Was Employed As an Electricity Market Expert During the Drafting Process.

L.-M. Mohr (✉)
Magnus Energy, Naarden, Netherlands
e-mail: lisam.mohr2@gmail.com

A. Estermann et al. (eds.), *European Electricity Market Coupling*, Future of Energy,
https://doi.org/10.1007/978-3-031-86315-8_13

13.1 The Energy Community

The Energy Community (EnC, 2024c) is an international organization bringing together the European Union and its Eastern neighbours based on a legally binding framework. It was founded by the Treaty establishing the Energy Community (EnC Treaty, 2005), in force since 2006, and aims at extending the European Union's *acquis* to create a single energy market. The founding members were the European Community on the one side and the so-called Contracting Parties on the other, namely Albania, Bulgaria, Bosnia and Herzegovina, Croatia, Kosovo[1], Montenegro, North Macedonia, Romania and Serbia. Meanwhile, Bulgaria, Croatia and Romania joined the European Union and Moldova (2010), Ukraine (2011) and Georgia (2017) acceded to the EnC Treaty. The European Union is a Party to the EnC Treaty, represented by the European Commission which serves as permanent Vice-President of the organization (Energy Community, 2024c). Armenia, Norway and Türkiye are observers to the Energy Community.

The Energy Community was established to pursue the following key objectives as defined in Article 2 of the EnC Treaty (Energy Community, 2024c):

- establish a stable regulatory and market framework capable of attracting investment in power generation and networks;
- create an integrated energy market allowing for cross-border energy trade and integration with the European Union's market;
- enhance the security of supply to ensure stable and continuous energy supply that is essential for economic development and social stability;
- improve the environmental situation in relation with energy supply in the region and foster the use of renewable energy and energy efficiency; and
- develop competition at regional level and exploit economies of scale.

13.1.1 Institutional Set-Up of the Energy Community

The Energy Community has different institutions supporting its processes as presented in Fig. 13.1 and described hereafter (Energy Community, 2024c; Karova, 2015b, pp. 30–31). The Ministerial Council,[2] the highest decision-making body, represents each Contracting Party at ministerial level as well as the European Community and meets once a year. It takes strategic decisions, steering the Energy Community and its development, with its presidency rotating annually. The work of the Ministerial Council is prepared by the Permanent High Level Group[3] (PHLG) which meets four times a year. The Energy Community Regulatory

[1] This Designation Is Without Prejudice to Positions On Status, and Is in Line With UNSC 1244 and the ICJ Opinion On the Kosovo Declaration of Independence.

[2] Article 47-52 EnC Treaty.

[3] Article 53-57 EnC Treaty.

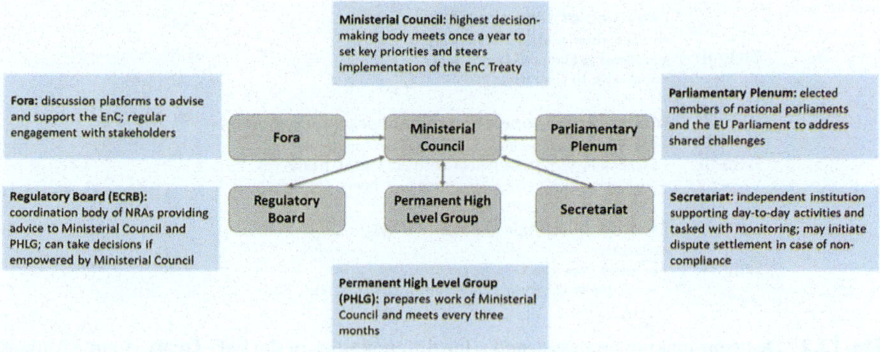

Fig. 13.1 Institutional set-up of the energy community. *Source* Author's own illustration

Board[4] (ECRB) is the coordination body of the national regulatory authorities (NRAs). The ECRB advises the Ministerial Council and the PHLG and is entitled to issue recommendations in case of cross-border disputes of two or more NRAs. Further, the ECRB may take legally binding decisions if empowered by the Ministerial Council. The only permanent institution of the Energy Community is the Energy Community Secretariat[5] based in Vienna. It is responsible for supporting the day-to-day activities of the Energy Community and obliged to monitor and report on the implementation of the EnC Treaty. To that end, the Energy Community Secretariat, upon complaint or on its own motion, may initiate dispute settlement procedures if a Party to the EnC Treaty does not comply with the *acquis* in force.[6] Decisions on the existence of a breach by a Party to the EnC Treaty are to be taken by the Ministerial Council.[7] Besides, the Parliamentary Plenum[8] brings together elected representatives of national parliaments and the European Parliament to address shared challenges while different Fora[9] (e.g. Electricity, Gas, Oil) support and advise the Energy Community.

13.1.2 Energy Community Acquis

The Contracting Parties committed to adopt parts of the European Union's legislation in the areas of energy, environment, competition and renewables.[10] New *acquis* of the Energy Community can be adopted, or existing can be amended,

[4] Article 58-62 EnC Treaty.
[5] Article 67-72 EnC Treaty.
[6] Article 90 EnC Treaty.
[7] Article 91 EnC Treaty.
[8] Ministerial Council Procedural Act 2015/05/MC-EnC.
[9] Article 63-66 EnC Treaty.
[10] Article 3(a) EnC Treaty.

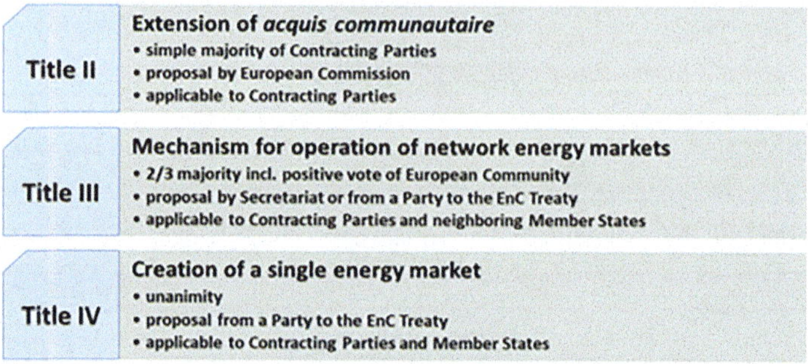

Fig. 13.2 Decision-making processes under the different titles of the EnC Treaty. *Source* Author's own illustration based on (EnC Treaty, 2005)

based on the European Union's legal acts with adaptations to the specific situation of each of the Contracting Parties and the Energy Community's institutional set-up.[11] This concept allows the Energy Community *acquis* to evolve over time and the Contracting Parties to stay aligned with the European Union's developments. To date, this has brought numerous updates and extensions of the Energy Community *acquis* to include new sectors or legislation newly adopted in the European Union.

Respective decisions are generally taken by a majority of the votes cast of the Ministerial Council (see Sect. 13.1.1 and Fig. 13.1) based on a European Commission's proposal (Energy Community, 2024a), with each Party to the EnC Treaty having one vote.[12] However, as summarized in Fig. 13.2, the exact decision-making process[13] required to adopt a proposal depends on its nature and the Title of the EnC Treaty under which it is proposed (Karova, 2019, pp. 280–286): Decisions under Title II (extension of *acquis communautaire*) require a proposal by the European Commission, simple majority of the Contracting Parties for adoption and are addressed to the Contracting Parties only, not covering the interface with Member States. Proposals under Title III (mechanism for operation of network energy markets) can be brought forward by the Energy Community Secretariat or any Party to the EnC Treaty and are accepted if a two-third majority, including a positive vote of the European Community, is in favour of the proposal. Measures under Title III are addressed to both the Contracting Parties and the neighbouring Member States, as defined in Article 27 of the EnC Treaty. The most extensive measures are governed by Title IV (creation of a single energy market), requiring a Party to the EnC Treaty to propose and unanimity of all Parties to decide as resulting in obligations towards the entire Energy Community, i.e. the Contracting Parties and all Member States.

[11] Article 5 EnC Treaty.

[12] Article 77 EnC Treaty.

[13] Article 76-85 EnC Treaty.

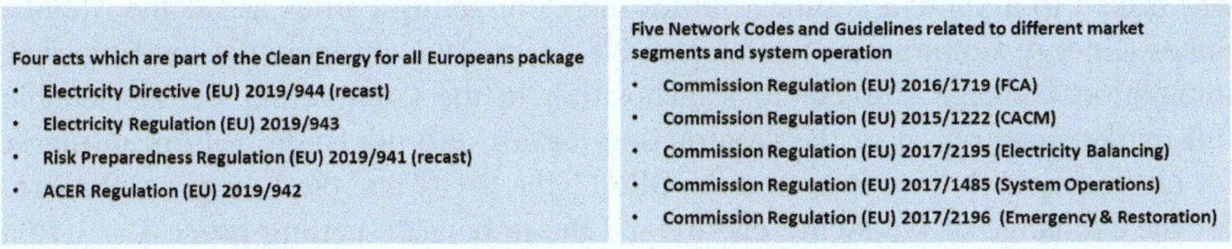

Fig. 13.3 Legal acts incorporated in the energy community *acquis* by the electricity integration package in 2022. *Source* Author's own illustration based on (Energy Community, 2024b)

13.1.3 Electricity Integration Package

The Ministerial Council Decision 2022/03/MC-EnC, adopted under Title III of the EnC Treaty on 15 December 2022, incorporated missing parts to fully align the Energy Community *acquis* with the, at that time,[14] applicable legal framework for the electricity sector in the European Union. The so-called Electricity Integration Package includes nine legal acts as described in Fig. 13.3[15] (Energy Community, 2024b).

By the Ministerial Council's decisions,[16] the Contracting Parties committed to bringing into force the laws, regulations and administrative provisions necessary to comply with the new provisions[17] by 31 December 2023. However, the legal deadline has passed and none of the Contracting Parties has transposed the new legal framework into national law (Energy Community Secretariat, 2024, p. 2).

The decision was accompanied by the Ministerial Council Procedural Act[18] 2022/PA/01/MC-EnC fostering regional market integration and ensuring reciprocity between energy sector stakeholders of the Member States and the Contracting Parties. As a novelty of the Electricity Integration Package, ACER (see also Chap. 15) was entitled to take decisions on cross-border issues concerning both Member States and Contracting Parties, and ENTSO-E (see also Chap. 1)

[14] Having in mind the electricity market design reform and the discussions with regards to future amendments of Network Codes and Guidelines in the European Union, it is important to note that, as explained in Sect. 13.1.2, newly adopted legislation in the European Union is not automatically incorporated in the Energy Community *acquis*.

[15] Directive (EU) 2019/944 and Regulation (EU) 2019/941 were already incorporated by the Ministerial Council Decision 2021/13/MC-EnC. However, the Ministerial Council Decision 2022/03/MC-EnC amended parts thereof to reflect and include references to Regulation (EU) 2019/943 ('recast').

[16] Article 2(1) Ministerial Council Decision 2021/13/MC-EnC and Article 2(1) Ministerial Council Decision 2022/03/MC-EnC.

[17] Different to the European Union, both Directives and Regulations incorporated in the Energy Community *acquis* need to be transposed into national law of the Contracting Parties (Karova, 2015a, p. 4).

[18] Besides recommendations and decisions, procedural acts are one of the measures the Energy Community can take, having binding force on the institutions of the Energy Community, and, if provided, on the Parties to the EnC Treaty (Article 86-88 EnC Treaty).

was tasked to assume a similar role for the Contracting Parties as for the Member States (Energy Community Secretariat, 2023b, p. 4). Consequently, the Electricity Integration Package, subject to transposition in the Contracting Parties, enables full market integration of the Contracting Parties into the single European market for electricity including inter alia the SDAC, the SIDC and the European platforms for the exchange of balancing energy and the imbalance netting process.

The European Network Codes and Guidelines in the electricity sector include subsequent terms and conditions or methodologies (TCMs) to be developed by relevant stakeholders such as TSOs and NEMOs (see also Chap. 2). TCMs define more detailed rules on specific topics such as capacity calculation, algorithms for operating the market coupling or products used therein. Their approval in the Energy Community differs, depending on their scope: pan-EU, regional or national. With the adoption of the Electricity Integration Package, the Contracting Parties committed to the following approach with regards to TCMs (Energy Community Secretariat, 2023b, pp. 6, 8). For the pan-EU TCMs, submitted by either all TSOs or all NEMOs in the European Union to ACER for decision, relevant stakeholders of the Contracting Parties need to implement them without adaptations including future amendments thereof. On the contrary, regional TCMs, developed by more than one TSO of Member States and/or Contracting Parties (e.g. within capacity calculation or system operation regions), are submitted to the NRAs concerned, for their approval. In case of disagreement or voluntary action of the relevant NRAs, the decision is referred to ACER in case both Member States and Contracting Parties are concerned. If only Contracting Parties are involved, the ECRB is tasked to issue a decision. National TCMs, concerning only one Contracting Party, are subject to the relevant NRAs' approval as is the case in the European Union. A high-level overview of the different concepts is provided in Fig. 13.4.

As described in Chap. 16, transparency and integrity are of utmost importance for the well-functioning of electricity markets. Even though not part of the Electricity Integration Package, but due to the strong link with the SDAC and the SIDC, it shall be noted that an adapted version of Regulation (EU) 1227/2011 (EU REMIT) was adopted in the Energy Community in 2018[19] (EnC REMIT). However, there are major differences between the principles of the EU REMIT and the adapted EnC REMIT. Most prominently, the EnC REMIT neither includes centralized data collection nor grants ACER any power related to data collection or surveillance in the Contracting Parties. As described in a guidance on REMIT published by the ECRB (2022, p. 4), at the time of adoption, it was considered more appropriate to amend the EnC REMIT at a later stage when there is clarity on the way forward in particular in relation to market integration between the Contracting Parties and the Member States. To date, even though the Electricity Integration Package was adopted in 2022, no such amendment has been incorporated in the Energy Community *acquis*.

[19] Ministerial Council Decision 2018/10/MC-EnC.

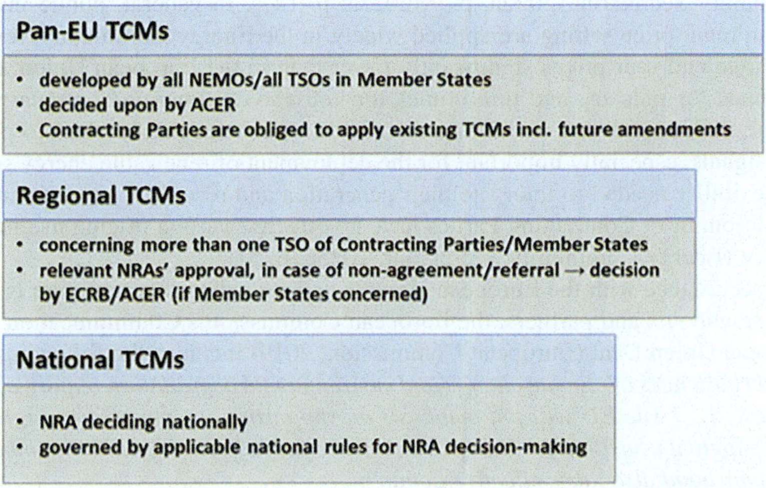

Fig. 13.4 Approach towards pan-EU, regional and national TCMs in the energy community. *Source* Adapted from Energy Community Secretariat (2023b, p. 6)

13.2 Extension of the SDAC and the SIDC to the Energy Community

As outlined in Sect. 13.1.3, the Electricity Integration Package for the first time enables the integration of the Contracting Parties into the European internal market for electricity. In particular, the adapted and adopted Commission Regulation (EU) 2015/1222[20] (EnC CACM) enables the integration into the SDAC and the SIDC.

13.2.1 Importance of Market Coupling for the Energy Community

When compared to the European Union, the electricity markets in the Contracting Parties are not yet as well-functioning. Short-term markets of Contracting Parties partially lack liquidity or are not at all operational, and a share of market segments is price regulated hindering competition to be developed (Energy Community Secretariat, 2023a). In addition, due to the energy crisis and the rise of prices for electricity in 2022, the Contracting Parties intervened in markets with emergency measures, some not yet phased out,[21] slowing down their development (Energy

[20] Commission Regulation (EU) 2015/1222 as adapted and adopted in the Energy Community by Ministerial Council Decision 2022/03/MC-EnC.

[21] According to (Energy Community Secretariat, 2023a), not yet phased out and concerning measures are, for example, Albania's non-market-based procurement of electricity for network losses

Community Secretariat, 2022a, p. 7, 2023a, p. 147). In general, public interventions in retail price setting are applied widely in the Energy Community, resulting in average end user prices significantly lower than in the European Union, almost two times for industry and three times for households (Energy Community Secretariat, 2023c, p. 14). Considering the above, short-term markets fail to provide price signals, especially important for the deployment of renewable energy sources and flexibility needed to integrate their generation and react to their intermittency. In addition, most Contracting Parties have no effective carbon pricing mechanisms in place (Energy Community Secretariat, 2023c, p. 5).

In accordance with the European Union's understanding that its future is linked to its neighbours and partners, the European Commission's Communication on the European Green Deal (European Commission, 2019) includes the following: "*The global challenges of climate change and environmental degradation require a global response. […] The EU will put emphasis on supporting its immediate neighbours. The ecological transition for Europe can only be fully effective if the EU's immediate neighbourhood also takes effective action*".

Consequently, the integration of the Contracting Parties into the European Union's internal electricity market, including into the SDAC and the SIDC, is highly important not only for the Energy Community but also for the European Union. The envisaged integration will provide market participants with access to well-functioning, liquid and competitive electricity markets and will create a stable regulatory framework for further investments in renewable energy sources across Europe. Market coupling is key for decarbonization and the successful implementation of the 2030 targets in the Energy Community[22] and, thus, will contribute to the above-mentioned ecological transition in the European Union's immediate neighbourhood. Finally, the Contracting Parties' adherence to the SDAC and the SIDC will ultimately support the creation of an integrated market leveraging synergies across Europe and provide affordable and secure (reliable) energy supply to end consumers enabling the transition to a clean energy system.

Electricity market integration and the European Union's efforts towards reducing carbon emissions are further linked by (EU, 2023/956) establishing a carbon border adjustment mechanism (CBAM), which was adopted in the European Union to address the so-called "carbon leakage" risk. According to the description of the European Union (2024), such risk occurs if polluting production is transferred to other countries with less stringent climate policies than the European Union or more carbon-intensive imports replace products in the European Union. Therefore, to create a system putting a fair price on carbon emitted during production, the CBAM imposes financial and administrative costs on importers of CBAM goods into the European Union from third countries, including the Contracting Parties. Consequently, the CBAM focuses on goods most at risk of carbon leakage such as

or the so-called equity fee imposed on all customers in Moldova that would like to switch their supplier.

[22] With Ministerial Council Decision 2022/02/MC-EnC, the Energy Community adopted the 2030 targets for greenhouse gas emissions, energy efficiency and renewable energy.

cement, iron and steel, aluminium, fertilizer, hydrogen as well as electricity. However, the CBAM Regulation (EU, 2023/956) allows for exemptions with regards to electricity if certain criteria are met. These criteria include inter alia the integration of the third country's electricity market with the European Union's (market coupling) and the implementation of an emissions trading system (Energy Community Secretariat, 2023c, p. 3).

13.2.2 Day-Ahead and Intraday Markets and Their Operators in the Energy Community

Any further integration with the European Union's market coupling requires short-term markets and their operators to be established—a step which still needs additional efforts in several Contracting Parties. At the beginning of 2024, all Contracting Parties except Bosnia and Herzegovina, Georgia[23] and Moldova have operational day-ahead markets, while only Serbia and Ukraine have established intraday markets (Energy Community Secretariat, 2024, p. 3). Three of the day-ahead markets (Albania, Montenegro, North Macedonia) and the Serbian intraday market opened only during 2023 (Energy Community Secretariat, 2023a, pp. 15, 84, 98, 112), and the day-ahead market in Kosovo* at the beginning of 2024[24] (ALPEX, 2024).

The respective operating power exchanges are the Albanian Power Exchange (ALPEX) for Albania and Kosovo*,[25] the Montenegrin Power Exchange (MEPX) for Montenegro, the national electricity market operator MEMO for North Macedonia, SEEPEX for Serbia and the joint-stock company Market Operator (JSC MO) for Ukraine (Energy Community Secretariat, 2024, p. 3). Bosnia and Herzegovina lacks the legal basis for the establishment of short-term markets (Energy Community Secretariat, 2023c, p. 12) while for Georgia, the opening of the short-term markets, to be operated by the Georgian Energy Exchange (GENEX), was postponed several times, latest to July 2024 (Energy Community Secretariat, 2023a, p. 42). Most recent information indicates that OPEM, a subsidiary of OPCOM,[26] will set up short-term markets as electricity market operator in Moldova (Gridina, 2024).

[23] Due to missing interconnection with any other Party to the EnC Treaty, Georgia is exempted from obligations under the EnC CACM in accordance with its Article 1(2). Therefore, the obligations explained in the following text only refer to the other eight Contracting Parties.

[24] The day-ahead markets were launched in April 2023 in Albania and Montenegro, in May 2023 in North Macedonia and in February 2024 in Kosovo*. The intraday market in Serbia opened in July 2023.

[25] The day-ahead markets of Albania and Kosovo* are coupled as of the start of operations in Kosovo*.

[26] OPCOM is the electricity (and gas) market operator (OPCOM, 2024) and one of the designated NEMOs (ACER, 2023) in Romania.

Aligned with the European Union's framework, the SDAC and the SIDC shall only be organized by market operators designated as NEMOs (see Chap. 2, Sect. 2.2.3), respecting the designation criteria as defined in Article 6 of the EnC CACM, which are the same as in Commission Regulation (EU) 2015/1222 (CACM Guideline). Article 4(1) of the EnC CACM obliged the Contracting Parties to designate at least one NEMO by six months after entry into force, i.e. 15 June 2023. Additionally, according to Article 5(1) of the EnC CACM, the Contracting Parties were required to notify the Energy Community Secretariat in case of a legal monopoly for day-ahead and intraday trading[27] by 15 February 2023. In such case, the competent authorities may refuse the provision of trading services by more than one NEMO in the Contracting Party and by NEMOs designated in other Member States or Contracting Parties ('passporting').

The Energy Community (2024b) publishes and regularly updates a list of the NEMOs in the Energy Community on its webpage based on the Contracting Parties' information provided. According to this list, at the beginning of 2024, only Albania (ALPEX), Kosovo* (ALPEX), North Macedonia (MEMO) and Serbia (SEEPEX) have informed about designated NEMOs while the existence of a legal monopoly for day-ahead and intraday trading services was notified for Moldova, Montenegro, North Macedonia, Serbia and Ukraine.[28] Nevertheless, a NEMO designation in a Contracting Party should only be considered equivalent to one in the European Union, if the respective Contracting Party has transposed all rights and obligations of a NEMO, defined in different legal acts of the Electricity Integration Package, into national law. Otherwise, such NEMO, even if compliantly designated based on the same criteria, would not perform its tasks on an equal footing with the NEMOs of Member States, which is particularly relevant in case of competitive NEMOs.

Figure 13.5 provides an overview of day-ahead and intraday markets and their operators in the Contracting Parties.

13.2.3 Capacity Calculation Regions and Possibilities for Cross-Zonal Trade in the Energy Community

With regards to cross-zonal capacity, the Ministerial Council Decision 2022/03/MC-EnC established three capacity calculation regions (CCRs) for the Energy Community (Fig. 13.6): Shadow South-East Europe (Shadow SEE), Italy-Montenegro (ITME) and Eastern Europe (EE). The EnC CCRs are defined by Annex I to the EnC CACM upon entry into force and include both bidding zone

[27] According to Article 5(2) of the EnC CACM, a national legal monopoly is deemed to exist where national law expressly provides that no more than one entity can carry out day-ahead and intraday trading services.

[28] According to the list, all Contracting Parties which have notified about a monopoly status missed the legal deadline.

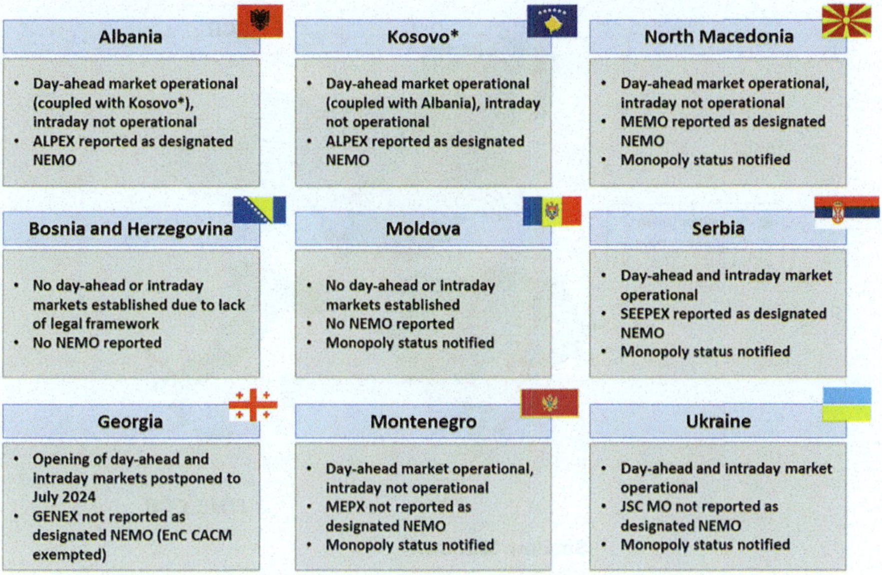

Fig. 13.5 Day-ahead and intraday markets and their operators in the Energy Community at the beginning of 2024. *Source* Adapted from (Energy Community Secretariat, 2024, p. 3) with flags reproduced from (Central Intelligence Agency, 2021)

borders between Contracting Parties, and bidding zone borders between Contracting Parties and Member States (Energy Community, 2023, pp. 3–4; Energy Community Secretariat, 2023a, p. 10).

According to Articles 3 and 5 of Annex I to the EnC CACM, the TSOs of the Shadow SEE and the EE CCR shall conclude cooperation agreements as a basis for the cooperation between TSOs of Contracting Parties and TSOs of Member States by 15 June 2023. At the same time, TSOs of the EnC CCRs should have submitted coordinated capacity calculation methodologies pursuant to Article 20(2) of the EnC CACM. However, as reported by the (Energy Community Secretariat, 2023a, pp. 10–11), discussions on a possible reconfiguration of the EnC CCRs are delaying their operationalization and neither cooperation agreements were concluded, nor capacity calculation methodologies submitted.

As in the European Union, the determination of CCRs can be amended. Importantly, amendments to the EnC CCRs are subject to an amendment proposal of all TSOs of the European Union pursuant to Article 15 of the CACM Guideline, defining the pan-EU CCR determination (see Chap. 3, Sect. 3.1), in consultation with TSOs of the Contracting Parties and decided upon by ACER as prescribed by Article 1(2) of Annex I to the EnC CACM.

Cross-border trade and effective market coupling further depends on available cross-zonal capacities as recognized by the applicable legal framework. Article 16 of Regulation (EU) 2019/943 (Electricity Regulation) requires certain amounts of cross-zonal capacity to be made available to market participants, commonly

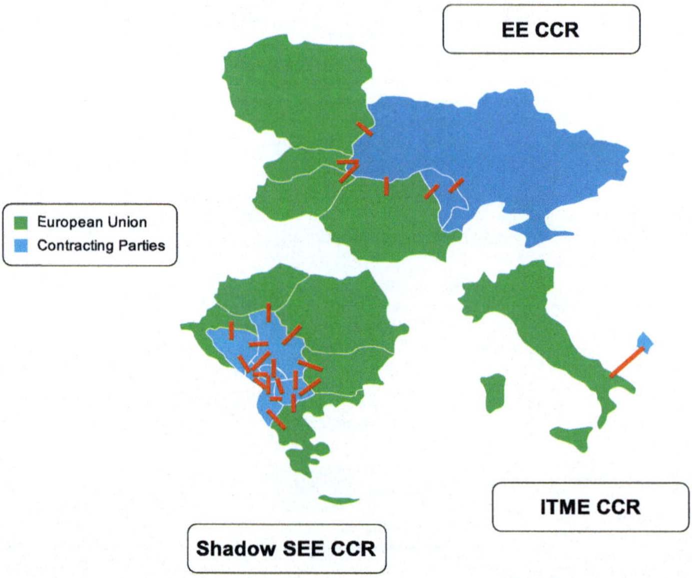

Fig. 13.6 Capacity calculation regions (CCRs) in the Energy Community. *Source* Adapted from ENTSO-E (2023a, p. 3)

referred to as the 70% target (see Chap. 3, Sect. 3.1.7). With the adoption of the adapted Electricity Regulation in the Energy Community,[29] the Contracting Parties obliged themselves to make available the same minimum levels of cross-zonal capacity. The results of a study, contracted by the Energy Community Secretariat in 2023, indicate that none of the Western Balkan 6 Contracting Parties[30] currently fulfils the 70% target and that the margin available for cross-zonal trade is low for a considerable amount of network elements (EIHP, 2023). Furthermore, the study identifies the lack of coordinated capacity calculation as one of the main reasons and concerns to be addressed and stresses the need to adopt action plans in case of structural congestions to achieve the 70% target by the end of 2027.[31]

[29] Regulation (EU) 2019/943 as adapted and adopted in the Energy Community by Ministerial Council Decision 2022/03/MC-EnC.

[30] The analysis for Ukraine and Moldova was simplified and focused only on the future time-frame (2028), while Georgia, due to the missing interconnections with other Contracting Parties or Member States, was not covered by the study.

[31] According to Article 15(2) of Regulation (EU) 2019/943 as adapted and adopted in the Energy Community by Ministerial Council Decision 2022/03/MC-EnC, Contracting Parties applying an action plan shall ensure that the minimum capacity as defined in Article 16(8) is reached by 31 December 2027.

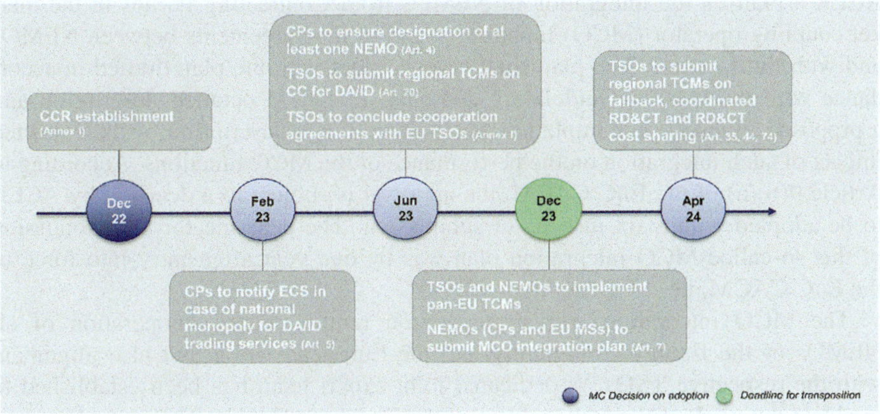

Fig. 13.7 Most important implementation milestones and their respective deadlines according to the EnC CACM.[32] *Source* Energy Community Secretariat (2023b, p. 12)

13.2.4 Implementation and Timeline of the Integration of the Contracting Parties into the SDAC and the SIDC

Figure 13.7 provides an overview of the most important implementation milestones and their respective deadlines according to the EnC CACM. This includes the designation of NEMOs and the first tasks within EnC CCRs until June 2023, as outlined in Sects. 13.2.2 and 13.2.3, as well as the requirement to implement the majority of the pan-EU TCMs by the end of 2023. As described in Chap. 2, Sect. 2.4, these pan-EU TCMs link, for example, to the common grid model with the respective data provision, the intraday cross-zonal gate opening and closure times, the maximum and minimum technical price limits, or the products to be used in the SDAC and the SIDC. Further regional TCMs are to be developed by the TSOs in the EnC CCRs and submitted by April 2024 (coordinated redispatching and countertrading and related cost sharing, fallback). These steps are to be completed to ensure the implementation of the EnC CACM and with that the harmonization required for the integration into the SDAC and the SIDC, whose successful operation is based on common principles and rules.

Most importantly, Article 7(3) of the EnC CACM requires the NEMOs from Contracting Parties and Member States to submit to all NRAs, the ECRB and

[32] In the figure, the articles refer to the EnC CACM and the abbreviations to the following: capacity calculation (CC), Contracting Party (CP), day-ahead (DA), Energy Community Secretariat (ECS), intraday (ID), Ministerial Council (MC), redispatching & countertrading (RD&CT).

ACER a plan on the integration of NEMOs from Contracting Parties in the market coupling operator (MCO) functions[33] and in the agreements between NEMOs and with third parties. The plan shall be consistent with the plan drafted in accordance with the CACM Guideline[34] and shall include a detailed description and a proposed timescale for implementation as well as a description of the expected impact of such integration on the performance of the MCO functions. According to Article 9(6)(a) of the EnC CACM, the approval is subject to a decision by ACER to be adopted within six months of submission. The deadline for the submission of this so-called MCO integration plan was by one year after entry into force of the EnC CACM, i.e. 15 December 2023.

The MCO integration plan's preparation requires close cooperation of all NEMOs of the Energy Community and the European Union but also alignment with the respective TSOs. A dedicated joint expert team has been established to facilitate the work of the stakeholders of the European Union and Energy Community to accomplish this challenging task (Energy Community Secretariat, 2023a, p. 148). Nevertheless, upon a joint message shared by the European Commission, the Energy Community Secretariat and ACER, the submission of the MCO integration plan was postponed with a newly envisaged submission date of 15 June 2024 (Energy Community Secretariat, 2024, p. 4; NEMO Committee, 2023). The postponement was recommended due to the delayed transposition in the Contracting Parties and is linked to the consideration that a NEMO designation is only valid if the respective Contracting Party has compliantly transposed the legal requirements for the SDAC and the SIDC, including the exhaustive definition of rights and obligations of NEMOs (Energy Community Secretariat, 2024, p. 4).

The final approved MCO integration plan should outline the process for the adherence of NEMOs of Contracting Parties to relevant contracts and required steps for the extension of the SDAC and the SIDC to the Contracting Parties in the upcoming years. This will serve as the basis for the inclusion of bidding zone borders into the SDAC and the SIDC, grouped in so-called local implementation projects (LIPs), as was the case for already completed extension projects in the European Union.

In practical terms, the adherence of the Contracting Parties to the SDAC and the SIDC requires, as for all extension projects, different steps to be completed.[35] Besides the required transposition of legal acts related to market coupling, the NEMOs and TSOs of the Contracting Parties need to adhere to the contractual framework in place. This includes contracts between NEMOs, between TSOs,

[33] According to Article 2(30) of the EnC CACM and the CACM Guideline, "market coupling operator (MCO) function" means "the task of matching orders from the day-ahead and intraday markets for different bidding zones and simultaneously allocating cross-zonal capacities".

[34] Article 7(3) of the EnC CACM refers to Article 7(3) of the CACM Guideline according to which the NEMOs in the Member States had to submit the so-called MCO plan. The MCO plan was approved by all EU NRAs in 2017 and not amended since then.

[35] It is not necessary to join both the SDAC and the SIDC at the same time. Instead, it can be done one after the other, for example, starting with day-ahead and later joining the intraday project.

joint NEMO-TSO contracts as well as contracts with third party service providers as further detailed in Chap. 14. Even prior to this, all parties involved need to sign the Single Day-Ahead and Intraday Coupling Observership and Non-Disclosure Agreement (CACM Global NDA, cf. Section 14.2.1) to protect the exchange of confidential information of NEMOs and TSOs when fulfilling their obligations. This preparatory step has not yet been completed for all Contracting Parties (ENTSO-E, 2023b, pp. 33–35).

From a technical perspective, the implementation of the EnC CACM and the different LIPs' readiness has to be ensured. This requires intensive regional cooperation and joint work streams within the LIPs dedicated to design and develop inter alia IT solutions and operational processes, including for pre- and post-coupling[36] activities (see Chap. 2, Sect. 2.3.1). Resulting requests for change towards the common pan-EU systems are to be submitted well in advance. In this regard, the NEMOs and TSOs of the Member States announced further assessments of the requirements and possible timelines for such requests for change and proposed a first SDAC go-live window for the Energy Community at the end of 2025 (MCSC, 2024, pp. 2, 4). Finally, as also described in Chap. 14, Sect. 14.4, the actual go-live is to be approved by the Market Coupling Steering Committee (MCSC) after successfully completed testing phases and involved parties having complied with all legal and technical requirements.[37]

13.3 The Energy Community's Electricity Markets and Their Characteristics

Five Contracting Parties are relatively small with electricity consumption lower than 10 TWh per year, while Bosnia and Herzegovina and Georgia range around 13 TWh (IEA, 2023). Serbia, with approximately 35 TWh per year (IEA, 2023), is roughly double in size, and Ukraine, by far the largest market, had around 150 TWh of electricity consumption in 2021 before the war of aggression of the Russian Federation against Ukraine started[38] (Energy Community Secretariat, 2023c, p. 6). In general, electricity consumption in the Energy Community increased by approximately 7% between 2020 and 2021 (Energy Community Secretariat, 2022b, p. 5) and decreased slightly for 2022 (Energy Community Secretariat, 2023c, p. 7).

[36] The clearing and settlement (post-coupling), as described in Chap. 9, poses challenges beyond the required harmonization of the electricity sector, for example, in case of non-alignment with the European Union's Value-Added-Tax (VAT) legislation.

[37] Even though the MCO integration plan is a NEMO-only TCM, the roadmaps for further developments of the SDAC/SIDC and decisions on go-live are approved by the MCSC, including NEMOs and TSOs, both being tasked with the day-to-day management of the SDAC/SIDC according to Article 10 of the CACM Guideline.

[38] Due to the ongoing war, the electricity consumption in Ukraine dropped significantly in 2022.

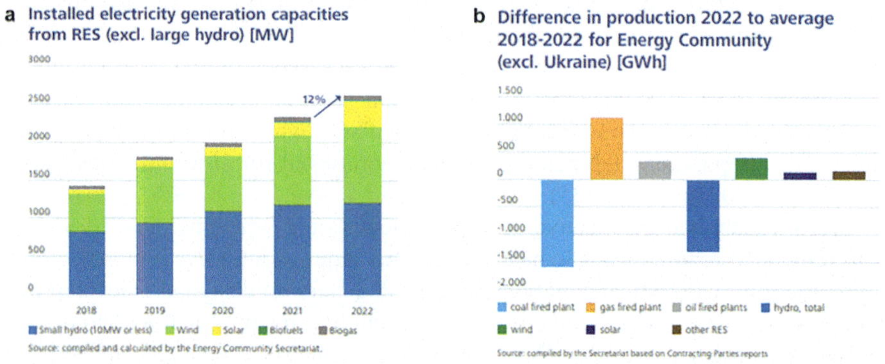

Fig. 13.8 **a** Installed electricity generation capacities from renewable energy sources (excluding large hydro) [MW], **b** Difference in production 2022 to average 2018–2022 for Energy Community (excl. Ukraine) [GWh]. *Source* Energy Community Secretariat (2023c, p. 7).

13.3.1 Generation Mix in the Energy Community

With regards to the generation mix in the Energy Community, as presented by the Energy Community Secretariat (2023c, p. 7), the following can be observed. The installed capacities in coal-fired plants remained unchanged with their share in total installed capacity decreasing slowly to approximately 35% in 2022, while coal-based electricity production covered close to 50% of total production in the same year.[39] However, the generation capacities of renewable energy sources (RES) were steadily growing over the last years as indicated in Fig. 13.8a (left). Solar, wind and biofuels increased from a share of 1% in 2017 to 5% of total installed capacities in 2022. Despite the stable electricity production from fossil-fuelled thermal power plants, the difference in production in 2022 compared to a five-year average (2018–2022), as presented in Fig. 13.8b (right), shows significant decline of electricity generation from coal, whereas production from natural gas, fuel oil and non-hydro renewable sources increased. Furthermore, generation is sensitive to hydrology due to the considerable amount of hydropower plants in the Contracting Parties, one of the reasons for a five-year record production in the Energy Community in 2021, followed by a five-year minimum in 2022.[40]

13.3.2 Day-Ahead Market Prices in the Energy Community

As described in Sect. 13.2.2, only Serbia (day-ahead) and Ukraine (day-ahead and intraday) had operational short-term electricity markets throughout 2022 and 2023. Albania, Montenegro and North Macedonia launched day-ahead markets only in

[39] The figures are for eight Contracting Parties only, excluding Ukraine.

[40] In 2022, the electricity production in Ukraine and Moldova fell due to the ongoing war.

spring 2023, and, in July 2023, SEEPEX's intraday market went live. In the first month of operation, the volumes at ALPEX, MEPX and MEMO amounted to around 20%, 17% and 8% of the respective Contracting Parties' total consumption (Energy Community Secretariat, 2023c, p. 12).

Based on transparency data from (ENTSO-E, 2024), the day-ahead market prices in Serbia (SEEPEX) mostly coincided with Croatia (CROPEX) and Hungary (HUPX) in 2023 with a correlation of around 90%, which decreased slightly compared to 2022. Importantly, the Serbian market rules are not yet aligned with the European Union's framework as prices both for day-ahead and intraday are downwards limited with zero (SEEPEX, 2023, pp. 31, 33), where on the contrary, markets in the European Union also reach negative prices. The same[41] is true for the already established short-term markets in Montenegro (MEPX, 2023, p. 1), North Macedonia (MEMO, 2023, p. 21) and Ukraine[42] (NEURC, 2023).

Figure 13.9 shows the average monthly day-ahead market prices based on the available data[43] from (ENTSO-E, 2024) for Montenegro, North Macedonia, Serbia and, additionally, for Croatia and Hungary. The prices follow a common trend—after 2022, being a year of high and fluctuating prices during the energy crisis, the monthly average day-ahead prices in all observed markets range around 100 EUR/MWh since spring 2023.

13.4 Outlook

The Contracting Parties' electricity markets have very different levels of development. For some, basic building blocks are missing[44] while others already started working towards the implementation of Commission Regulation (EU) 2015/1222 already before its official adoption in the Energy Community (Energy Community Secretariat, 2022b, p. 12). Currently, the Contracting Parties are in the process of transposing and implementing the entire Electricity Integration Package "at once" with ambitious deadlines. However, at the end of February 2024, the legal deadline has passed and none of the Contracting Parties has transposed the new legal framework into national law (Energy Community Secretariat, 2024, p. 2). Even if preparations can start in parallel, delayed transposition in the Contracting Parties is a major risk to the extension of the SDAC and the SIDC. Therefore, first and foremost, the Contracting Parties need to speed up the compliant transposition of

[41] The current minimum price limit in North Macedonia and Ukraine is not exactly zero but less than 0,5 EUR/MWh (0,1 MKD and 10 UAH respectively).

[42] In Albania, the permitted minimum price is −500€/MWh (ERE, 2023, Annex 2).

[43] For Montenegro and North Macedonia data is included as of the first full month of operation being May and June 2023 respectively. For Albania and Ukraine, no data was available.

[44] As reported by the Energy Community Secretariat (2023a), the Contracting Parties have not yet finalized the transposition and implementation of the Third Energy Package (e.g. some do not comply with unbundling provisions).

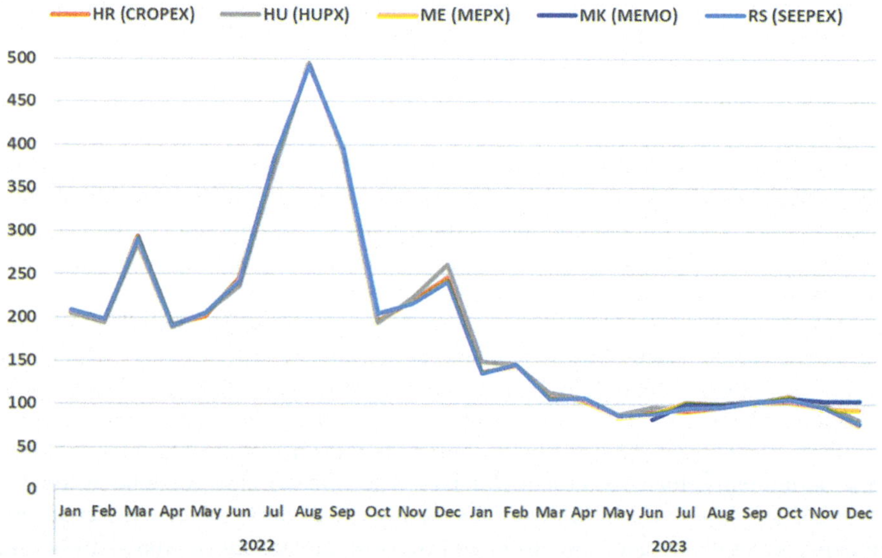

Fig. 13.9 Monthly average day-ahead market prices in Croatia (HR, CROPEX), Hungary (HU, HUPX), Montenegro (ME, MEPX), North Macedonia (MK, MEMO) and Serbia (RS, SEEPEX) for 2022 and 2023 in EUR/MWh. *Source* Author's own illustration

the Electricity Integration Package and increase their efforts to proceed with the required reforms of their national electricity markets.

As the projects in the European Union have shown, regional cooperation among key stakeholders is essential. This aspect of the Electricity Integration Package needs to be strengthened in the Energy Community. Regional governance structures to cope with the newly assigned tasks must be set up urgently, even more so considering that more than one year has passed without notable progress in this regard. It is recommended to use the structures established in the European Union as blueprints to avoid lengthy discussions and further delays jeopardizing the implementation of the EnC CACM.

Furthermore, the Electricity Integration Package for the first time expanded the concept of TCMs to the Energy Community, a complex system which is based on direct applicability of ACER's individual decisions on pan-EU, and, if applicable, regional TCMs in the European Union. When transposing, the Contracting Parties are demanded to mirror this concept and ensure that relevant ACER decisions are applicable as part of their national legislation without time lags. In addition, the ECRB is newly mandated to take decisions on regional TCMs in case of non-agreement or referral of Contracting Parties' NRAs (without Member States being affected). Different from ACER, a permanent Agency of the European Union, the ECRB has no staff itself but is composed of representatives of the NRAs, supported by the Energy Community Secretariat, and may therefore face challenges when implementing the Electricity Integration Package. The Contracting

Parties and respective NRAs need to acknowledge that additional human resources and new expertise are required due to the new competences and address this by respective measures. As mentioned above, the Contracting Parties and in particular the TSOs and the NRAs need to increase their cooperation to develop, submit or approve regional TCMs such as the capacity calculation methodologies. Importantly, this partly involves the neighbouring Member States and their stakeholders which are expected to take a lead based on their already acquired knowledge and experience.

With regards to one of the main prerequisites for market coupling—available cross-zonal capacity—it is recommended to accelerate the EnC CCRs' operationalisation and to introduce a coordinated approach for calculating cross-zonal capacities for all timeframes without further delay. This requires the TSOs of the Contracting Parties and the neighbouring Member States to submit joint proposals to their respective NRAs and to improve cooperation to allow for a better usage of existing infrastructure. Additionally, the Contracting Parties are obliged to tackle (structural) congestion to increase cross-zonal capacities and, hence, ensure sufficient possibilities for trade. In this regard, the Contracting Parties should follow up on the study conducted by the Energy Community Secretariat and perform detailed assessments to conclude on their status quo and, if needed, define respective measures to achieve the 70% target.

On the other hand, the stakeholders of the European Union can support this process by sharing knowledge and are required to prepare for the Contracting Parties' adherence in their planning and processes based on experience gained with already completed extension projects. In this context, the submission of the MCO integration plan by 15 December 2023 should have been a first important milestone and a key tool to ensure predictability and joint commitment as a basis for both sides' planning and development. Unfortunately, due to the delayed transposition, which resulted in a lack of alignment of the Contracting Parties' national legislation with the legal framework for market coupling, the MCO integration plan has not yet been submitted. Consequently, not only the transposition of relevant legal acts related to the SDAC and the SIDC but also the designation of at least one NEMO in each Contracting Party has to be completed urgently.

With regards to market transparency and integrity, the differences in REMIT principles in the European Union and the Energy Community, as described in Sect. 13.1.3, should be addressed by means of an amendment to the EnC REMIT to allow for centralized data collection and surveillance by ACER for the entire future scope of the SDAC and the SIDC. In addition, due to the electricity market design reform in the European Union, also addressing market integrity and transparency, any amendments of the EnC REMIT should be aligned with the to be adopted recast of the EU REMIT. It is further worth noting that the Energy Community *acquis* is sector specific and, hence, does not cover tax legislation. This will require additional effort in some of the Contracting Parties to address the challenges that arise for post-coupling processes, for example, due to non-aligned VAT rules.

The experience in the European Union shows that the implementation of the legal framework and to achieve technical readiness for both NEMOs and TSOs on a national and regional level is challenging even where markets and associated operators are already established. However, the Contracting Parties can build on experience, already existing solutions and partly on established governance structures in the European Union. Nevertheless, the implementation of this extension project is dependent on both sides' commitment and effort. Thus, for the upcoming years, it will be crucial that all involved parties proceed fully coordinated to successfully enlarge the SDAC and the SIDC across Europe.

Bibliography

ACER. (2023) *NEMO list*. Retrieved February 2024, from https://www.acer.europa.eu/sites/default/files/documents/en/Electricity/MARKET-CODES/CAPACITY-ALLOCATION-AND-CONGESTION-MANAGEMENT/Documents/NEMO_list.pdf

ALPEX. (2024). *Market notice*. Retrieved February 06, 2024, from https://alpex.al/wp-content/uploads/2024/01/The-Gate-Opening-for-the-Day-Ahead-Market-of-Kosovo-Bidding-Zone.pdf

Central Intelligence Agency. (2021). *The World Factbook 2021*. Retrieved February 06, 2024, from https://www.cia.gov/the-world-factbook/

Commission Regulation (EU) 2015/1222 of 24 July 2015 establishing a guideline on capacity allocation and congestion management, OJ L 197, 25.7.2015, pp. 24–72. Retrieved February 06, 2024, from https://eur-lex.europa.eu/legal-content/EN/TXT/?uri=CELEX%3A32015R1222

Commission Regulation (EU) 2016/1719 of 26 September 2016 establishing a guideline on forward capacity allocation, OJ L 259, 27.9.2016, pp. 42–68. Retrieved February 06, 2024, from https://eur-lex.europa.eu/legal-content/EN/TXT/?uri=uriserv%3AOJ.L_.2016.259.01.0042.01.ENG

Commission Regulation (EU) 2017/1485 of 2 August 2017 establishing a guideline on electricity transmission system operation, OJ L 220, 25.8.2017, pp. 1–120. Retrieved February 06, 2024, from https://eur-lex.europa.eu/legal-content/EN/TXT/?uri=CELEX%3A32017R1485

Commission Regulation (EU) 2017/2195 of 23 November 2017 establishing a guideline on electricity balancing, OJ L 312, 28.11.2017, pp. 6–53. Retrieved February 06, 2024, from https://eur-lex.europa.eu/eli/reg/2017/2195/oj

Commission Regulation (EU) 2017/2196 of 24 November 2017 establishing a network code on electricity emergency and restoration, OJ L 312, 28.11.2017, pp. 54–85. Retrieved February 06, 2024, from https://eur-lex.europa.eu/eli/reg/2017/2196/oj

Directive (EU) 2019/944 of the European Parliament and of the Council of 5 June 2019 on common rules for the internal market for electricity and amending Directive 2012/27/EU (recast), OJ L 158, 14.6.2019, pp. 125–199. Retrieved February 06, 2024, from https://eur-lex.europa.eu/legal-content/EN/TXT/?uri=CELEX:32019L0944

ECRB. (2022). *ECRB guidance on REMIT regulation*. Retrieved February 06, 2024, from https://www.energy-community.org/dam/jcr:04da54ae-178c-4924-98f4-ba3ea363668e/ECRB_REMIT_guidance_0422.pdf

EIHP. (2023). *Final report on the study on 70% target for electricity interconnection capacities to be made available to market participants*. Retrieved February 06, 2024, from https://www.energy-community.org/dam/jcr:66eda2aa-67cc-4e05-bf45-6e4b6a274e2a/EIHP_70target_study_102023.pdf

Energy Community. (2023). *Conclusions of the 28th energy community electricity forum*. Retrieved February 06, 2024, from https://www.energy-community.org/dam/jcr:918e954a-084b-484d-bc7c-ff729f2bf8cd/AF_2023_Conclusions%20Final_clean.pdf

Energy Community. (2024a). *Energy community acquis*. Retrieved February 06, 2024, from https://www.energy-community.org/legal/acquis.html

Energy Community. (2024b). *The electricity integration package*. Retrieved February 06, 2024, from https://www.energy-community.org/implementation/package/EL.html

Energy Community. (2024c). *Who we are*. Retrieved February 06, 2024, from https://www.energy-community.org/aboutus/whoweare.html

Energy Community Secretariat. (2022a). *Annual implementation report*. Retrieved February 06, 2024, from https://www.energy-community.org/dam/jcr:3706068e-9a3d-47d5-ae22-f7af0bbd1097/EnC_IR2022.pdf

Energy Community Secretariat. (2022b). *Energy transition tracker*. Retrieved February 06, 2024, from https://www.energy-community.org/dam/jcr:a09255dc-ac8a-47b1-b664-3463705906de/EnC_Tracker_07_2022.pdf

Energy Community Secretariat. (2023a). *Annual implementation report 2023*. Retrieved February 06, 2024, from https://www.energy-community.org/dam/jcr:3da7c4f8-ea23-4169-b1e9-66b0ed05fcb7/EnC_IR2023.pdf

Energy Community Secretariat. (2023b). *CACM and its implementation in the EnC*. Retrieved February 06, 2024, from https://www.energy-community.org/dam/jcr:98b603c6-7326-4c63-98c2-bd644e53e50c/20230516%20WS%20on%20MC-Session1-EnCS.pdf

Energy Community Secretariat. (2023c). *CBAM-readiness tracker*. Retrieved February 06, 2024, from https://www.energy-community.org/dam/jcr:d6e80d5e-9290-4e8b-ac7e-5170ec59808a/EnC%20Tracker%2006_2023_final.pdf

Energy Community Secretariat. (2024). *The electricity integration package in the energy community—Update*. Retrieved February 28, 2024, from https://eepublicdownloads.blob.core.windows.net/public-cdn-container/clean-documents/Network%20codes%20documents/MESC/2024/240228_MESC_TOP_2.2_EnCS_Update.pdf

ENTSO-E. (2023a). ENTSO-E's EnC Task Force—Update. Retrieved February 06, 2024, from https://www.energy-community.org/dam/jcr:385745bd-7b0a-4aa1-985e-3501a133986f/230706_Session%205_Status%20quo%20in%20EnC%20CPs_EE%20EnC%20TF.pdf

ENTSO-E. (2023b). *Market report 2023*. Retrieved February 06, 2024, from https://eepublicdownloads.entsoe.eu/clean-documents/nc-tasks/ENTSO-E_Market_Report_2023.pdf

ENTSO-E. (2024). *Transparency platform*. Retrieved February 06, 2024, from https://transparency.entsoe.eu/dashboard/show

ERE. (2023). *Board Decision No. 106, dated 23.03.2023, on approving some additions in the electricity market rules approved with ERE Board Decision No. 347, dated 27.12.2022*. Retrieved February 06, 2024, from https://alpex.al/wp-content/uploads/2023/10/Decision_No_106_2023_On_approving_some_additions_in_the_electricity_market_rules_approved_with_ERE_Board_decision_no.347_dated_27.12.2022.pdf

European Commission. (2019). *Communication from the Commission COM(2019) 640 on The European Green Deal*. Retrieved February 06, 2024, from https://eur-lex.europa.eu/legal-content/EN/TXT/HTML/?uri=CELEX:52019DC0640

European Union. (2024). *Carbon border adjustment mechanism*. Retrieved February 06, 2024, from https://taxation-customs.ec.europa.eu/carbon-border-adjustment-mechanism_en

Gridina, M. (2024) *Moldovan government appointed a new electricity market operator*. Retrieved February 28, 2024, from https://moldovalive.md/moldovan-government-appointed-a-new-electricity-market-operator/

IEA. (2023). *Energy statistics data browser*. Retrieved February 06, 2024, from https://www.iea.org/data-and-statistics/data-tools/energy-statistics-data-browser

Karova, R. (2015). The application of the EU constitutional principles in the energy community. *European Networks Law & Regulation Quarterly, 3*(3), 176–186.

Karova, R. (2015b). The dispute settlement system of the energy community: Testing its effectiveness. In D. Buschle & K. Talus (Eds.), *The energy community: A new energy governance system* (pp. 25–67), Intersentia.

Karova, R. (2019). Energy community in its second decade: Where does it stand and where is it heading. In L. De Almeida, M. Cantero Gamito, M. Djurovic, & K. Purnhagen (Eds.), *The transformation of economic law—Essays in Honour of Hans Micklitz* (pp. 271–294). Hart Publishing.

MCSC. (2024). *MESC 3.1 Update on SDAC & SIDC*. Retrieved February 28, 2024, from https://eepublicdownloads.blob.core.windows.net/public-cdn-container/clean-documents/Network%20codes%20documents/MESC/2024/240228_MESC_TOP_3.3_Update_on_SDAC___SIDC.pdf

MEMO. (2023). *MEMO—Operational rules*. Retrieved February 06, 2024, from https://dam.memo.mk/wp-content/uploads/2023/04/2_Operational-Rules-_ENG.pdf

MEPX. (2023). *Product specification*. Retrieved February 06, 2024, from https://belen-spot.me/wp-content/uploads/2023/03/II.3.-Product-Specification.pdf

Ministerial Council Decision 2018/10/MC-EnC: implementing Regulation (EU) No 1227/2011 of the European Parliament and of the Council on wholesale energy market integrity and transparency. Retrieved February 06, 2024, from https://www.energy-community.org/dam/jcr:aec217c0-51fa-426a-a986-f36455546f51/Decision_2018_10_MC-EnC_Reg_1227-2011.pdf

Ministerial Council Decision 2021/13/MC-EnC: amending Annex I to the Treaty Establishing the Energy Community and incorporating Directive (EU) 2019/944 and Regulation (EU) 2019/941 in the Energy Community acquis communautaire. Retrieved February 06, 2024, from https://www.energy-community.org/dam/jcr:3304cadf-c63b-433f-9636-79d9ec63b186/Decision_2021-13-MC-EnC.pdf

Ministerial Council Decision 2022/02/MC-EnC: on amending Ministerial Council Decision No 2021/14/MC- EnC amending Annex I to the Treaty Establishing the Energy Community and incorporating Directive (EU) 2018/2001, Directive (EU) 2018/2002, Regulation (EU) 2018/1999, Delegated Regulation (EU) 2020/1044, and Implementing Regulation (EU) 2020/1208 in the Energy Community acquis communautaire. Retrieved February 06, 2024, from https://www.energy-community.org/dam/jcr:421f0dca-1b16-4bb5-af86-067bc35fe073/Decision_02-2022-MC_CEP_2030targets_15122022.pdf

Ministerial Council Decision 2022/03/MC-EnC: Decision on the incorporation of Regulation (EU) 2019/942, Regulation (EU) 2019/943, Regulation (EU) 2015/1222, Regulation (EU) 2016/1719, Regulation (EU) 2017/2195, Regulation (EU) 2017/2196, Regulation (EU) 2017/1485 in the Energy Community acquis, amending Annex I of the Energy Community Treaty, and on the amendments of the Ministerial Council Decisions 2021/13/MC-EnC and 2011/02/MC-EnC. Retrieved February 06, 2024, from https://www.energy-community.org/dam/jcr:d5a1a894-88db-4326-818b-f2c648bd237e/Decision03-2022-MC_newELacquis_15-12-2022.pdf

Ministerial Council Procedural Act 2015/05/MC-EnC: on the establishment of an Energy Community Parliamentary Plenum meetings. Retrieved February 06, 2024, from https://www.energy-community.org/dam/jcr:db8a2dff-5757-4dd4-8ccd-a4b71803747b/PA_2015_05_MC_parliamentary.pdf

Ministerial Council Procedural Act 2022/PA/01/MC-EnC: on Regional Market Integration. Retrieved February 06, 2024, from https://www.energy-community.org/dam/jcr:2e32c79c-7a00-4f85-8481-e377218a7bb2/MCPA202201_%20REM_15-12-2022.pdf

NEMO Committee. (2023). *The All NEMO Committee Decisions*. Retrieved February 06, 2024, from https://www.nemo-committee.eu/assets/files/nemo-committee-meeting-minutes-summary-q4-2023.pdf

NEURC. (2023). *Resolution No.2099*. Retrieved February 06, 2024, from https://www.nerc.gov.ua/acts/pro-granichni-cini-na-rinku-na-dobu-napered-vnutrishnodobovomu-rinku-ta-balansuyuchomu-rinku

OPCOM. (2024). *Home*. Retrieved February 06, 2024, from https://www.opcom.ro/acasa/en

Regulation (EU) No 1227/2011 of the European Parliament and of the Council of 25 October 2011 on wholesale energy market integrity and transparency, OJ L 326, 8.12.2011, pp. 1–16. Retrieved February 06, 2024, from https://eur-lex.europa.eu/legal-content/EN/TXT/?uri=CELEX%3A32011R1227

Regulation (EU) 2019/941 of the European Parliament and of the Council of 5 June 2019 on risk-preparedness in the electricity sector and repealing Directive 2005/89/EC, OJ L 158, 14.6.2019, p. 1–21. Retrieved February 06, 2024, from https://eur-lex.europa.eu/legal-content/EN/TXT/?uri=CELEX:32019R0941

Regulation (EU) 2019/942 of the European Parliament and of the Council of 5 June 2019 establishing a European Union Agency for the Cooperation of Energy Regulators (recast), OJ L 158, 14.6.2019, pp. 22–53. Retrieved February 06, 2024, from https://eur-lex.europa.eu/legal-content/EN/TXT/?uri=celex%3A32019R0942

Regulation (EU) 2019/943 of the European Parliament and of the Council of 5 June 2019 on the internal market for electricity (recast), OJ L 158, 14.6.2019, pp. 22–53. Retrieved February 06, 2024, from https://eur-lex.europa.eu/legal-content/EN/TXT/?uri=celex%3A32019R0943

Regulation (EU) 2023/956 of the European Parliament and of the Council of 10 May 2023 establishing a carbon border adjustment mechanism, OJ L 130, 16.5.2023, pp. 52–104. Retrieved February 06, 2024, from https://eur-lex.europa.eu/eli/reg/2023/956/oj

SEEPEX. (2023). *Operational rules*. Retrieved February 06, 2024, from https://seepex-spot.rs/wp-content/uploads/2023/07/OPERATIONAL-RULES.pdf

Treaty establishing the Energy Community. (2005). Retrieved February 06, 2024, from https://www.energy-community.org/legal/treaty.html

Lisa-Marie Mohr is a Managing Consultant at Magnus Energy. She holds a master's degree in Technical Mathematics from the Vienna University of Technology. Previously, as an Electricity Market Expert at the Energy Community Secretariat in Vienna, she focused on the assistance to Contracting Parties during the transposition and implementation of the Energy Community *acquis* related to electricity. Before joining the Energy Community Secretariat, she worked as a Technical Expert for the German regulatory authority Bundesnetzagentur, where she was tasked with projects related to electricity markets and cross-border issues, in particular the implementation of the CACM and Electricity Regulation. During a secondment to ACER, she contributed to ACER's Recommendation on reasoned amendments to the CACM Regulation. Having acquired both experience in the European Union and the Energy Community, she is dedicated to facilitating the further integration of the European internal market for electricity.

Part V

Legal Framework of Market Coupling

The Contractual Framework from a TSO Point of View

14

Vicky Blana

Abstract

This chapter introduces procedures and tasks associated with the CACM Guideline and the day-to-day management of the Single Day-Ahead and Intra-Day Coupling. The chapter also highlights that well-structured collaboration across stakeholders is so essential to the success of the European internal energy market that a contractual framework for cooperation among NEMOs and TSOs on a pan-European level has been created. This market coupling contractual framework is described in detail, whereas other elements such as the TSO-only contracts are described at a higher level. Moreover, cost sharing for Nominated Electricity Market Operators and TSOs is described. This chapter closes with some illustrative examples and a recommended approach for TSOs willing to join the contractual structure.

14.1 Introduction

The operation of the Single Day-Ahead Coupling (referred to as the "SDAC") and the Single Intraday Coupling (referred to as the "SIDC"), as regulated under Articles 38 et seq. and 51 et seq. respectively of Commission Regulation (EU) 2015/1222 of 24 July 2015 establishing a guideline on capacity allocation and congestion management (referred to as the "CACM Guideline") (EU, 2015), comprises of several layers of complex procedures and tasks, applying several pertinent methodologies. Each of these procedures and tasks should be performed by different subsets of entities, including NEMOs, TSOs, and as the case may be, third

V. Blana (✉)
Independent Transmission Operator SA (IPTO), Athens, Greece
e-mail: v.blana@admie.gr

A. Estermann et al. (eds.), *European Electricity Market Coupling*, Future of Energy,
https://doi.org/10.1007/978-3-031-86315-8_14

parties (e.g. JAO) in a coordinated and precise manner, within specific predetermined time windows. Subsequently, it is crucial that the cooperation among all involved parties is well-structured and organized in a robust manner, safeguarding that each one of them is abiding by the agreed procedures and is performing their respective duties and responsibilities diligently, to ensure effective and efficient operation of the market, and subsequently delivery of market results for the benefit of all European consumers.

In this context, Article 10 of the CACM Guideline addresses the day-to-day management of the single day-ahead and intraday market coupling, by providing that "TSOs and NEMOs shall jointly organize the day-to-day management of the single day-ahead and intraday coupling. They shall meet regularly to discuss and decide on day-to-day operational issues. TSOs and NEMOs shall invite the Agency and the Commission as observers to these meetings and shall publish summary minutes of the meetings". Article 10 of the CACM Guideline is reflected in the contractual framework for cooperation among NEMOs and TSOs on a pan-European level. This framework is also enhanced by NEMO and TSO cooperation agreements and arrangements on a regional, as well as on a local level. More illustratively, the market coupling contractual framework could be described as a pyramid, comprised of several layers:

(a) On the pan-European level, the cooperation is organized under two Europe-wide agreements, one for the SDAC and one for the SIDC, where all NEMOs and TSOs participating in the SDAC and SIDC activities are parties, namely, for the day-ahead timeframe, the DAOA (ENTSO-E, 2020a, 2020b, 2020c, 2020d), and for the intraday timeframe, the IDOA (ENTSO-E, 2020a, 2020b, 2020c, 2020d). In parallel, pan-European cooperation among TSOs-only is organized under the TCDA agreement for the day-ahead timeframe (ENTSO-E, 2020a, 2020b, 2020c, 2020d) and the TCID (ENTSO-E, 2020a, 2020b, 2020c, 2020d) agreement for the intraday timeframe, merged in one TCMC agreement, while pan-European cooperation among NEMOs-only is organized under the ANDOA (NEMO Committee, 2024a, 2024b) for the day-ahead timeframe and the ANIDOA (NEMO Committee, 2024a, 2024b) for the intraday timeframe.
(b) On a regional level, the cooperation is organized under several regional cooperation agreements where only NEMOs and TSOs participating in the SDAC and SIDC in the specific regions are parties.
(c) On a local level, the above-mentioned pan-European and regional agreements are supplemented as necessary, by local agreements or local arrangements between individual NEMOs and TSOs operating in a bidding zone or NEMOs and TSOs operating on either side of a bidding zone border.
(d) The overall contractual structure is also complemented by agreements among NEMOs and/or TSOs on the one side and third entities (e.g. ENTSO-e, JAO) on the other side, as the case may be, to assign the performance of specific tasks relevant to the SDAC and SIDC processes.

The determination of parties to each agreement is contingent on its relevance to the specific phase of the market coupling to which the agreement pertains. Notably, both NEMOs and TSOs are parties to agreements that encompass procedures for the pre- and post-coupling phases. These agreements include the procedures followed by the NEMOs and TSOs for the submission of the inputs of the market coupling algorithm, such as available cross-zonal network capacities and allocation constraints (pre-coupling) and the procedures followed by the NEMOs and the TSOs for verification of the market coupling results, transfer of the market coupling results and scheduled exchanges, as well as collection and distribution of the congestion income (post-coupling). Conversely, agreements governing procedures linked to the coupling phase, where the sole responsibility lies with the NEMOs, such as the operation of the market coupling algorithm, involve exclusively NEMOs as parties. See Chapter 2, Sect. 2.3.1 for a description of the three phases of market coupling. See also Chapter 4 for the role of TSOs in the SDAC and SIDC.

The structure of the market coupling contractual framework is built on the principle of decentralization, in the sense that overarching agreements only regulate issues that require uniform application and seamless coordination among all involved parties. It is understood that this is the case for purely operational matters, which by definition may affect the overall performance of the SDAC and SIDC. For such purely operational issues, compliance with high-level pan-European procedures is mandatory both at regional as well as at a local level. However, overarching agreements do not require compliance with issues that can be agreed upon on a regional or local level. In other words, regional and local agreements can define regional or local issues without restrictions by the provisions of the overarching agreements, except for operational matters where coordination needs to take place centrally. Nevertheless, it should be noted that overarching agreements provide the only contractual basis and thus, a contractual bridge for parties among which no regional or local agreement needs to be concluded, as for instance parties with no common bidding zone borders.

14.2 Overview of Individual Agreements

This section provides an overview and brief description of the most important individual agreements entered into by the TSOs in order to perform their activities under the SDAC and SIDC.

14.2.1 The CACM Global NDA

Confidentiality is of paramount importance when it comes to sharing sensitive commercial information among NEMOs and TSOs within the context of SDAC and SIDC operations. Additionally, when NEMOs, power exchanges or

TSOs intend to join the SDAC and SIDC activities by becoming project or non-operational parties in preparation for their go-live, or are attributed the status of observer within the context of any of the pan-European or regional SDAC or SIDC agreements, the need for robust non-disclosure provisions becomes even more evident.

In order to safeguard confidentiality of the information to be exchanged among concerned parties within the context of the SDAC and SIDC, the signing of a non-disclosure agreement (NDA) is a fundamental prerequisite before acceding to any agreement or attaining the status of observer. This crucial condition is addressed through the endorsement of the CACM Global NDA, which is overseen by ENTSO-e. In particular, instead of managing several individual NDAs with different sets of parties, each potentially having overlapping scopes and durations, NEMOs and TSOs, in close collaboration with ENTSO-e, have chosen to streamline the process by adopting the "Single Day-Ahead and Intra-Day Coupling Observership and Non-Disclosure Agreement", commonly referred to as the "CACM Global NDA". This agreement came into force on 23 February 2016 with several parties having adhered ever since. As of March 2023, a total of 20 NEMOs and 51 TSOs have become parties to the CACM Global NDA (ENTSO-E, 2023). Evidently, adherence of new parties to the CACM Global NDA is possible for any NEMO, power exchange or TSO, upon consent of the other CACM Global NDA parties.

The CACM Global NDA encompasses standard non-disclosure provisions, ensuring not only the efficient and secure exchange of data, as a vital part for market coupling operations but also enabling its flexible expansion to include new parties. By consolidating these agreements into a single comprehensive framework, the process becomes more straightforward and adaptable, promoting broader participation and enhancing the overall effectiveness of the market coupling mechanism.

14.2.2 The DAOA and IDOA Agreements

The cornerstone of the market coupling contractual framework revolves around two pivotal agreements: the Single Day-Ahead Coupling Operations Agreement (referred to as the 'DAOA') and the Single Intraday Operations Agreement (referred to as the 'IDOA'). These agreements have been collaboratively established by TSOs and NEMOs across all market areas participating in the SDAC and SIDC processes, respectively. The DAOA and the IDOA outline the high-level processes that NEMOs and TSOs engage in during the pre- and post-coupling phases, including normal, as well as backup and fallback processes, i.e. processes to be triggered in the event of circumstances that prevent normal operation. Such processes are dealt with in high-level operational procedures, which are attached to the agreements as annexes. For the time being, the IDOA only covers the continuous intraday trading (referred to as the "XBID"), and future amendments are anticipated to encompass also intraday auctions (referred to as the "IDAs").

In line with Article 20, Paragraph 7, of ACER Decision No 4/2020 on the algorithm methodology for day-ahead and intraday coupling, ACER mandated that all NEMOs, in coordination with TSOs, must publish and must continuously update operational contracts and procedures, including change control procedures, monitoring procedures and fallback and back up procedures. In compliance with these ACER requirements, current non-confidential versions of the DAOA and the IDOA along with their annexes have been published in the websites of ENTSO-e and the NEMO Committee.

The DAOA and the IDOA, as overarching agreements, follow the already mentioned principle of decentralization, thus only regulating operational issues that require the overall alignment and compliance by all parties on a European level and that cannot be dealt with on a regional or local level.

The DAOA and the IDOA have been amended in January 2022 to implement a joint governance structure that would include all NEMOs and TSOs, signatories to each of the DAOA and the IDOA, with the aim to increase synergies among the two parallel projects. This restructuring led to the creation of the Market Coupling Steering Committee (referred to as the "MCSC"), a central governing body responsible for discussing and agreeing on joint SDAC and SIDC matters. Besides, the competence of the MCSC, comprised by representatives of the parties to the DAOA or the IDOA only, covers matters related to the operation and adaptations of the SDAC-only or SIDC-only respectively. With reference to the MCSC, Article 10 of the CACM Guideline states that '[TSOs and NEMOs] shall meet regularly to discuss and decide on day-to-day operational issues. TSOs and NEMOs shall invite the Agency and the Commission as observers to these meetings and shall publish summary minutes of the meetings'. The scope of observership has been broadened since the adoption of the CACM Guideline by Decision No 2023/03/MC-EnC of the Ministerial Council of the Energy Community (Energy Community, 2022a, 2022b), by extending it to include the Energy Community Regulatory Board as well as the Energy Community Secretariat, in addition to ACER and the Commission already provided for by Article 10 of the CACM Guideline. In particular, Article 10 of the adapted CACM Guideline states that "TSOs and NEMOs from Member States acting in accordance with Regulation (EU) 2015/1222 shall invite the Energy Community Regulatory Board and the Energy Community Secretariat as observers to these meetings and shall publish summary minutes of the meetings". See Chapter 13 for more information on the Energy Community.

The governance of the SDAC and SIDC is further enhanced by subcommittees and working groups created by the MCSC to which the MCSC may delegate specific tasks and powers. The structure of subcommittees and working groups under the DAOA and the IDOA is, as far as possible, joint or mirroring for both projects, in order to achieve better and more efficient organization and leverage the synergies of SDAC and SIDC, while at the same time taking into account the individual needs and specificities of each project. Notable among the subcommittees with delegated powers by the MCSC and mentioned directly within the DAOA and the IDOA are the Operational Committees (referred to as the "OPSCOM") and Incident Committees. The respective SDAC and SIDC OPSCOMs are responsible for monitoring daily operations, offering operational advice to the MCSC, proposing operational enhancements, and managing day-ahead and intraday operations. Meanwhile, the respective SDAC and SIDC Incident Committees handle the management of critical or major events during market coupling operations.

The DAOA and the IDOA have been further amended in September 2022 to implement a qualified majority voting (referred to as the "QMV"), in order to speed up decision-making and avoid escalations when NEMOs and TSOs are deciding on non-operational issues. However, decisions on operational matters can only be taken validly if both NEMO and TSO votes (as pre-aligned under their respective NEMO-only and TSO-only pan-European agreements) converge.

As the SDAC and SIDC cooperation aims, among others, to be further expanded to new borders, both the DAOA and the IDOA contain provisions to allow enlargement of the cooperation to new parties or observers. In particular, the agreements provide for different levels and requirements of expansion, depending on the legal framework within which the prospective entities interested in joining the cooperation are established and operate.

In case of NEMOs and TSOs within the scope of application of the CACM Guideline, new parties can join the DAOA or the IDOA by accession to the respective agreements, as long as they have paid their share to the historical costs and under the condition that they have put in place all necessary regional and local agreements and arrangements required to perform their obligations within the SDAC and SIDC.

In case of NEMOs, power exchanges and TSOs outside the scope of application of the CACM Guideline and established in countries members of the European Economic Area, or the European Free Trade Association and Energy Community, or in a country which has exited the European Union or the European Economic Area (e.g. such as in the case of "Brexit"), or in any other country covered by a relevant European Commission decision, the NEMO, power exchange or TSO can become a party to the DAOA or the IDOA, contingent upon an intergovernmental agreement enabling such NEMO, power exchange or TSO to participate to the SDAC or SIDC and a MCSC decision allowing such accession. If the above conditions are met, new parties can join the cooperation by accession to the respective agreement, as long as they comply with any applicable legal and regulatory requirements to enter such agreement (e.g. intergovernmental agreements), they have paid their share to the historical costs and they have put in place all necessary regional and local agreements and arrangements required to perform their obligations under the SDAC and SIDC. In addition, the new parties should obtain regulatory comfort for their accession to the agreements, either in the form of a regulatory approval or lack of regulatory objection by their respective national regulatory authorities, depending on their applicable legal and regulatory framework. The accession to the SDAC or SIDC operations takes place according to a roadmap agreed among the acceding party and existing parties on a midterm basis.

Any other power exchanges and TSOs cannot become parties to the above agreements, but can only apply to become observers.

New parties acceding the cooperation initially start as non-operational parties. Both the DAOA and the IDOA contain provisions on the procedure to be followed by non-operational parties in order to become operational. In particular, a non-operational party can become operational only if they have complied with all legal and technical criteria set forth in the agreements, have successfully performed all necessary testing and their technical readiness is confirmed by a MCSC decision. In addition, non-operational parties must have in place all necessary regional and local agreements or arrangements that enable them to perform their obligations under the respective agreements and have regulatory comfort for the launch of their go-live, either in the form of a regulatory approval or lack of regulatory objection, depending on their applicable legal and regulatory framework. For clarity, it should be mentioned that a NEMO, power exchange or TSO doesn't necessarily need to join both agreements at the same time.

The different status of operational and non-operational parties is also reflected in the governance, by categorizing decisions as either operational or governance and development decisions and defining the parties with voting powers on each decision category. It should be noted, that while as a general rule, non-operational parties should not have any voting powers as regards to operational decisions, some operational decisions may impact their go-live. Thus non-operational parties should be kept informed properly and may have the right to raise to operational parties potential concerns about certain operational decisions which may have material adverse effect on their interests.

The DAOA and the IDOA also outline similar provisions for obtaining an observer status, available to any NEMO, power exchange, and TSO irrespective of their jurisdiction. The observer status may be granted to a party upon a MCSC decision, which in case of observers not falling within the scope of the CACM Guideline can be subject to additional conditions indicated by the MCSC. Observers should sign the CACM Global NDA to protect confidentiality of exchanged information, do not have voting rights and can only access limited documentation and attend limited meetings of the cooperation as defined by the MCSC decision.

14.2.3 The TSO Cooperation Agreements for the Market Coupling

While the DAOA and the IDOA, having both NEMOs and TSOs as contracting parties, tackle joint NEMO and TSO duties and responsibilities in the performance of the SDAC and SIDC respectively, they do not cover NEMO-only or TSO-only tasks. In fact, NEMOs and TSOs need a further contractual framework to define their individual NEMO-only and TSO-only duties and responsibilities and coordinate their respective tasks and deliverables with regard to the operation and further development of the SDAC and SIDC. To that end, they have put in place separate agreements, where either only NEMOs or only TSOs-only are parties. On the one hand, on the NEMO-only side, NEMOs have concluded the All NEMO

Day-Ahead Operational Agreement (referred to as the "ANDOA") with respect to the SDAC and the All NEMO Intraday Operational Agreement (referred to as the "ANIDOA") with respect to the SIDC. On the TSO-only side, the TSOs have concluded the TSO Cooperation Agreement for the Single Day-Ahead Coupling (referred to as the "TCDA agreement") and the TSO Cooperation Agreement for the Single Intraday Coupling (referred to as the "TCID agreement"). Further, in 2023, the TSOs have decided to merge the two agreements into one single TSO Cooperation Agreement for the Market Coupling (referred to as the "TCMC agreement") to cover both the day-ahead and intraday timeframes.

According to Article 20, Paragraph 7 of ACER Decision No 4/2020 approving the algorithm methodology for the day-ahead and intraday coupling, ACER mandated all NEMOs in coordination with the TSOs to publish non-confidential versions and then continuously update operational contracts and procedures, including change control procedures, monitoring procedures and fallback and back up procedures. In compliance with the above ACER requirement, current non-confidential versions of the ANDOA and ANIDOA, along with their annexes have been published on the NEMO Committee website, while current non-confidential versions of the TCDA and the TCID agreements along with their annexes have been published on the website of ENTSO-e.

Following the same principle of decentralization as for the DAOA and the IDOA, the TCDA and the TCID agreements, as merged into a single TCMC agreement, as overarching TSO-only agreements, regulate only operational issues that require the alignment and compliance by all TSO parties and that cannot be dealt with on a regional or local level.

The TSOs have also opted for joint SDAC and SIDC governance for their cooperation, via the TSO Market Coupling Steering Committee (referred to as the "TSO MCSC"), with competence to decide on all matters related to the TSO-only management and operations, including the pre-alignment of the TSOs vote in light of common NEMO and TSO decisions to be taken under the decision-making rules of the DAOA and the IDOA. It should be noted that the TSO MCSC is a different body from the "All TSOs". The "All TSOs", operating under the auspices of the ENTSO-e Market Committee, is comprised by all TSOs within the scope of the CACM Guideline and responsible for the CACM deliverables, while the TSO MCSC is comprised by representatives of all TSOs participating in the SDAC or SIDC activities. For the intraday timeframe, the TCID also contains high-level procedures with TSO-only relevance for the XBID (for the time being).

In the likeness of the DAOA and the IDAO, both TCDA and TCID agreements include provisions that allow the further expansion of the cooperation to new parties or observers. New parties may join the cooperation by acceding to any of the agreements, contingent to a decision of the TSO MCSC approving the accession of the new party and under the conditions that the acceding party is a TSO. In addition, in order to accede to the agreements, the acceding party should have previously signed the DAOA or the IDOA respectively, as well as become a party to a set of other service level agreements necessary for the performance of TSOs duties and responsibilities under the SDAC and SIDC. At the time being, such agreements include the CSE SLA concluded among the TSOs and JAO for the day-ahead timeframe (a list of the SLAs is included as Appendix F to the TCDA Agreement) and framework agreements among the TSOs and third party providers for project management and testing services for the intraday timeframe (a list of the SLAs is included as Appendix H to the TCID Agreement).

As regards the observership, any TSO may be granted the status of an observer upon a TSO MCSC decision, which, in case of observers not falling within the scope of the CACM Guideline, can be subject to additional conditions indicated by the TSO MCSC. Observers should sign the CACM Global NDA to protect confidentiality of exchanged information, do not have voting rights and can only access limited documentation and attend limited meetings of the cooperation as defined by the TSO MCSC decision.

14.2.4 The CSE SLA

In order to increase efficiency of their cooperation and simplify the process of allocating common costs, the NEMOs and TSOs, parties to the DAOA, have decided to delegate the task of allocating their SDAC joint NEMOs and TSOs common costs and joint NEMOs and TSOs regional costs of operating the day-ahead market coupling to a Central Settlement Entity. Such Central Settlement Entity shall be responsible for settling the above-mentioned costs with third party service providers and distributing them to the respective NEMOs and TSOs according to the agreed upon sharing keys (Annex 6 to the DAOA). In addition, the TCDA TSOs have also decided to extend the delegated tasks to the Central Settlement Entity to further include the task of handling their TSO-only common costs and TSOs regional costs for operating the day-ahead market coupling (Appendix E to the TCDA). Currently the role of the Central Settlement Entity has been assigned to JAO, by concluding the Central Settlement Entity Service Level Agreement (referred to as the "CSE SLA") among NEMOs and TSOs on the one side and JAO on the other side.

Under the CSE SLA, JAO is acting as a central contracting entity for the above-mentioned categories of common costs by entering into agreements with external service providers and receiving and paying the relevant invoices, on behalf of SDAC NEMOs and TSOs. Subsequently, JAO distributes the amounts paid to such external service provides among NEMOs and TSOs by invoicing each NEMO and TSO on a regular basis with its share of the costs by using the sharing keys as agreed among NEMOs and TSOs under the DAOA and the TCDA. JAO has also undertaken the task of supporting the budgeting and monitoring of costs in cooperation with the appropriate governance bodies under the DAOA and the TCDA. Lastly, JAO has been assigned with the additional task of calculating the historical costs to be paid by any new parties acceding the cooperation.

Due to the importance of the Central Settlement Entity tasks for the well-functioning of the cooperation under both the DAOA and the TCDA agreement, accession to the CSE SLA is a prerequisite to the accession of a new TSO to the

DAOA and the TCDA agreement. Resembling the requirements for adhering to the DAOA and the TCDA agreement, accession to the CSE SLA is subject to the new TSO paying its share in the historical costs.

It should be noted that a similar agreement may be concluded to expand the tasks of the Central Settlement Entity to NEMO and TSO common costs under the SIDC.

14.2.5 Regional Cooperation of NEMOs and TSOs Within SDAC and SIDC

For the day-ahead timeframe, the SDAC Agreement is complemented on a regional level by regional cooperation agreements among NEMOs and TSOs operating in the respective regions. Such regional cooperation agreements have replaced pre-existing agreements among power exchanges and TSOs for the design and implementation of the day-ahead market coupling. The regional cooperation agreements regulate in further detail the procedures that should be jointly performed by the NEMOs and TSOs in the pre- and post-coupling phases. Such agreements include detailed common operational procedures, assigning specific roles and responsibilities to the parties, and cover, not only normal operation of the day-ahead market coupling, but also back up and fall back procedures triggered in the event of circumstances that prevent normal operation. It should be noted that regions have been created before the European guidelines and network codes entered into force, as early implementation and pilot projects. Consequently, their composition is not necessarily corresponding to Capacity Calculation Regions but mostly derives from historical cooperation groups among the power exchanges and TSOs.

Regional cooperation for the SDAC is governed via the steering bodies described in the respective regional cooperation agreements. Similar to the DAOA, regional cooperation agreements follow the principle of decentralization and regulate issues that require, at a minimum, the cooperation of all parties at a regional level and that cannot be dealt with at a local level.

Furthermore, regional cooperation agreements may also contain specific provisions that allow the extension of the regional cooperation to new parties. Such extension can take effect with adherence to the regional cooperation agreements and may be contingent upon compliance with specific conditions. Such conditions may include the approval of the adherence by existing parties, the support by the adhering party's national regulatory authority expressed, depending on its national regulatory framework, either explicitly or by lack of regulatory objection, and the payment by the new party of its share in historical costs of the cooperation.

As a matter of fact, a party joining a regional cooperation agreement shall not by default and as of its adherence, be considered as a party operating the day-ahead market coupling processes (operational party). On the contrary, a new party shall join the cooperation as not yet operational, with the aim of preparing for its go-live within a targeted timeframe agreed upon with the other parties of the cooperation and approved by the competent steering bodies (such party assuming the status of a "project party"). Within the agreed timeframe, the project party should set in place all legal, business and technical requirements that are necessary for the performance of its tasks under the SDAC when it eventually becomes operational. More precisely, from a legal point of view, the project party should put in place all necessary local agreements and introduce all necessary local arrangements, as the case may be. For example, TSOs may need to enter into bilateral agreements with NEMOs operating in the respective bidding zone, bilateral agreements with neighbouring TSOs, or SLAs for delegation of specific tasks to JAO. A list of all local agreements and local arrangements may need to be submitted to the other parties for information purposes. On the business and technical side, the project party should procure all IT systems and implement all operational procedures that are necessary for its operation after its go-live. The regional operational procedures may also need to be updated to include details relevant to the project party or the new borders. The project phase also includes testing sessions with the other parties for several scenarios as agreed upon by the parties and deemed appropriate.

When all of the above have been achieved, the competent regional steering bodies may confirm the technical readiness of the party and indicate the go-live date. Technical readiness should be understood as the confirmation that all legal agreements and arrangements are in effect, all IT systems are in place, all regional operational procedures are updated as necessary and implemented by the new party and all testing has been completed successfully. Confirmation of the Go-live may be conditional to regulatory approval or lack of regulatory objection, as the case may be, as well as verification that the project party has paid its share in the historical costs of the cooperation. After the successful go-live, the project party becomes an operational party.

The different status of parties within the same regional cooperation, with some parties being operational and potentially some parties being project parties, may also be reflected in the governance structure introduced by the regional cooperation agreements, by categorizing decisions as either common, or operational, or project decisions and defining the parties with voting powers on each decision category.

For the intraday timeframe and with regard to XBID, there is no need for regional cooperation agreements among NEMOs and TSOs already operating the SDIC/XBID, as all aspects of their cooperation are included in the SIDC Agreement. Regional cooperation agreements for the design and implementation of the intraday continuous market coupling are generally terminated after the successful go-live of the respective parties. That being said, regional cooperation agreements may need to be set in place in case of Local Implementation Projects (referred to as "LIPs") as part of the roadmap for extension of the XBID to new parties or borders under the IDOA.

As for complementary regional intraday auctions (referred to as "CRIDAs"), an approach similar to the day-ahead market is followed, for parties that have opted for regional intraday auctions, as an intermediate step until intraday auctions (referred to as the "IDAs") under the IDOA are available. Those agreements will be terminated once the IDAs on the respective borders go-live.

As regards regulatory approval of regional cooperation agreements it should be noted that even though there is no provision in the CACM Guideline requiring the approval of regional cooperation agreements as a whole by the respective national regulatory authorities, the cost sharing mechanism agreed among the parties under such agreements is explicitly subject to individual approval by the competent national regulatory authority of each Member State, according to the provisions of Article 80, Paragraph 4 of the CACM Guideline. In addition, national regulatory frameworks applicable to specific NEMOs or TSOs may also require approval of the regional cooperation agreements, as the case may be.

14.2.6 Local Agreements and Local Arrangements Within SDAC and SIDC

Beyond the overarching pan-European and regional cooperation agreements, local provisions are often necessary to accommodate a more detailed framework for coordination among individual NEMOs and TSOs involved in SDAC and SIDC (both for XBID & CRIDAs streams and subsequently, for IDAs), particularly concerning data exchanges. These provisions may take the form of legal contracts between the relevant parties or, depending on the jurisdiction, may be governed by the applicable national legislation, such as national laws, regulatory decisions, or network codes. From the TSOs perspective, such local agreements or arrangements may be necessary between neighbouring TSOs and between TSOs and NEMOs.

In more detail, TSOs may need to enhance the cooperation with their neighbouring TSOs, operating on the opposite side of their bidding zone borders. This collaboration aims to address specific aspects related to the execution of TSO-only pre- and post-coupling tasks. Such aspects may include scheduling, calculation of available cross-zonal transmission capacity, matching processes for market participants' nominations of usage of capacities allocated on the bidding zone borders, the sharing of the congestion income, full and partial decoupling arrangements, specifications, procedures and timings for relevant necessary data exchanges, among other elements related to the pre- and post-coupling processes.

Respectively, TSOs may also need to cooperate further with the NEMOs operating within their biding zones, in order to determine the specifics, procedures, and timing for essential data exchanges. The exchanged data between local NEMOs and TSOs may encompass technical data related to market participants, notifications of available transmission capacity from TSOs to NEMOs, notification of market results and market schedules from NEMOs to TSOs to be used as inputs in their balancing markets, and communication among the parties in case of market disruptions.

14.2.7 Agreements Among TSOs and JAO

According to Articles 8 and 44 of the CACM Guideline, the establishment and operation of fallback procedures (see Chapter 4 for more information) to ensure efficient, transparent and non-discriminatory capacity allocation in the event that the single day-ahead coupling process is unable to produce results is included among the TSO tasks. Such fallback procedures should be developed by the TSOs of each Capacity Calculation Region and be approved by their respective NRAs. For several of the existing Capacity Calculation Regions, the TSOs have opted for the performance of shadow explicit day-ahead auctions, as the preferred fallback procedure to allocate daily capacity, in case of full or partial decoupling in the day-ahead timeframe, i.e. processes that kick in, in case an incident in the single day-ahead processes prevents the production of market results. In the shadow auctions, only capacity rights are auctioned, while electricity is to be purchased directly through power exchanges. To enable the performance of shadow explicit auctions, the TSOs have decided to delegate this task to JAO, as the already designated Single Allocation Platform for allocation of long-term transmission rights under Article 48 of Commission Regulation (EU) 2016/1719 of 26 September 2016 establishing a guideline on forward capacity allocation (referred to as the "FCA Regulation").

Furthermore, according to Article 68, Paragraphs 8 and 9 of the CACM Guideline, all central counter parties or shipping agents should collect congestion incomes (see Chapter 10 for congestion income and congestion income distribution) arising from the single day-ahead and intraday market coupling and ensure its transfer to the TSOs. Consequently, according to Article 73, Paragraph 3 of the CACM Guideline, the TSOs shall distribute such congestion incomes among them in accordance with the congestion income distribution methodology approved by ACER. The application of the above-mentioned legal provisions outlined the need for a congestion income distribution system and the appointment of a congestion income distributor, responsible for receiving the congestion income created from the day-ahead market coupling, as well as the CRIDAs (in the borders where complementary regional intraday auctions take place) and eventually, the IDAs from congestion income collectors (central counter parties or shipping agents) and distributing it to the relevant TSOs using appropriate sharing keys. Several TSOs have opted for delegating the task of congestion income distributor to JAO, as external service provider.

In addition, some TSOs, depending on the Region where they operate or the borders under their control, have also opted to mandate JAO with other tasks relevant to the performance of TSO responsibilities under the SDAC or SIDC, such as of publication, on its website and on the transparency platform operated by ENTSO-e according to Regulation (EC) 543/2013, of certain market data on behalf of the TSOs, etc. Please see Chapter 16 for a discussion on transparency.

The delegation of the above-mentioned tasks or any other additional tasks, as the case may be, by the TSOs to JAO takes place by the conclusion of service level agreements (referred to as "SLAs") among relevant TSOs and JAO. According to such SLAs, JAO has undertaken the task to perform the shadow auctions for explicit daily capacity allocation in case the implicit allocation under SDAC fails, by applying the Shadow Allocation Rules (JAO, 2023), as in force and approved from time to time, as well as to perform all other relevant sub-tasks (such as entering into participation agreements with market participants, notifying and updating market participants in case of high risk of decoupling in affected borders etc.). JAO has also undertaken the role of congestion income distributor, under which JAO shall operate the congestion revenue distribution system (referred to as the "CRDS"), collect the congestion income from the congestion income collectors and distribute it to the TSOs according to the respective applicable sharing keys, by invoicing and settling relevant transactions.

14.2.8 Cost Sharing

Article 80 of the CACM Guideline introduces the principles of cost sharing among NEMOs and TSOs for the performance of the day-ahead and intraday market coupling, defining an important aspect of their cooperation. It should also be noted that Article 80 of the CACM Guideline has also been adapted by Decision No 2022/03/MC-EnC of the Ministerial Council of the Energy Community (Energy Community, 2022a, 2022b), so that it extends and covers also cost sharing with NEMOs and TSOs established in non-EU countries, members of the Energy Community.

Costs to be shared by NEMOs and TSOs that derive from establishing, amending and operating single day-ahead and intraday coupling should be broken down in three main categories:

(a) Common costs resulting from coordinated activities of all NEMOs or TSOs participating in the single day-ahead and intraday coupling. Especially for NEMOs and TSOs established in non-EU countries, members of the Energy Community, (see Chapter 13 for more detail on the Energy Community) those costs refer only to their interconnections with Energy Community Contracting Parties. According to Article 80, Paragraph 3 of the CACM Guideline, these costs shall be shared among the TSOs and NEMOs in the Member States and third countries participating in the single day-ahead and intraday coupling. To calculate the amount to be paid by the TSOs and NEMOs in each Member State and, if applicable, third countries, one eighth of the common cost shall be divided equally between each Member State and third country, five eighths shall be divided between each Member State and third country proportionally to their consumption, and two eighths shall be divided equally between the participating NEMOs;

(b) Regional costs resulting from activities of NEMOs or TSOs cooperating in a certain region. According to Article 80, Paragraph 4 of the CACM Guideline, NEMOs and TSOs cooperating in a certain region shall jointly agree on a proposal for the sharing of regional costs. The proposal shall then be individually approved by the competent national authorities of each of the Member States and Energy Community Contracting Parties in the region. NEMOs and TSOs cooperating in a certain region may alternatively use the cost sharing arrangements set out in point (b) above (common pan-European sharing key); and

(c) National costs resulting from activities of the NEMOs or TSOs in one Member State or Energy Community Contracting Party. The approval of costs are subject to the national regulatory framework of the respective NEMOs and TSOs.

Details on cost sharing, applying the above principles of Article 80 of the CACM Guideline are further detailed in NEMOs and TSOs agreements. Decentralization of these agreements indicates that each agreement only manages costs arising from the cooperation in the respective level. Subsequently, common pan-European NEMO and TSO costs fall under the scope of the DAOA and the IDOA, pan-European NEMO-only or TSOs-only costs fall under the scope of the ANDOA/ANIDOA or TCMC agreement respectively, regional costs fall under the scope of regional cooperation agreements and should be approved by the regional NRAs, while national costs are undertaken either in local agreements or local arrangements, as the case may be.

14.3 Illustrative Examples

14.3.1 TSO Participation in SDAC and SIDC Agreements

Figure 14.1 shows the agreements to which a TSO currently participating in the SDAC and SIDC is a party, at pan-European, regional and local level. The TSO may also need to be a party to additional agreements (e.g. with JAO) to delegate additional tasks and roles, depending on each party's individual circumstances and legal and regulatory framework.

14.3.2 Overview of the Pan-European SDAC and SIDC Governance

Figure 14.2 shows the current market coupling organizational structure for joint NEMO and TSO activities under the DAOA and IDOA. Subcommittees and working groups are as much as possible joint or mirroring to exploit synergies among the two parallel projects.

	SDAC	SIDC
Pan-European	Single Day-Ahead and Intra-Day Coupling Observership and Non-Disclosure Agreement (CACM Global NDA)	
	Single Day-Ahead Coupling Operations Agreement (DAOA)	Single Intraday Operations Agreement (IDOA)
	TSO Cooperation Agreement for the Market Coupling (TCMC Agreement)	
	Central Settlement Entity Service Level Agreement (CSE SLA)	PMO Services Agreement & Testing Services Agreement
Regional	Regional Cooperation Agreements for Day-Ahead Market Coupling	Regional Cooperation Agreements for LIPs and CRIDAs (until IDAs)
Local	TSO-TSO Agreement	TSO-TSO Agreement
	TSO-NEMO Agreement	TSO-NEMO Agreement

Fig. 14.1 Overview of market coupling agreements. *Source* Author's Own Illustration

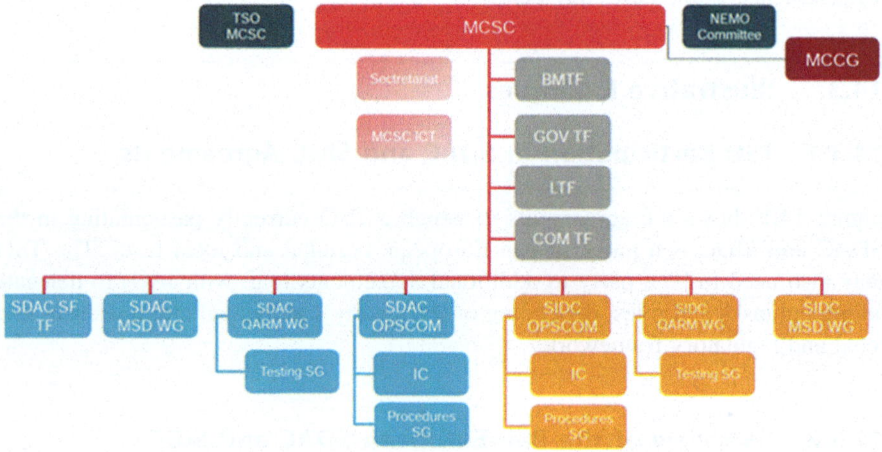

Fig. 14.2 Market coupling organizational chart. *Source* (Market Coupling Steering Committee, 2022)

14.4 Recommended Approach for New TSOs

It is undeniable that the contractual structure for the day-ahead and intraday market coupling, although necessary, is very complex. This complexity makes it essential for NEMOs, power exchanges and TSOs who are interested in joining the SDAC or SIDC to have a clear roadmap and step wise approach for their accession. For

TSOs, this approach can be summarized as follows, but may need to be adapted depending on the specific circumstances pertinent to each TSO:

First Step: As a first step, a TSO interested in joining the SDAC or SIDC, should adhere to the CACM Global NDA, by referring a request for adherence to ENTSO-e. When all existing CACM Global NDA parties have provided their consent, the adhering party should sign the relevant adherence form.

Second Step: As a next step, the interested TSO should request to become an observer under the DAOA or the IDOA and the TCMC agreement. Approval of the observership takes place by a MCSC decision for the DAOA and the IDOA and by TSO MCSC decision for the TCMC agreement. In case of TSOs not operating within the scope of the CACM Guideline, such decisions may be conditional to additional requirements indicated by the Steering Committees. At this point it should be noted that TSOs operating outside the scope of the CACM Guideline or outside the European Economic Area, or the European Free Trade Association or Energy Community, or in a country which has exited the European Union or the European Economic Area, or in any other country not covered by a relevant European Commission decision, can only come this far and cannot proceed beyond this step.

Third Step: Subsequently, the interested TSO should request to become a full party to the DAOA or the IDOA and the TCMC agreement. Upon fulfilment of all requirements (necessary agreements set in place, payment of pan-European pro-quota historical costs and as the case may be, compliance with legal and regulatory requirements and regulatory comfort) and approval by MCSC and TSO MCSC decisions, the interested TSO can become full party by signing an accession form. Upon adhering to the pan-European agreements, the party is considered as non-operational. At this stage, adhering to other pan-European agreements, as indicated by the relevant agreements, such as the CSE SLA for the day-ahead timeframe and the PMO services agreement and Testing Services agreement for the intraday timeframe, is also required.

Fourth Step: In parallel, the interested TSO should request to become party to a regional cooperation agreement. Upon fulfilment of all requirements (such as payment of pro-quota regional costs and regulatory approval) and approval by the competent governance body under such regional cooperation agreement, the interested TSO can become a project party by signing an adherence form.

Fifth Step: When the interested TSO has become non-operational/project party to pan-European and regional cooperation agreements respectively, the TSO should start working together with the other parties to create a roadmap and targeted timeframe for its go-live. This process can be enabled by the creation of dedicated working groups in addition to already existing permanent working groups of each cooperation.

Sixth Step: The project phase is the most demanding part of the process, as it entails both legal and technical actions to be taken. The interested TSO should negotiate and put in place all local agreements and arrangements that are necessary for the performance of its tasks under the SDAC or SIDC tasks. Although an exhaustive list cannot be compiled, as it depends on the circumstances of the particular TSO, these may include bilateral TSO-TSO agreements among neighbouring TSOs and agreements between TSOs and NEMO operating in the same bidding zone. The TSO may also need to join the appropriate SLAs with JAO depending on the tasks that may need to be delegated (e.g. congestion income distribution, performance of shadow auctions as fallback procedures) or other agreements with third parties as the case may be. In parallel, the TSO should procure the appropriate IT systems and implement the applicable operational procedures. It should also cooperate with the other parties in order to identify the potential need for update of any operational procedures with details relevant to such TSO or its borders. Finally, the TSO must successfully complete all indicated testing sessions for all appropriate scenarios.

Last step: Upon successful completion of the sixth step and regulatory approval, the MCSC in parallel with the competent governance body of the regional cooperation can confirm technical readiness and approve the Go-live date. On the approved date, the TSO, along with the corresponding NEMO shall become operational and launch its participation in the SDAC or SIDC. It is evident that even though initiating the process and moving forward by taking further steps is an individual TSO process, elaboration of regional and local operational procedures, testing and go-live can only be achieved by coordination between the TSO and NEMOs active within its bidding zones. A TSO can only go-live at the same time as the corresponding NEMO and vice-versa.

Besides, regulatory comfort should be considered by the TSO in every step to be taken, depending on the requirements of the legal and regulatory framework applicable to such TSO. As already mentioned, the CACM Guideline only foresees the need for individual approval of regional costs by the competent national authorities of each of the Member States and Energy Community Contracting Parties in the region. Nevertheless, applicable legal and regulatory frameworks may impose further obligations for notification or approval of contracts by the competent national regulatory authorities.

Bibliography

Energy Community. (2022a). Commission Regulation (EU) 2015/1222 of 24 July 2015 establishing a guideline on capacity allocation and congestion management, as incorporated and adapted by the Ministerial Council Decision 2022/03/MC-EnC of 15 December 2022. Retrieved October 30, 2024, from https://www.energy-community.org/dam/jcr:4e533934-4a9f-4a25-87a8-9853e146a4ef/REGULATION_EU20151222.pdf

Energy Community. (2022b). Decision of the Ministerial Council of the Energy Community D/2022/03/MC-EnC of 15 December 2022. Retrieved October 30, 2024, from https://www.energy-community.org/dam/jcr:d5a1a894-88db-4326-818b-f2c648bd237e/Decision03-2022-MC_newELacquis_15-12-2022.pdf

ENTSO-E. (2020a). Single Day-Ahead Operational Agreement (DAOA). Retrieved October 30, 2024, from https://www.entsoe.eu/network_codes/cacm/implementation/sdac/

ENTSO-E. (2020b). Single Intraday Operational Agreement (IDOA) . Retrieved October 30, 2024, from https://www.entsoe.eu/network_codes/cacm/implementation/sidc/

ENTSO-E. (2020c). TSO Cooperation Agreement for Single day-ahead coupling (TCDA) . Retrieved from https://www.entsoe.eu/network_codes/cacm/implementation/sdac/

ENTSO-E. (2020d). TSO cooperation agreement for single intraday coupling (TCID). Retrieved October 30, 2024, from https://www.entsoe.eu/network_codes/cacm/implementation/sidc/

ENTSO-E. (2023). ENTSO-e market report 2023. Retrieved October 30, 2024, from https://eepublicdownloads.entsoe.eu/clean-documents/nc-tasks/ENTSO-E_Market_Report_2023.pdf

EU. (2015). Commission Regulation (EU) 2015/1222 of 24 July 2015 establishing a guideline on capacity allocation and congestion management. Retrieved October 30, 2024, from https://eur-lex.europa.eu/legal-content/EN/TXT/?uri=CELEX%3A32015R1222#d1e3225-24-1

JAO. (2023). Shadow allocation rules. Retrieved October 30, 2024, from https://www.jao.eu/sites/default/files/2023-02/Shadow%20Allocation%20Rules_Merged.pdf

Market Coupling Steering Committee. (2022). Market coupling consultative group meeting presentation of 1st December 2022. Retrieved October 30, 2024, from www.nemo-committee.eu/assets/files/mccg-presentation.pdf

NEMO Committee. (2024a). All NEMO Day Ahead Operational Agreement (ANDOA). Retrieved October 30, 2024, from NEMO Committee website https://www.nemo-committee.eu/sdac

NEMO Committee. (2024b). All NEMO Intraday Operational Agreement (ANIDOA). Retrieved October 30, 2024, from https://www.nemo-committee.eu/sidc

Vicky Blana is an Attorney at Law, member of the Athens Bar Association and licensed to practice before the Supreme Courts of Greece. Currently, she is the Director of the Legal Regulatory Issues Branch within Independent Transmission Operator SA (IPTO), the Greek Transmission System Operator, and has been a member of the company's legal team since 2012. With a focus on energy law, she has substantial legal experience representing IPTO in various legal working groups and committees, as well as in energy litigation. Since 2014, she has been actively involved in market coupling activities. She graduated from the Faculty of Law of the Aristotle University of Thessaloniki and holds two master's degrees, in Civil Procedure Law and in Commercial Law, both from the National and Kapodistrian University of Athens. She is also an accredited mediator and holds a certification as a Compliance Officer, with a particular interest in energy-related issues. She has also published several articles in legal journals and contributed to legal handbooks.

The Role of National Regulatory Authorities in Market Coupling 15

Nico Schoutteet

Abstract

This chapter introduces the role of the National Regulatory Authorities in European market coupling in line with the Electricity Directive (EU) 2019/944. National Regulatory Authorities are assigned a crucial role to adopt and enforce the market coupling mechanisms in Europe. These tasks and duties are assigned to them by the legislative energy package in Europe, against the background of the electricity market liberalization process. This chapter highlights some key challenges which may need to be addressed to truly deliver on a harmonized, transparent, and efficient European Internal Electricity Market: national interests, institutional and governance barriers, enforcement mechanisms and information asymmetry. Where these aspects are not dealt with properly, suboptimal market design choices risk being implemented.

15.1 Introduction

National regulatory authorities ('NRAs') perform a crucial task in European market coupling initiatives. Established and strengthened in their duties by the European energy packages, their role is to ensure that markets are designed and organized efficiently, that the actors involved comply with the applicable regulation and that these markets function in consumers' and market participants' interest.

The liberalization process of the European electricity sector has been elaborated in previous chapters of this Handbook. This project, initiated in the 1990s, started with the unbundling of formerly vertically integrated utilities into different (legal) entities, performing different functions across the value chain. This involved the

N. Schoutteet (✉)
Zedelgem, Belgium
e-mail: n.schoutteet@gmail.com

A. Estermann et al. (eds.), *European Electricity Market Coupling*, Future of Energy,
https://doi.org/10.1007/978-3-031-86315-8_15

separation of these elements of the value chain into monopolistic and competitive activities—both of which require specific rules and regulations to ensure that they are performed in a manner which is in line with the applicable European and national policy objectives. On a European level, these policy objectives include the promotion of consumers' interest (linked to affordability), the safeguarding of security of supply and environmental sustainability.

To ensure that these policy objectives are pursued by enforcing the applicable legislation, NRAs were established by Member States as a direct consequence of the adoption and implementation of the Second Energy Package[1] (in 2003). The Third Energy Package (in 2009) established, furthermore, a European Union Agency for the Cooperation of Energy Regulators ('ACER').

The functioning and governance of the relationship between NRAs, and ACER, reflects the interconnected nature of the European electricity system and the relevant market coupling activities.

15.2 European Regulatory Scheme

15.2.1 Establishing and Mandating NRAs to Regulate Electricity Markets

Liberalizing (electricity) markets in Europe has been a gradual, stepwise process. From a legal perspective, this process consisted of the adoption and implementation of different "Energy Packages". To date, five such packages can be identified: the First Energy Package (1996), the Second Energy Package (2003), the Third Energy Package (2009), the Clean Energy Package (2019) and the Fit for 55 Package (2021). Each of these packages added to the roles and responsibilities of regulatory authorities, starting from the Second Energy Package in 2003 which laid the basis for the NRAs' foundation in all European Member States.

Today, the tasks and duties assigned to NRAs are laid down in the Electricity Directive (EU) 2019/944. Being a Directive, this legal provision requires national transposition, so no fully harmonized approach exists between Member States on exactly how NRAs are organized, which tasks they perform and to whom they are accountable. Nevertheless, a minimum list of duties and powers assigned to NRAs can be found in Article 59 of the Electricity Directive.

Among others, these tasks relate to ensuring compliance of energy undertakings, including TSOs and DSOs, overseeing the implementation of regulation

[1] Article 23 of the Directive 2003/54/EC stipulates that all Member States shall designate one or more regulatory authorities, independently from the interests of the electricity industry. The article furthermore expresses a number of guidelines for the functioning of these NRAs, as well as principles to uphold in performing their tasks (e.g. promoting non-discrimination, competition and efficient market functioning). The provisions in this Directive were subsequently transposed into national legislation of the Member States in the following years.

(including network codes and guidelines), enforcing in case of non-compliance, monitoring investment plans, and fixing or approving tariffs.

One of these tasks, particularly relevant for the focus of this Handbook, relates to the development of competitive and properly functioning regional cross-border markets, to achieve a competitive, flexible, secure, and environmentally sustainable Internal Energy Market.

Given the European and regional scope of this objective, additional requirements on the interaction and cooperation between regulatory authorities are detailed in the Electricity Directive, the Electricity Regulation (EU) 2019/943 and the ACER Regulation (EU) 2019/942. This aspect will be further explored in the next section.

One of the key enablers of successfully mandating NRAs to regulate the electricity sector, is their independence. While the Electricity Directive and Regulation sometimes leave considerable room for discretion to the Member States on how to establish and issue their national energy policies, the independence of NRAs from public and private interests must be ensured.

This independence is particularly important to allow NRAs to, among others:

issue objective decisions, based on technical expertise and economic analysis, in the interests of the consumers and the market as a whole;
ensure a consistent regulatory framework, contributing positively to investments by ensuring consistent and predictable decisions over time; and
promote transparency and efficiency.

Different elements could, in theory, impinge on this independence. Political interference, insufficient financial or human resources, lack of transparency, conflicts of interests or the absence of parliamentary or judicial scrutiny could have a negative impact on the effectiveness and fairness of the regulatory framework.

The Electricity Directive prescribes a clear set of principles, to which Member States should adhere, in order to safeguard the NRAs' independence. The degree to which these principles are adhered to, by Member States and regulatory authorities, is the subject of a regular report from the European Commission to the European Parliament and Council.

15.2.2 Cooperation Between NRAs

Following the adoption of the First and Second Energy Package, bilateral agreements between TSOs and newly established peer exchanges and bottom-up voluntary market coupling processes emerged. Some relevant and noteworthy initiatives include the Trilateral Coupling (TLC) between Belgium, France and the Netherlands in 2006, the Price Coupling of Regions (PCR) between Central Western Europe (CWE) countries, Nordics, Baltics, Poland and Great Britain) in 2010 and the CWE Flow-Based Market Coupling in 2015.

Table 15.1 Comparison of ACER and CEER. *Source* Author's Own Table

ACER (European Union Agency for the Cooperation of Energy Regulators)	CEER (Council of European Energy Regulators)
Official EU body (decentralized agency)	Not-for-profit association, including a Secretariat
Existence mandated by the Electricity and ACER Regulation	Voluntarily formed by European energy regulators
Based in Ljubljana, Slovenia	Based in Brussels, Belgium
Created in 2011	Created in 2000
Supports cooperation between NRAs, assigned specific tasks and responsibilities through legislation	Acts as platform for cooperation, information exchange and assistance between NRAs
Staff: ~ 100	Staff: ~ 10

Given the requirement for NRAs to approve cross-border capacity calculation and allocation mechanisms, at the time mostly embedded in national legislative frameworks, bottom-up coordination between the involved NRAs has emerged. Based on this coordination, clearer and enforceable rules for cooperation would be established in subsequent energy packages.

Different bodies have been established to foster the coordination between NRAs, with varying degrees of formalism and legal basis. The two main ones are CEER (the Council for European Energy Regulators) and ACER (the Agency for the Cooperation of Energy Regulators). Their main similarities and differences are summarized in Table 15.1.

15.2.2.1 Organizational Structure

In addition to its permanent staff, ACER consists of an organizational structure, of which we recall the most important bodies here:

ACER 's Board of Regulators (BoR) steers the regulatory policy of ACER. It plays a key role in the cooperation between ACER and NRAs, and provides opinions and guidance to the ACER Director. ACER 's BoR is established by senior representatives of the 27 European NRAs, as well as observers from the EEA EFTA States, the European Commission and the EFTA Surveillance Authority.

Its functioning is governed by its rules of procedure, a document which describes, among others, the representation and the decision-making procedure. One important element of these procedures is the fact that opinions and guidance are adopted through a qualified majority of its present members. This stands in contrast to decisions taken by (regional) groups of NRAs, where the unanimity principle is applied.

ACER 's Working Groups and Task Forces provide support to the Board of Regulators. Currently, 4 Working Groups (WGs) are operational:

The Electricity Working Group (EWG)
The Gas Working Group
The REMIT Working Group
The Retail Markets Working Group

In turn, these working groups are supported by Task Forces (TFs), composed of expert representatives from the European NRAs and ACER. Historically, the EWG TFs were established according to corresponding network codes and/or guidelines, for example the CACM TF, etc. Within these TFs, NRAs and ACER align on common positions regarding the adoption of market coupling rules and conditions or methodologies, proposed by stakeholders such as TSOs, NEMOs, or others.

Other relevant ACER bodies include the **Board of Appeal** (where complaints against ACER decisions can be lodged) and the **Administrative Board** (mainly tasked with exercising budgetary powers and the appointment of ACER 's bodies and Director). ACER itself is managed by its **Director**, who is appointed for a five-year term and supported by the internal departments and staff of ACER.

15.2.3 Procedures for Interaction

The roles and rules of procedure for the interactions between NRAs and with ACER are inscribed in different (regulatory) instruments. These range from formal legislative texts (such as the Electricity Directive and the ACER Regulation, network codes and guidelines) to enforceable rules of procedure or even informal memoranda of understanding between (regional groups of) NRAs and ACER. Depending on the (geographical or contextual) scope for cooperation, different rules may apply.

With the entry into force of the Capacity Allocation and Congestion Management (CACM) Guideline in 2015 and the need to adopt, by unanimous decision of large groups of NRAs, a significant number of terms and conditions or methodologies (TCMs), an informal decision-making process has been elaborated by the members of ACER 's Working Groups and Task Forces. Several "routes" for regulatory adoption of these methodologies can be distinguished, depending on the need to amend proposals or the (in)ability to identify unanimous positions among NRAs.

The **Public Consultation** is usually performed by TSOs or NEMOs, in order to seek the input from key stakeholders, ACER and NRAs before the formal submission of a methodology for approval.

The **submission of a methodology** follows the requirements and deadlines in network codes and guidelines, and is done—depending on the context—to ACER, all European NRAs, regional groups or even individual NRAs. The obligation to submit Europe-wide terms and conditions or methodologies to all European NRAs has been replaced, in the Clean Energy Package, by a requirement to submit these directly to ACER for approval. Regional methodologies, however, remain to be submitted to the regional groups of NRAs.

Requests for amendments are issued by NRAs or ACER when they judge that the original submission of a methodology is not in line with the spirit or the requirements of the applicable regulation.

A **referral to ACER** is agreed when disagreements prevent (regional groups of) NRAs from finding a common position on the approval. ACER subsequently issues a decision, to be adopted by the qualified majority of Board of Regulators ' members.

When NRAs agree among themselves, methodologies can be **approved**. This informal agreement among NRAs forms the basis for national approval decisions, issued in accordance with individual NRAs ' decision-making rules and procedures.

15.3 The Regulation of Market Coupling Initiatives in Europe

Over the last decades, several packages have introduced legislation to foster the integration of formerly national electricity markets in Europe, into the Single Electricity Market. As explored in the introduction, these packages have shaped the way in which markets are organized, and which tasks are assigned to the different key stakeholders involved.

European electricity markets as we know them today are governed mostly by the provisions in the Third and Fourth (Clean) Energy Package. These include different legal instruments, of which the most relevant ones are:

The Electricity Directive (EU) 2019/944
The Electricity Regulation (EU) 2019/943
The ACER Regulation (EU) 2019/942

Aside from these legal instruments themselves, they form the basis for the adoption and the implementation of network codes and guidelines. These network codes and guidelines lay out different rules—on different topics—for harmonization of, among others, the organization of European electricity markets. For market coupling, the three most important guidelines are:

The CACM Guideline (EU) 2015/1222, dealing with day-ahead and intraday markets
The Forward Capacity Allocation (FCA) Regulation (EU) 2016/1719, dealing with long-term markets
The Electricity Balancing (EB) Regulation (EU) 2017/2195, dealing with balancing markets

The CACM Guideline sets out minimum harmonized rules for single day-ahead and intraday market coupling, including methods for calculating and allocating cross-zonal transmission capacity. In these guidelines, the roles of TSOs and NEMOs are defined, as well as the framework for cooperation between them and the (regional) deliverables which are subject to regulatory scrutiny. Key principles include fair and transparent price formation and non-discriminatory access to the transmission network.

The FCA Regulation establishes rules for operating long-term (forward) markets. This allows market participants to secure capacity well before the day-ahead and intraday timeframes, allowing them to effectively hedge their business needs.

The EB Regulation intends to create a sequence of markets where reserves can be shared or exchanged by TSOs in order to continuously and immediately balance electricity supply and demand.

Other network codes (NC) and guidelines (GL) exist as well, but these deal less with markets and more with the operation of the electricity network, or requirements for connection of generation, demand or high voltage direct current units to the transmission networks.

The requirements in these network codes and guidelines have shaped the ways in which NRAs and ACER cooperate today. The entry into force of the first of these instruments, the CACM Guideline in 2015, has proven to be the first formal requirement for all European NRAs to align their positions and approval processes, in order to make possible the adoption of the CACM Terms and Conditions or Methodologies. Over the course of the following years, these rules and procedures have been further refined, based on advancing experience and changing legal requirements, such as the entry into force of the recast ACER Regulation in the Clean Energy Package.

Without explicitly governing the functioning of the Single Day-Ahead and Intraday Coupling mechanisms (SDAC & SIDC), the NRAs and ACER play an important role in the adoption of the terms and conditions or methodologies which these projects have to implement. These adoption processes mostly relate to the content of the rules and procedures by which cross-zonal exchanges are made possible (capacity calculation, allocation, clearing, etc.), rather than defining the internal governance of these projects. The latter typically involve contractual arrangements between the involved parties: transmission system operators and nominated electricity market operators, and—where relevant—other entities such as regional coordination centres.

Taking the CACM Guideline as an example, when approving these terms and conditions or methodologies, NRAs scrutinize the proposals based on their compliance with different elements:

- the general objectives of the CACM Guideline (e.g. those listed in Art. 3, such as promoting effective competition, ensuring optimal use of the transmission infrastructure, ensuring non-discriminatory access to transmission capacity, etc.);
- the specific content-wise requirements assigned to the methodologies;

their timescale for implementation, i.e. the deadlines by which the rules should be in place; and

where relevant, national (legal) considerations not reflected in the CACM Guideline

Other guidelines (such as the FCA Regulation, EB Regulation and System Operation (SO) Regulation) have very similar requirements, and proposals for terms and conditions or methodologies are approved in a similar way.

While these requirements may seem clear and enforceable, differences in views on their interpretation arise frequently, not only between the regulators and the regulated entities, but also between NRAs themselves. As shown later, this frequently leads to disagreements on the implementation and subsequently, referrals of the decision-making powers to ACER.

The state of play regarding the adoption of the TCMs under the CACM Guideline paints an ambiguous picture. On the one hand, a lot has been achieved in terms of adopting different methodologies, streamlining rules and enabling intense cooperation among all parties involved. Most—if not all—of the methodologies foreseen in Art. 9 of the CACM Guideline have been developed and adopted (i.e. approved). This can be considered a success.

On the other hand, however, implementation of these methodologies is often a complex and lengthy project. As evidenced by the European Court of Auditors (ECA, 2023), delays have piled up, and the disagreements among parties have been significant. These issues must be resolved so as not to further hamper the achievement of the Internal Electricity Market.

For an overview of the different methodologies to be approved under the CACM Guideline, we refer to the analyses in Chapter 2 regarding the framework for market coupling.

15.4 Key Regulatory Challenges

The effective regulation of cross-zonal electricity markets, as well as the undertakings that operate within them, can be made more difficult by several key challenges. This section explores a couple of these: national interests, barriers related to the institutional framework, enforcement, and information asymmetry. Each of these, when not mitigated or addressed consequently, risk leading to suboptimal outcomes when designing and operating electricity markets.

15.4.1 National Interests

While the efficiency of market coupling initiatives is, most commonly, measured against increases or decreases in socio-economic welfare, the benefits may be very different between different Member States. These differences result from aspects such as different set-ups of national markets, with different levels of integration and liquidity already established.

One aspect which is often overlooked in market coupling and integration activities, is the inherent solidarity mechanism which applies, by which efforts to reduce overall energy costs and increase resiliency against external (price) shocks or safeguard operational security and security of supply, are addressed together, rather than on an individual basis. In other words: elements of market design which tend to have a positive impact on one individual aspect (such as, for example, lowering wholesale prices in one country) might have a negative impact on other aspects (such as, for example, increasing exchanges to neighbouring countries). This is particularly relevant considering many of the individual proposals for redesigning European electricity markets (see previously).

Aside from these inherently different objectives, other national interests (regulatory or imposed externally by other interests) may render the harmonization of rules across countries difficult. This touches upon the core objective of the network codes and guidelines: overcoming individual approaches by harmonizing the rules in a way that maximizes value for the European market as a whole.

15.4.2 Institutional and Governance Barriers

The aim of market coupling regulation is to streamline the previous patchwork of national legislation governing European electricity markets. These include different rules for adopting, and enforcing, desired behaviour from TSOs and NEMOs. Despite the paradigm shift, whereby NRAs currently work under a coordinated governance framework to decide on matters for cross-border relevance, there remains room for improvement.

Here, we can refer again to the ECA report mentioned previously (ECA, 2023): the choice of instruments to govern market coupling (network codes and guidelines), leads, in the opinion of the ECA, to "a complex legal architecture of cross-border trade rules and delays in implementation".

This challenge must be addressed and resolved in order to progress with the delivery of the Internal Electricity Market.

15.4.3 Absence of Efficient Enforcement Mechanisms

Most national legislative frameworks prescribe the processes and means assigned to NRAs to enforce behaviour from regulated entities, in case of non-compliance with the legal provisions. Given the cross-border scope of market coupling initiatives, legal questions related to the responsibility of these entities, and who should enforce compliance, have frequently hampered effective regulation in the past years.

The legal framework does not, today, provide ACER with enforcement tools, nor does it provide clear rules for joint enforcement by(groups of) NRAs towards groups of regulated entities, be they TSOs, power exchanges or other stakeholders.

The absence of these rules may put at risk the ability of NRAs to effectively regulate the functioning of electricity markets and their operators. (ECA, 2023).

15.4.4 Information Asymmetry

In a liberalized market, the role of information is crucial. This has several aspects: availability, relevance, transparency, precision, etc. This section explains that several elements related to these aspects of information risk endangering the implementation of the most appropriate market design.

In particular, information asymmetry between regulators (NRAs) and regulated entities (TSOs, or NEMOs, or RCCs…) may impede the delivery of the most efficient version of the Internal Electricity Market.

Information asymmetry may relate to the knowledge of governance, processes, and tools for market coupling on the one hand, but also on the general functioning of markets and the impact of specific design choices. In order to overcome this, the increased push for transparency on these procedures as well as on market (coupling) data is to be understood. The availability of reliable information for all stakeholders, but NRAs in particular, is a prerequisite for accountability and enforcement.

In basic economic theory, information asymmetry is an externality whereby the knowledge imbalance between economic actors may lead to suboptimal outcomes or even market failures. The problem of information asymmetry is broader than just designing the most appropriate electricity market mechanisms: it is widespread in the regulation of the electricity sector. Some examples include the Regulation on Wholesale Energy Market Integrity and Transparency (REMIT), price formation and bidding strategies, mitigation of market power, resource adequacy and security of supply, consumer protection and the design of regulatory oversight.

15.5 Outlook

Section 2.1 already highlighted the importance of NRA independence. Indeed, independence from political, public or private interests is crucial to enable regulatory authorities to decide on market coupling arrangements in a way that best serves society as a whole, but consumers and market participants in particular (in line with their legal mandate).

For NRAs to properly achieve said independence, several elements are imperative. They should be granted, by law, the mandate to issue directly applicable and enforceable decisions without the confirmation (or invalidation) from other (governmental or political) bodies.

Appropriate financial and personnel resources should allow the efficient functioning of NRAs. This is all the more relevant in the context of the considerable resources that undertakings in the regulated sector dispose of, often at least partly funded by the consumers through network tariffs. The independence of regulators is achieved, not only via the resources allocated to them, but also through their human resource management and expertise on subject matters. Avoiding regulatory capture entails implementing human resource management schemes aimed at, among others, transparent recruitment and selection processes, training and development of employees and the adoption of clear and transparent codes of conduct and ethical standards.

Typically, NRAs are accountable to their legislator (i.e. the parliamentary institutions in a Member State), but also to consumers and other stakeholders. While the former is typically enshrined within legal provisions that establish them, the latter should be included in the "corporate culture" of the NRAs: the interests of consumers and market participants are to be considered the guiding principle when deciding on appropriate market arrangements.

Bibliography

Commission Regulation (EU) 2015/1222 of 24 July 2015 establishing a guideline on capacity allocation and congestion management

Commission Regulation (EU) 2016/1719 of 26 September 2016 establishing a guideline on forward capacity allocation

Commission Regulation (EU) 2017/2195 of 23 November 2017 establishing a guideline on electricity balancing

Commission Regulation (EU) 2017/1485 of 2 August 2017 establishing a guideline on electricity transmission system operation

Directive (EU) 2019/944 of the European Parliament and of the Council of 5 June 2019 on common rules for the internal market for electricity and amending Directive 2012/27/EU (recast)

European Court of Auditors (2023) Internal electricity market integration. https://www.eca.europa.eu/.03. European Union. https://doi.org/10.2865/578602

Regulation (EU) 2019/943 of the European Parliament and of the Council of 5 June 2019 on the internal market for electricity (recast)

Regulation (EU) 2019/942 of the European Parliament and of the Council of 5 June 2019 establishing a European Union Agency for the Cooperation of Energy Regulators (recast)

Nico Schoutteet is an economic advisor to the Director for the Technical Operation of Electricity and Gas Markets at the Belgian national regulatory authority CREG. His work focuses on the functioning of electricity markets, from a national, regional and European perspective. For the past decade, he has been active in all regulatory aspects related to the adoption, implementation and enforcement of different guidelines (in particular the CACM and FCA Regulations) as well as the Electricity Regulation. He is an active member of ACER′s task forces, working groups and regional fora for the cooperation among European national regulatory authorities. As co-chair of several of these task forces, he has been closely involved in the development of mechanisms and procedures to streamline the work between regulators, and their interaction with TSOs, NEMOs and other stakeholders. Nico is very passionate about the positive role that electricity markets play to enable the energy transition, and strongly committed to contribute to an increased understanding of these markets, through effective monitoring and visualization of market data.

Part VI
Importance of Data

Transparency in European Electricity Markets and ENTSO-E Transparency Platform

16

Nikolaj Nåbo Andersen and Jayaram Anandha

Abstract

In this chapter introduces market transparency and its importance in maintaining market integrity, enabling participants to foster open and fair competition in wholesale markets for the benefit of energy consumers. This chapter describes the Regulation on Wholesale Energy Market Integrity and Transparency (REMIT) (EU. (2011). Regulation (EU) No 1227/2011 of the European Parliament and of the Council of 25 October 2011 on wholesale energy market integrity and transparency.) and the four key principles on which it is based: transparency; integrity; monitoring and cooperation. Moreover, this chapter introduces the ENTSO-E Transparency Platform, the central electricity market data platform of Europe, from its governance and processes to an overview of users and the data items handled within. Finally, this chapter closes with the future outlook of market transparency and the need for the continuous evolution of transparency regulations.

16.1 Introduction

A well-functioning market shall be free from any market abuses, which can be market manipulation or insider trading. It is important that all the market players have a level playing field and equal access to all the information.

N. N. Andersen (✉)
Energinet Systemansvar A/S, Fredericia, Denmark
e-mail: nae@energinet.dk

J. Anandha
Accenture Industry X, Kronberg im Taunus, Germany
e-mail: jayaram.anandha@accenture.com

A. Estermann et al. (eds.), *European Electricity Market Coupling*, Future of Energy,
https://doi.org/10.1007/978-3-031-86315-8_16

The EU electricity market model is about the integration of diverse national markets into a European framework via Market Network codes. Transparency is key when diverse market participants across Europe, divided by nationality or language come together to make cross-border or internal transactions.

A transparent market reflects on the electricity sector as a whole. Market transparency maintains the trust and confidence in the integrity of markets, allowing participants to foster open and fair competition in wholesale markets for the benefit of final consumers of energy.

In general, the publication of market information improves transparency. However, making the monitoring process long and resource-intensive is inefficient. That's why identifying the appropriate market indicators, having diverse toolsets, agile processes, and proper governance and coordination among market parties are equally important for transparency in electricity markets.

Transparency is key for all stakeholders:

- **Regulators**: There is a regulatory interest in ensuring that wholesale prices are not distorted by abusive market practices or lack of transparent rules.
- **Investors**: Transparency helps investors understand the market adequately, while access to available data such as prices and quantities equip them with necessary investment signals.
- **Market participants**: Information disclosure is a useful instrument for creating a level playing field for market participants so small players have no disadvantages over big players.
- **Politicians**: Transparency helps politicians engage in informed debates and decision-making on electricity market developments.
- **Consumers**: Transparency protects consumers from volatile energy prices, supports sustainable energy choices, allows making informed decisions, and enhances overall predictability in their energy usage and costs.

16.2 The Need for an EU Framework in Market Transparency

The creation of an EU Internal Energy Market is not just about combining diverse national markets but also about creating new regulations and new entities such as ACER, ENTSO-E, RCCs, NEMOs, Allocation Platforms, etc. With new rules and players in the EU electricity market, previous transparency regulations from finance families, voluntary regulations from different market parties or even national regulations proved insufficient to achieve necessary transparency. The risk is that market abuse in one Member State can affect the price of energy in other Member States as Energy markets in Europe are interlinked.

Therefore, the requirement was clear: A unique and sector specific set of transparency regulations to ensure all the market parties can benefit from common European provisions.

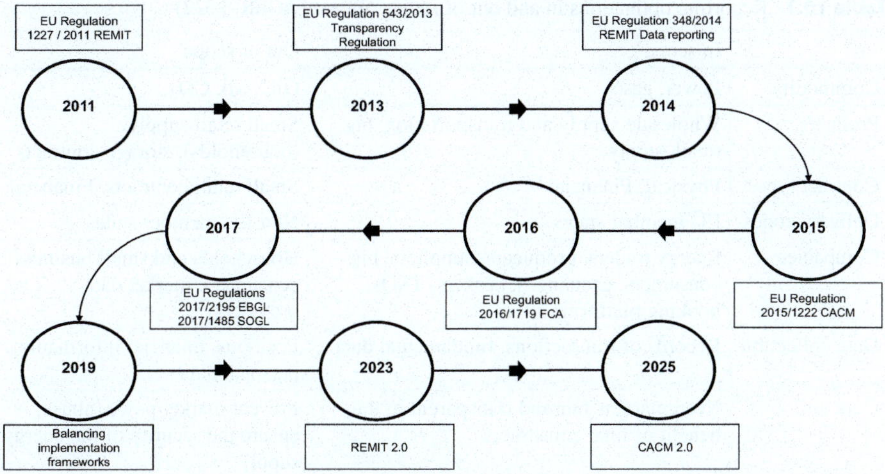

Fig. 16.1 Timeline of new regulations including transparency requirements. *Source* Author's own illustration

The transparency requirements are continuously evolving and being improved via new regulations or related implementation frameworks to ensure that the need for new market information is met and a level playing field is maintained. This is illustrated in Fig. 16.1.

16.3 Introduction of REMIT

Maintaining open and fair competition in the internal markets for electricity and gas requires a commitment to integrity and transparency in wholesale energy markets. To address this, the Regulation on Wholesale Energy Market Integrity and Transparency (REMIT) was enacted on December 28, 2011. REMIT specifically aims to enhance competition in European wholesale energy markets by enforcing regulations that prevent trading based on insider information and deterring market manipulation.

Since its implementation, REMIT has proven to be an effective tool in supporting transparent and competitive practices in these markets. It equips regulators (ACER and NRAs) with various tools to monitor market activities and identify any irregularities. Additionally, market participants are required to comply with specific obligations under REMIT, including the reporting of transactions and public disclosure of relevant information, to help ensure a level playing field for all. Table 16.1 provides an overview of the different reporting obligations.

REMIT applies to wholesale energy markets, encompassing any market within the Union where wholesale energy products are traded. Wholesale energy markets include both commodity markets and derivative markets. Importantly, all entities

Table 16.1 Reporting obligations in and out of scope. *Source* (ACER, 2022)

	In scope	Out of scope
Commodity	Power, gas	Oil, coal, CO_2,...
Products	Wholesale supply and transportation, big retail supply	Small retail supply (households), storage contracts
Contract types	Physical, Financial	Small retail contracts, Financial
Delivery zones	EU member states	Non-EU member states
Companies	Energy traders, producers, suppliers, big consumers, exchanges, brokers, TSOs, booking platforms,.	Households and small business consumers, market data providers
Data collection	Records of transactions, fundamental data	Positions, financial information, personal data
Core aim	Keep markets fair and transparent to the benefit of final consumers	Prevent market power abuse, ensure the security of the energy supply

and persons who participate in, or whose conduct affects, wholesale energy markets within the Union should be compliant with REMIT. It makes no difference whether the person is a resident of the EU or whether he or she is a professional investor.

16.3.1 Four Key Principles of REMIT

Transparency: Market participants have obligations to disclose inside information effectively and promptly, as well as other reporting responsibilities. Market parties refer to any person, including TSOs, who enter transactions, including the placing of orders to trade, in one or more wholesale energy markets. Examples include energy trading companies, generators, shippers of natural gas, balance responsible entities, wholesale customers, final customers with a consumption capacity > 600GWh/per year, TSOs, investment firms, etc. The REMIT Implementing Regulation (IR) clearly defines what data Market participants need to report, and how and when to report it.

Integrity: The primary prohibitions against abusive practices in wholesale energy markets, aiming to prevent actions that artificially drive prices to levels not justified by market forces of supply and demand, are (1) market manipulation and (2) insider trading. The concept of Market Manipulation includes performing false or misleading transactions, price positioning which secures or attempts to secure the price at an artificial level, as well as transactions involving fictitious devices or deception, and the dissemination of false and misleading information. Trading on inside information creates information asymmetry among market parties. To prevent insider trading, REMIT requires Market participants to share inside information with all other market participants via central platforms making the information available to a broader audience.

Monitoring: REMIT also proposes several obligations to establish a comprehensive and effective monitoring framework with a central reporting point. ACER and NRAs are given a wide range of power and toolsets for efficient market monitoring. ACER oversees the entire European-wide energy market, while NRAs largely focus on national transactions. Since ACER has a wider cross-border overview, it can provide assistance in identifying suspicious behaviours, but only NRAs are authorized to impose fines on market participants found guilty of breaching REMIT provisions.

Cooperation: The EU electricity market system involves several National/European entities and thousands of Market participants. Therefore, it is a no-brainer that Cooperation is one of the four principles of the REMIT. A clear cooperation framework and ownership are essential to perform EU-wide market surveillance. NRAs and EU agencies are responsible for cooperation to ensure REMIT is applied in a coordinated manner. In case the need to improve the cooperation was identified, ACER may request cooperation and coordination among EU Agencies, NRAs or Market party groups. As the responsible party for promoting cooperation, ACER has issued guidance documents on the application of REMIT for NRAs to carry out their tasks in a coordinated and consistent way and for Market parties on reporting procedures. It is worth highlighting that ACER has entered a cooperation agreement with different Market Parties or groups (ENTSO-E, ESMA, etc.) to further improve information exchange and market transparency.

16.3.2 Revised REMIT

In the wake of Russia's military invasion of Ukraine in February 2022, an unprecedented energy crisis broke out in Europe and globally. As a response, the European Commission has proposed reforms to the EU's electricity market design, which includes the REMIT regulation. This revised REMIT, taking effect as of 7 May 2024, introduces new measures to better protect EU citizens and businesses from energy market abuse. As a recognition of its importance, the scope of REMIT has now expanded to also cover energy storage (electricity and gas), and the market abuse provisions under REMIT now apply to financial instruments as well. Additionally, the scope of ACER's decision-making powers and regular surveillance has been further expanded. ACER has been granted investigatory powers in cross-border cases (involving two or more Member States); however, enforcement continues to be at the national level. Ultimately, the revised REMIT aims to bring the EU rules on transparency and integrity of energy markets into closer alignment with those in the financial markets as per (ACER, 2024).

16.4 Transparency Regulation

REMIT took the first step in improving market transparency by requiring the public disclosure of inside information and reporting information to the regulators. However, it was shortly after being recognized that these measures were not fully satisfying as relevant market information. Therefore, a second step towards reaching the necessary level of transparency was needed. The Transparency Regulation (EU, 2013), or Regulation 543/2013 was developed to ensure that market participants, regardless of being new or well established, would have equal access to timely and relevant information. The regulation lays down the minimum common set of data relating to the electricity market to be made available to market participants free of charge.

Under Regulation 543/2013 on submission and publication of data in electricity markets (EU, 2013), ENTSO-E shall establish and operate a central information transparency platform for publishing electricity data. The ENTSO-E Transparency Platform was launched on 5th January 2015 and published the Manual of Procedures, which provides information necessary to submit or extract data to or from the Platform.

Through this Regulation, it has now become mandatory for European Member State data providers and owners to submit fundamental information related to electricity generation, load, transmission and balancing for publication through the ENTSO-E Transparency Platform.

With the dissemination of the data to the public and market participants free of charge, the Transparency Regulation aims to:

- Enhance market participants' ability to make efficient production, consumption, and trading decisions.
- Facilitate deeper market integration and the rapid development of intermittent renewable energy generation sources, which require the disclosure of complete, timely available, high quality and easily digestible information relating to supply and demand fundamentals.
- Improve the security of energy supplies by the timely availability of complete sets of data and allow market parties to match supply and demand, reducing the risk of blackouts.
- Support the new market participants or participants without their own assets at a disadvantage.

16.4.1 Relationship Between Transparency Regulations and REMIT

Even though REMIT entered into force in 2011, its reporting requirements only became effective in 2015. The transparency regulation, established in 2013, enables all market participants to access the same information easily and free of charge, contributing not only to shared market knowledge but also to complementing REMIT's objectives by aiding market surveillance and preventing market manipulation.

Transparency regulations require market operators, energy companies, and relevant authorities to disclose certain information to the public and regulatory bodies. This transparency helps maintain the integrity of the energy markets and enhances investor confidence. REMIT requires market participants, including energy companies, traders, and transmission system operators, to report certain data related to their trading activities and fundamental market information to the relevant regulatory authorities. This data is then used by the authorities to monitor the markets for potential irregularities and manipulation.

In summary, although both transparency regulation and REMIT regulation are related to energy markets, the transparency regulation aims to make market information accessible to the public and participants. In contrast, REMIT specifically targets preventing market abuse and manipulation through mandatory reporting and regulatory oversight.

That being said, the REMIT and the Transparency regulation have certain overlaps. More specifically, the overlaps are related to the so-called "fundamental data", which includes data related to transmission of electricity (available transmission capacity, scheduled exchange, etc.) generation and load and the unavailability of transmission, production or consumption capacity.

As this information is required by the Transparency regulation and REMIT, the ENTSO-E Transparency platform is submitting the relevant data directly to ACER on behalf of TSOs and relevant market participants.

16.5 Transparency Requirements in Network Codes

Network codes are a set of rules developed to facilitate the harmonization, integration and efficiency of the European electricity market. There are three Network Codes from the Markets family: Capacity Allocation and congestion Management (CACM), Electricity Balancing Guidelines (EB GL) and Forward Capacity Allocation (FCA). These three network codes are an integral part of the drive towards completion of the internal energy market. Within these network codes and their related methodologies, certain obligations and tasks concerning Transparency apply to market parties such as TSOs, RCCs and ENTSO-E. These Network Codes and methodologies usually do not explicitly specify the platform for data publication. TSOs have the right to decide how and where this data needs to be published.

16.5.1 The Benefit of a Central Publication Platform

Publication requirements that are uniform for a group of TSOs often prompt TSOs to opt for centralized publication on the ENTSO-E Transparency Platform thus

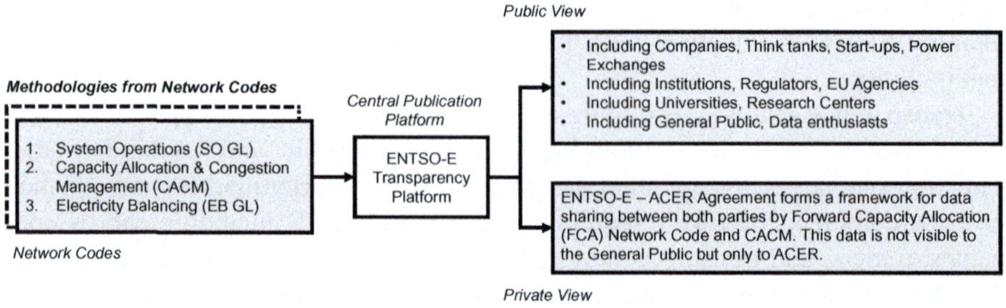

Fig. 16.2 A high-level overview of the fulfilment of data publication requirements from network codes using the ENTSO-E transparency platform. *Source* Author's own illustration

helping the market participants to have all the data in one place. Additionally, publishing centrally is more efficient than publishing on individual TSO websites in terms of quality control, operation, and maintenance. Figure 16.2 provides a high-level overview of this process.

16.6 ENTSO-E Transparency Platform: Mandate and Operations

EU Regulation 543/2013 of 14th June 2013 mandates the European Member State data providers and owners to submit fundamental information related to electricity markets to the ENTSO-E Transparency platform. The platform serves as a central information platform by publishing market-related data for all the market participants across Europe on a non-discriminatory basis.

As per the legal mandate, ENTSO-E is responsible for the development and maintenance of the central Transparency Platform. The ENTSO-E Transparency Platform was launched on 5 January 2015. Figure 16.3 illustrates the ENTSO-E Transparency platform development timeline.

16.6.1 Transparency Platform Governance

Operating and maintaining a central transparency platform requires a lot of effort. Specialists from all TSOs are involved in all the stages of Transparency Platform development and operation; from interpreting the legal requirements to ensuring data quality and operations of the platform. Furthermore, the platform is at all times expected to deliver stable operation and the needed functionalities without compromising the integrity of the data. Figure 16.4 illustrates the governance of the Transparency Platform.

ENTSO-E is the legal entity responsible for the Transparency Platform. So naturally, the Platform follows the established governance structure of ENTSO-E. Transparency Platform activities are overseen by the Markets Committee, and

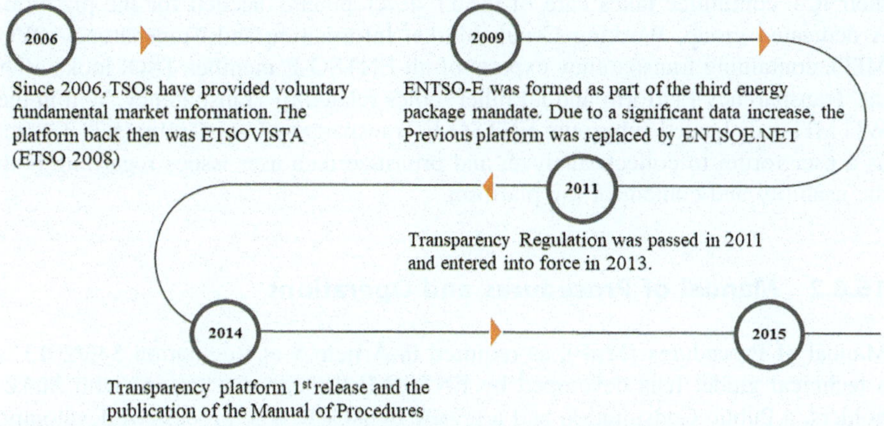

Fig. 16.3 ENTSO-E Transparency development timeline. *Source* (ENTSO-E, 2024)

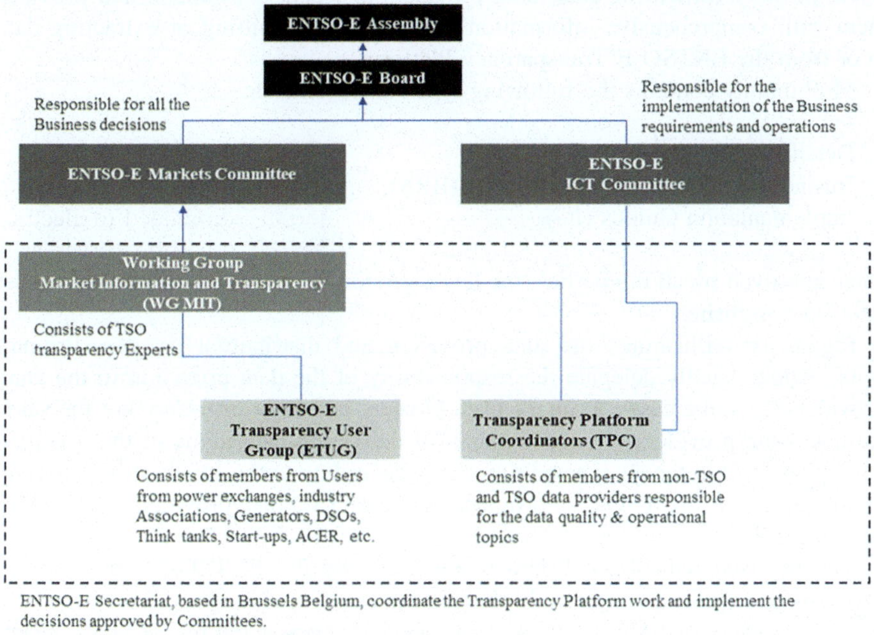

Fig. 16.4 Governance. *Source* Author's own illustration

their ICT committee takes care of the IT developments needed for the platform. A dedicated group, Working Group Market Information and Transparency (WG MIT), containing transparency experts of all ENTSO-E member TSO, looks after the Transparency Platform and all other topics related to Transparency. Within the WG MIT a dedicated subgroup, ENTSO-E Transparency User Group (ETUG), act as a user forum to collect, analyze, and prioritize data user issues regarding both the usability and content of the platform.

16.6.2 Manual of Procedures and Operations

Manual of Procedures (MoP), as required in Article 5 of Regulation 543/2013 is a technical guide. It is developed by ENTSO-E through discussions with Stakeholders, a Public Consultation and a review by ACER. The process of developing and updating the MoP is highly transparent, allowing all market stakeholders to express their opinions. MoP addresses all parties that are associated with ENTSO-E TP, primary owners of data, data providers and data consumers, and provides them with comprehensive information required for submitting or extracting data to or from the ENTSO-E Transparency Platform.

MoP mainly contains the following types of documents:

1. Detailed Data Description (DDD)
2. Business requirements specification (BRS)
3. Implementation Guides (IG).

Once approved by all the parties, the latest version of the MoP is published on the ENTSO-E website.

Figure 16.5 illustrates the data provision and distribution process. Primary Data owners usually delegate the responsibility of the data provision to the Data Providers. In some cases, Primary Data Owners and Data providers are the same entities. Data providers are accountable for delivering data items of the specified quality and within the designated timeframe according to the MoPs. It is important to note that there are TSO data providers and non-TSO data providers who submit data to the TP.

The received data is published to the users by the ENTSO-E Transparency Platform, accessible through multiple channels.

The electricity market has undergone many changes in the recent years, where markets are coupled, new markets are developed, and the granularity of the market time unit is lowered. In addition, the market coupling increases the competition among European market participants. This seems to drive the market participants towards relying more on data, thus increasing their need for data. In addition, the energy crisis also seems to have further increased the interest in Market Coupling data. At least the data consumption from the transparency platform has increased dramatically since 2018. The number of registered users has increased from approx. 11.000 users in 2018 to approx. 50.000 registered users in 2022.

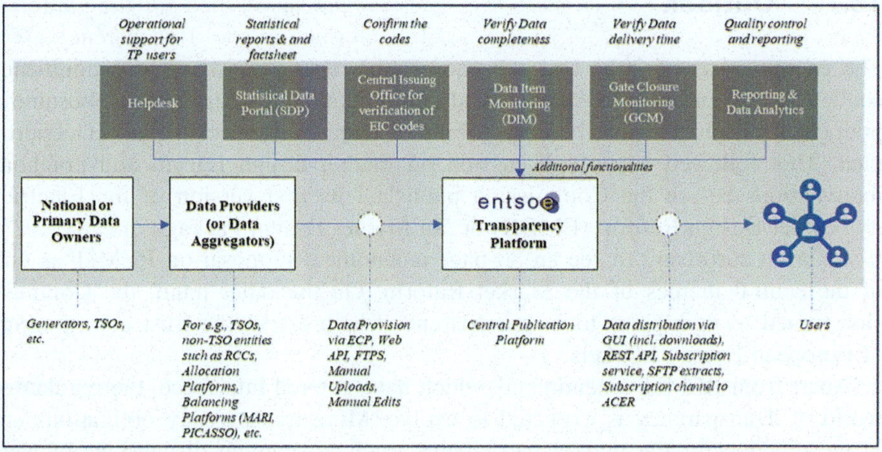

Fig. 16.5 High-level overview of ENTSO-E Transparency Platform. *Source* Author's own illustration

Also, in 2022 more than 6.000 users were active daily. In 2018, this number was around 2.000 daily users active on the Transparency platform.

8.000 + daily users may seem like a moderate usage of transparency platform. However, most users are machine-to-machine users who consume all data items on the platform repeatedly during the day. This resulted in more than 325 million application programming interface (API) requests in 2022. Many stakeholders are very interested in market coupling data, including politicians, officials, trading companies, universities and media organizations. All data items handled by the transparency platform are shown in Fig. 16.6.

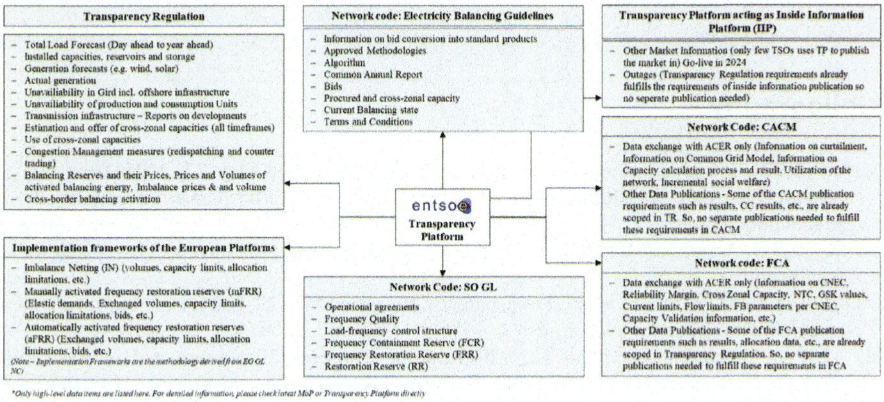

Fig. 16.6 Data items handled in ENTSO-E transparency platform. *Source* Author's own illustration

16.7 Outlook

The energy crisis of 2022 and subsequent high energy prices posed significant challenges for European businesses and consumers. High energy prices became a political focal point when the energy bills of many consumers tripled or quadrupled. This followed a swift discussion on market design reform and potential interventions before the Commission published its first version of the Electricity Market Design Reform (EMDR) or 5th Market Design Package in early 2023. It was a bit surprising to see an 80-page amendment proposal on REMIT as one of the central themes of the Market Reform. On the other hand, the Commission turned to its biggest tool—transparency—to preserve the trust and integrity of European Energy Markets.

Apart from REMIT amendments which have entered into force, the regulatory world of Transparency is expected to evolve. More transparency obligations are in the pipeline for the market participants, such as ongoing discussions of Network Code amendments (CACM, EB GL and later SO GL), upcoming go-live of the various capacity calculation processes across Europe. In October 2023, the EU Council adopted the amended Renewable Energy Directive (RED III) (EU, 2023), part of the "Fit for 55" package, which also includes new transparency requirements.

Additionally, as the energy transition picks up the pace and rapid digitalization transforms the energy sector, there is a burgeoning demand for market data. Earlier, electricity prices were only followed regularly by data enthusiasts or industry insiders, but since the recent energy crisis, the public (non-experts) has also started keenly following price forecasts.

In terms of transparency and market data, the trajectory is quite predictable: The demand for market data will exponentially increase, accompanied by the ongoing evolution of new regulatory requirements. To respond to the trend, Market participants shall adopt a proactive outlook towards Transparency by prioritizing the topic high on their agenda.

Bibliography

ACER. (2022). Introduction and scope of REMIT data. Retrieved May 27, 2024, from https://acer.europa.eu/sites/default/files/events/documents/2022-10/REMIT_Crash%20course_all%20Final.pdf

ACER. (2024). About REMIT. Retrieved June 10, 2024, from https://www.acer.europa.eu/remit/about-remit

ENTSO-E. (2024). Electricity market transparency. Retrieved May 27, 2024, from https://www.entsoe.eu/data/transparency-platform/

EU. (2011). Regulation (EU) No 1227/2011 of the European Parliament and of the Council of 25 October 2011 on wholesale energy market integrity and transparency. Retrieved November 01, 2024, from https://eur-lex.europa.eu/legal-content/EN/TXT/?uri=CELEX%3A02011R1227-20240507

EU. (2013). Commission Regulation (EU) No 543/2013 of 14 June 2013 on submission and publication of data in electricity markets. Retrieved November 01, 2024, from https://eur-lex.europa.eu/legal-content/EN/TXT/?uri=CELEX%3A32013R0543

EU. (2023). Directive (EU) 2023/2413 of the European Parliament and of the Council of 18 October 2023 as regards the promotion of energy from renewable sources. Retrieved November 01, 2024, from https://eur-lex.europa.eu/legal-content/EN/TXT/?uri=CELEX%3A32023L2413&qid=1699364355105

Nikolaj Nåbo Andersen is a Senior Manager for the Market Coupling and Transparency department at the Danish Transmission System Operator Energinet. He graduated from the University of Southern Denmark in 2015 with an MSc in Economics and Business Administration. Professionally, Nikolaj has worked with various projects within market coupling development and transparency including convening the ENTSO-E Working Group Market Information and Transparency from 2020 to 2022.

Jayaram Anandha is a Senior Consultant at Accenture Industry X Division, where he focuses on Utilities and Renewables. His work lies at the interface between technology and energy regulation/policy for the clean energy transition. Currently, he supports entities in integrating into the European Energy Market. Previously, he served as a Market Advisor for implementing Transparency Regulation and operating Market Platforms. Before joining Accenture, he worked at one of the Regional Coordination Centres (RCCs) in Europe, TSCNET, facilitating the implementation of European Network Codes. He led the ENTSO-E group of European TSOs and RCCs in transforming the diverse Outage Planning practices into a harmonized approach with the development of a common IT platform. Jayaram holds a master's degree in Power Engineering from the Technical University of Cottbus, Germany, and a bachelor's degree in Mechanical Engineering from Calicut University, India. He resides in Munich, Germany, with his wife.

A Market Player View

17

Thomas König and Sarah Pischler

Abstract

This chapter provides the reader with a perspective from a market participant on market coupling and why it matters. This chapter highlights the importance of accurate, real-time data provision and discusses why liquidity is so vital for market activities. Moreover, this chapter describes how a well-functioning wholesale market can create opportunities for trading and the importance of continued innovation from demand-side flexibility to algorithmic trading to drive forward the ways in which participants can engage in the markets. Finally, this chapter closes by highlighting the overall importance of market coupling for European consumers.

17.1 Why Market Coupling Matters, Especially for Intraday

Europe proudly boasts the world's largest interconnected power grid, with more than 400 interconnectors (Ember, 2023). This in itself is quite an achievement, but the story does not end there. Market coupling across the region allows new companies with innovative technologies to find traction in the European arena. The magnitude of algorithmic trades has risen tremendously within the last few years, which creates a lot of liquidity and makes the market more efficient. Market coupling cannot replace the technology needed for this, but it makes it possible

T. König (✉) · S. Pischler
enspired GmbH, Vienna, Austria
e-mail: thomas.koenig@enspired-trading.com

S. Pischler
e-mail: sarah.pischler@enspired-trading.com

A. Estermann et al. (eds.), *European Electricity Market Coupling*, Future of Energy,
https://doi.org/10.1007/978-3-031-86315-8_17

for the technology to be converted into use cases. Contrary to popular belief, it is not a coincidence that the roll-out of AI trading and market coupling, especially for intraday, happened at the same time—they are mutually dependent.

Consulting the statistics of social welfare gains, which are vital for reaching the intraday market coupling end goal of jointly securing supply for the population with the limited resources available, the question arises whether these gains were achieved *only* due to the enhanced trading efficiency. Naturally, the increase in cross-border capacity leads to an increase in grid utilization, causing prices to converge at lower levels between bidding zones. This is the measurable part of the equation. But one of the biggest advantages of market coupling lies in the lowering of market entrance barriers, which is of paramount importance for geographical expansion and serves as the backbone of automated trading in the system. Once a company is active in one market, the pathway to the next market within the coupled area is cut in half.

But let us dig a little deeper and take *enspired* as an example. Intraday markets started locally with hourly products in 2003. A user interface promptly emerged, and large volumes of energy were traded. In 2012, exchanges introduced an API (Application Programming Interface), and the market started to integrate simple automated trading systems. Flow-based capacity calculation and bigger capacity calculation regions brought about an increase in cross-border capacity calculation, causing the European energy market structure to move towards a system with small bids and high liquidity—a perfect playground for disruptive technologies such as algorithmic trading. Enspired was founded with a drive to exploit this new market situation by releasing as much flexibility as possible into it. Part of this goal was to leverage algorithms to maximize asset utilization at a low cost. Enspired's fully automated and cost-effective approach changed the way the industry navigates the market. In the old system, strong trading demands would have required a high degree of additional staffing, while the new set-up eliminates the need for human intervention and manually operated trading screens altogether. It is important to emphasize that none of these advantages would exist without market coupling.

17.2 Why Are We so Keen on Cross-Border Data?

Our world runs on data—the energy business is no exception. Accurate data analysis is key for any successful enterprise, and market coupling brings several advantages in this regard as information extracted from one market can be applied to others. For instance, an outage of a big power plant in France does not only impact the French power ecosystem but that of other bidding zones as well. Market coupling entails a significant reduction of workload because different use cases consider the same data within the same infrastructure. However, this only applies if enough physical capacity remains for a particular hour at a particular point in time. Once the markets are separated and prices no longer converge, data from a neighbouring bidding zone does not necessarily influence the current bidding zone. Real-time data provision is crucial for evaluating the effects of fundamental data on the internal market and the ability to act on them. Insight into open market data lowers the barriers for exploring trading strategies and validating business ventures.

17.3 Why is Liquidity Vital for Our Market Activities?

Liquidity ensures the ability to buy and sell products without having to pay significant spreads, which secures a robust service and solid supply to customers. Algorithmic trading not only generates liquidity but also relies on it to develop. Algorithms increase speed and responsiveness to changes in the order book and fundamental data. This is especially crucial when prices move quickly. Algorithms provide liquidity where it is needed: at the top of the continuous trading order book, which eventually leads to a reduction in price volatility. In markets with low liquidity and a rather static order book, large orders are needed to increase transaction volumes. Little trading activity creates a market situation in which algorithmic strategies have no significant advantage over a manual approach. Algorithms are most effective when they can leverage trading opportunities arising from a dynamic market environment. A rationally behaving order book is the feeding ground for algorithmic profit exploitation. Paradoxically, liquidity constitutes a major prerequisite for putting effective algorithms in place. This causality dilemma, more commonly known as a chicken-and-egg problem, was solved by the coupling of markets as it kickstarted a liquidity upward spiral of sorts, allowing neighbouring countries to profit from each other's liquidity.

17.4 Why Do We Innovate in Algorithmic Trading?

With the onset of market coupling emerged a common forum where TSOs, NEMOs, regulators and market participants can work together. Despite the different interests of these parties, workflows were established in a coordinated manner so that data provision is aligned, which in turn creates pressure to publish the right information at the right time. When it comes to data transparency, there is still room for improvement, but the progress made in that direction cannot be ignored.

The question now is whether market coupling has trading benefits other than a simple increase in trading. In short, yes! A well-functioning wholesale market opens many opportunities for trading—in particular with the help of algorithmic trading. As reported in S&P Global (S&P Global, 2023), experts estimate that around 50% of all intraday continuous traded power volume all over Europe in 2022 could have been executed via algorithms. The increasing familiarity with algorithmic trading establishes an ideal foundation for capturing flexibility and unleashing its full monetary potential. Without market coupling, the liquidity of the order book would be lower, and assets would earn a much smaller profit. Such developments can be observed in less developed markets. After the closure of the SIDC order book, locally available orders are no longer providing adequate liquidity. And that is not all. The environmental implications could be dire as well. It is a chain reaction. Less flexibility in the market means fewer renewables and more grid instabilities. Why? Because even with a smaller ratio of renewables, flexibility is needed to balance them out and avoid fluctuations in the power supply. Market coupling multiplies the availability of flexible sources by unprecedented levels.

17.5 How Do We Create New Business Opportunities?

How do we keep innovation afloat throughout the ongoing rise of intraday volume and API-based trading? It appears that the low-hanging fruits, mostly coming from the supply side, have been picked. Generators have put intelligence aplenty into the market design. The demand side, too, is experiencing an upswing as price pressure rises. Unleashing demand-side flexibility will continue to overhaul regulating policies, resulting in more opportunities to profit from flexibility.

This brings forth many new business models. Battery storage systems are seeing massive investments, and market coupling directly aids their trading volumes in the wholesale market. At enspired, we observe the explosive demand for combined battery storage marketing offers (covering wholesale, control reserve and ancillary service markets) on a daily basis. To reach the best commercial performance for a battery, we optimize the proportions of wholesale markets and frequency services based on price forecasts and high-speed backtesting. Batteries are indispensable for the energy transition, not least because they ease price volatility.

For future business cases, the commercial optimization of batteries is crucial. Trading strategies must maximize battery revenue by reaching a balance between asset profitability and asset lifetime. A safe approach is to prolong battery life, but if price spreads are extremely high or low, the profits can outweigh the potential downside of a shorter lifespan. Batteries bring grid stability, and active SoC (State of Charge) management ensures a well-functioning market.

Algorithmic trading is a key prerequisite for short-term battery marketing as storage optimization is not comparable to the dispatching of flexibility along a price curve. With quarter-hour products, a maximum of 132 quarter hours (the current and following day) can be optimized, taking into account restrictions like asset degradation and SoC (State of Charge). Therefore, a new trading schedule must be calculated for the whole day whenever the order book changes. This explains why the final trading schedule of any given day is not always indicative of the total volume traded on that day. Due to the increase in market liquidity, a dynamic structure can support optimization strategies much better than a static one.

Imagine a market where liquidity is so low that trading possibilities cannot be secured. This would result in a massive risk buffer and negatively impact the profitability of storage systems. Algorithmic trading would become next to impossible at this point. The commercial roll-out of batteries would be stalled or limited by the capacity of frequency maintenance markets. That is why well-functioning coupled markets are so important—they give the right incentives to integrate more technologies that foster a carbon-free power supply.

Aside from storage optimization, the integration of renewables also relies on a well-functioning market. Through close to real-time trading, even small forecast deviations can be traded, and in a future world that runs on 100% renewables, market coupling secures the ability to trade large volumes between bidding zones to keep local weather conditions from influencing the market disproportionately.

17.6 Would All This Be Possible Without Market Coupling?

In short, the answer is once again yes. BUT: Not to the extent we have now. From a geographical viewpoint, the same data and the same use cases have different footings, different potentials and different outcomes in different bidding zones. Market coupling gives assets larger areas for profitability. With all this in mind, it is necessary to keep making strides in common projects by enhancing transparency and working more closely with the actual market users. From an outside perspective, market coupling operates silently behind the scenes. In reality, it is the headquarters of innovative new business models that turn the energy market into a facilitator for a greener world.

Bibliography

Ember. (2023). *Breaking borders: The future of Europe's electricity is in interconnectors*. Retrieved December 15, 2024, from EMBER - Latest Insight: https://ember-energy.org/latest-insights/breaking-borders-europe-electricity-interconnectors/

S&P Global. (2023). *Power markets boost algorithmic trading amid changing sector dynamics*. Retrieved December 15, 2024, from S&P Global - Commodity Insights: https://www.spglobal.com/commodity-insights/en/news-research/latest-news/electric-power/083123-power-markets-boost-algorithmic-trading-amid-changing-sector-dynamics

Thomas König is Senior Regulatory Expert at enspired. His past thirteen years in the Energy Branch circled around short term markets but from different viewpoints. Starting very operatively with power plant dispatch and Intraday trading at Iqony (formerly Steag) he later focused on the conditions to make that happen: engaging in the regulatory world being an internal advisor and representative in associations like BDEW and Energy Traders Germany. Another angle was taken during his work for Amprion in International Regulations before entering enspired. He holds a Diploma in Business Engineering focusing on International Grid- and Energy Management from the Technical University in Dortmund.

Sarah Pischler has a background in linguistic studies and works as content expert at enspired. She has published articles, white papers and case studies on a broad range of energy market, energy technology and energy trading topics, including the power market structure in Europe, flexibility monetization, cross-market power asset optimization, and the various project phases of BESS development and marketing.

Glossary

Agency for the Cooperation of Energy Regulators (ACER) In order to help the different national regulators cooperate and ensure the smooth functioning of the internal energy market, the European Union Agency for the Cooperation of Energy Regulators was established in 2011 by the Third Energy Package legislation (EU Regulation 2009a, 2009b/713, 2009) now replaced by (EU Regulation 2019a, 2019b, 2019c/942, 2019) as an independent body to foster the integration and completion of the European Internal Energy Market for electricity and natural gas.

Available Transfer Capacity (ATC) ATC is the transfer capacity remaining available between two interconnected areas for further commercial activity over and above already committed utilization of the transmission networks. Also sometimes referred to as Available Transmission Capacity. (ENTSO-E, Pre-2015, 2015)

Balance Responsible Party (BRP) A market participant or its chosen representative responsible for its imbalances. (EU Regulation 2017/2195, 2017).

Balancing Balancing means all actions and processes, in all timelines, through which transmission system operators ensure, in an ongoing manner, maintenance of the system frequency within a predefined stability range and compliance with the amount of reserves needed with respect to the required quality. (EU Regulation 2019/943, 2019a, 2019b, 2019c)

Bidding Zone (BZ) A Bidding Zone is the largest geographical area within which market participants are able to exchange energy without capacity allocation. (EU Regulation 2019/943, 2019a, 2019b, 2019c)

Capacity Allocation and Congestion Management Guideline (or Regulation) The Capacity Allocation and Congestion Management Guideline lays down detailed guidelines on cross-zonal capacity allocation and congestion management in the day-ahead and intraday markets, including the requirements for the establishment of common methodologies for determining the volumes of

A. Estermann et al. (eds.), *European Electricity Market Coupling*, Future of Energy,
https://doi.org/10.1007/978-3-031-86315-8

capacity simultaneously available between bidding zones, criteria to assess efficiency and a review process for defining bidding zones. (EU Regulation 2015/1222, 2015)

Capacity Allocation Capacity allocation means the attribution of cross-zonal capacity. (EU Regulation 2019/943, 2019a, 2019b, 2019c)

Capacity Calculation Regions (CCR) A geographic area in which coordinated capacity calculation is applied. (EU Regulation 2015/1222, 2015)

Capacity Module (or Capacity Management Module) A system containing up-to-date information on available cross-zonal capacity for the purpose of allocating intraday cross-zonal capacity. (EU Regulation 2015/1222, 2015)

Carbon Border Adjustment Mechanism (CBAM) The initiative for a carbon border adjustment mechanism (the "CBAM") is part of the "Fit for 55" legislative package. The CBAM is to serve as an essential element of the Union's toolbox for meeting the objective of a climate-neutral Union at the latest by 2050 in line with the Paris Agreement by addressing the risk of carbon leakage that results from the Union's increased climate ambition. The CBAM is expected to also contribute to promoting decarbonization in third countries. (EU Regulation 2023/956, 2023)

Central Counter Party (CCP) The entity or entities with the task of entering into contracts with market participants, by novation of the contracts resulting from the matching process, and of organizing the transfer of net positions resulting from capacity allocation with other central counter parties or shipping agents. One of their main functions is to eliminate the counterparty credit risk arising from an organization's trade with a bilateral counterparty. (EU Regulation 2015/1222, 2015)

Clean Energy Package (or the Fourth Energy Package) In 2019 the EU overhauled its energy policy framework to help the bloc move away from fossil fuels towards cleaner energy—and, more specifically, to deliver on the EU's Paris Agreement commitments for reducing greenhouse gas emissions. The package consists of 8 new laws. Following entry into force of the different EU rules, EU countries have 1–2 years to convert the new directives into national law. (European Commission, 2019)

Clearing Price The price determined by matching the highest accepted selling order and the lowest accepted buying order in the electricity market. (EU Regulation 2015/1222, 2015)

Comitology Comitology is a process by which EU law is implemented, modified, or adjusted. It takes place within "comitology committees" chaired by the European Commission. The official term for the process is committee procedure. These committees assist the European Commission in exercising the implementing powers conferred on it by the legislative branch (i.e. the European Parliament and the Council of the European Union) with the assistance of committees consisting of Member State representatives. (Eurofound, 2011)

Common Grid Model (CGM) The common grid model means a Union-wide data set agreed between various TSOs describing the main characteristic of the power system (generation, loads and grid topology) and rules for changing these characteristics during the capacity calculation process. (EU Regulation 2015/1222, 2015)

Congestion Congestion is a situation in which an interconnection linking national transmission networks, cannot accommodate all physical flows resulting from international trade requested by market participants, because of a lack of capacity of the interconnectors and/or the national transmission systems concerned. (EU Regulation 2019/943, 2019a, 2019b, 2019c)

Congestion Income (CI) The revenue received as a result of capacity allocation. Congestion income arises in the electricity market when the transmission capacity between bidding zones is too low to even out the difference between supply and demand in the market areas. The market areas become separate price areas, and a buyer in one area pays a different price than someone in another area. The power exchange accrues the price difference in the form of congestion income, which the power exchange pays to the transmission system operators on both sides of the price area. (EU Regulation 2015/1222, 2015)

Contingency Contingency means the identified and possible or already occurred fault of an element, including not only the transmission system elements, but also significant grid users and distribution network elements if relevant for the transmission system operational security. (EU Regulation 2015/1222, 2015)

Continuous Trading Process for carrying out instructions for brokers and dealers to buy and sell securities as soon as the orders are received or as soon as the desired price becomes obtainable. (IATE, 2024)

Continuous Trading Matching Algorithm (CTMA) Continuous trading matching algorithm means the algorithm used in single intraday coupling for matching orders and allocating cross-zonal capacities continuously. (EU Regulation 2015/1222, 2015)

Control Area A control area means a coherent part of the interconnected system, operated by a single system operator and shall include connected physical loads and/or generation units if any. (EU Regulation 2019/943, 2019a, 2019b, 2019c)

Coordinated Capacity Calculator (CCC) The coordinated capacity calculator means the entity or entities with the task of calculating transmission capacity, at regional level or above. (EU Regulation 2015/1222, 2015)

Coordinated Net Transmission Capacity (or Coordinated Net Transfer Capacity) (CNTC) The CNTC approach is the capacity calculation method based on the principle of assessing and defining ex-ante a maximum energy exchange between adjacent bidding zones according to Article 2 of Regulation EU 2015/1222. In practical terms, CNTC involves calculating the net transfer capacity (NTC) for electricity transmission between different bidding zones. NTC is the maximum amount of power that can flow across interconnectors connecting bidding zones. The coordinated approach ensures that this capacity is determined collaboratively and transparently, considering factors like network constraints, operational security limits, and critical network elements. (Glowacki Law Firm, 2015)

Critical Network Element A critical network element means a network element either within a bidding zone or between bidding zones taken into account in the capacity calculation process, limiting the amount of power that can be exchanged. (EU Regulation 2019/943, 2019a, 2019b, 2019c)

Critical Network Element and Contingency (CNEC) The CNEC is a Critical Network Element limiting the amount of power that can be exchanged, potentially associated to a contingency. (ENTSO-E, 2016)

Cross Border Flow A cross-border flow means a physical flow of electricity on a transmission network of a Member State that results from the impact of the activity of producers, customers, or both, outside that Member State on its transmission network. (EU Regulation 2019/943, 2019a, 2019b, 2019c)

Cross-zonal Capacity The capability of the interconnected system to accommodate energy transfer between bidding zones. (EU Regulation 2019/943, 2019) There are two permissible approaches when calculating cross-zonal capacity: flow-based or based on coordinated net transmission capacity. The flow-based approach should be used as a primary approach for day-ahead and intraday capacity calculation where cross-zonal capacity between bidding zones is highly interdependent. The coordinated net transmission capacity approach should only be applied in regions where cross-zonal capacity is less interdependent, and it can be shown that the flow-based approach would not bring added value. (EU Regulation 2015/1222, 2019a, 2019b, 2019c)

Day-Ahead Operational Agreement (DAOA) According to Article 8 (1) of the CACM Guideline, all TSOs in EU Member States electrically connected to another Member State must participate in the single day-ahead (and intraday) coupling. The TSOs and NEMOs, which participate in the SDAC, cooperate under the day-ahead operational agreement (DAOA), which entered into force on 28 March 2019. (EU Regulation 2015/1222, 2015)

Electricity Markets Electricity markets means markets for electricity, including over-the-counter markets and electricity exchanges, markets for the trading of energy, capacity, balancing and ancillary services in all timeframes, including forward, day-ahead and intraday markets. (EU Regulation 2019/944, 2019a, 2019b, 2019c)

Energy Community Treaty The Energy Community Treaty provides for the creation of an integrated market in natural gas and electricity in South-East Europe which will create a stable regulatory and market framework capable of attracting investment in gas networks, power generation and transmission networks, so that all Parties have access to the stable and continuous gas and electricity supply that is essential for economic development and social stability. (Council Decision 2006/500/EC, 2006)

European Network of Transmission System Operators – Electricity (ENTSO-E) The Third Energy Package is a set of two European directives and three regulations from 2009. The Regulation that stipulates ENTSO-E's tasks and responsibilities is Regulation (EC) 714/2009 on conditions for access to the network for cross-border exchanges in electricity. The regulation sets out ENTSO-E's responsibilities in enhancing the cooperation between its 40 member TSOs across the EU to assist in the development of a pan-European electricity transmission network in line with European Union energy policy goals. (EU Regulation 2009/714, 2009a, 2009b)

EU Emissions Trading Scheme The EU Emissions Trading System was established by (EU Directive 2003/87/EC, 2003). It is a cornerstone of the Union's climate policy and constitutes its key tool for reducing greenhouse gas emissions in a cost-effective way. An update entered into force in 2023—(EU Directive 2023/959, 2023)

European Market Infrastructure Regulation (EMIR) The EMIR lays down clearing and bilateral risk management requirements for over-the-counter ('OTC') derivative contracts, reporting requirements for derivative contracts and uniform requirements for the performance of activities of central counterparties ('CCPs') and trade repositories. (EU Regulation 2012/648, 2012)

Explicit Allocation (or Explicit Capacity Allocation) Allocation of cross-border transmission capacity separate from the trade of electricity. (EU Regulation 2015/1222, 2015) Alternatively, explicit allocation means the allocation of cross-zonal capacity only, without the energy transfer. (EU Regulation 2013/543, 2013)

Flow-Based Approach This is a capacity calculation method in which energy exchanges between bidding zones are limited by power transfer distribution factors and available margins on critical network elements. (EU Regulation 2015/1222, 2015)

Generation Shift Keys A generation shift key means a method of translating a net position change of a given bidding zone into estimated specific injection increases or decreases in the common grid model. (EU Regulation 2015/1222, 2015)

Implicit Allocation Implicit allocation means a congestion management method in which energy is obtained at the same time as cross-zonal capacity. (EU Regulation 2013/543, 2013)

Individual Grid Model An individual grid model means a data set describing power system characteristics (generation, load and grid topology) and related rules to change these characteristics during capacity calculation, prepared by the responsible TSOs, to be merged with other individual grid model components in order to create the common grid model. (EU Regulation 2015/1222, 2015)

Intraday Auction (IDA) means the implicit intraday auction trading session for simultaneously matching orders from different bidding zones and allocating the available intraday cross-zonal capacity at the bidding zone borders by applying a market coupling mechanism. (ACER, 2019)

Intraday Power Trading Intraday power trading refers to continuous buying and selling of power at a power exchange that takes place on the same day as the power delivery. (Next Kraftwerke, 2019)

Market Coupling Method for integrating electricity markets in different areas, in which the daily cross-border transmission capacity between the various areas is not explicitly auctioned among the market parties but is implicitly made available via energy transactions on the power exchanges on either side of the border. (IATE, 2024)

Market Coupling Operator Function The market coupling operator (MCO) function means the task of matching orders from the day-ahead and intraday markets for different bidding zones and simultaneously allocating cross-zonal capacities. (EU Regulation 2015/1222, 2015)

National Regulatory Authorities (NRA) NRAs have a key role to play in ensuring that each European country meets its targets for energy markets and implements the relevant EU regulatory policy. In order to maintain the proper functioning of the single European market in gas and electricity, ACER supports NRAs in performing their regulatory function at European level and coordinates their contributions. (ACER, 2024)

Net Position The net position means the netted sum of electricity exports and imports for each market time unit for a bidding zone. (EU Regulation 2015/1222, 2015)

Net Transfer Capacity (or Net Transmission Capacity) The Net Transfer Capacity can be defined as the expected maximum volume of generation that can be wheeled through the interface between the two systems, which does not lead to network constraints in either system, respecting some technical uncertainties on future network conditions. (ENTSO-E, 2000)

Nominated Electricity Market Operator (NEMO) A NEMO means an entity designated by the competent authority to perform tasks related to single day-ahead or single intraday coupling. (EU Regulation 2015/1222, 2015)

Power Transfer Distribution Factors A power transfer distribution factor is the representation of the physical flow on a critical network element induced by the variation of the net position of a bidding zone. (IATE, 2024)

Price Coupling Algorithm Price coupling algorithm means the algorithm used in single day-ahead coupling for simultaneously matching orders and allocating cross-zonal capacities. (EU Regulation 2015/1222, 2015)

Price Coupling System A coupling system which in one step establishes both prices and volumes for each coupled market, based on all orders from all markets that are coupled. Market areas are considered in an anonymous manner in the coupling system. Market splitting is a form of price coupling, where all orders; pricing per bidding area and settlement is handled by one power exchange. A price coupling system can be placed in a unique legal entity or can be a unique system that is shared by the local power exchanges. (Nord Pool, NASDAQ OMX Commodities, 2013)

Regional Coordination Centres (RCC) The RCCs are established by the Electricity Regulation 2019/943 and as of 2022 replace the regional security coordinators (RSCs) foreseen by the System Operation Guideline. Their tasks include supporting the consistency assessment of transmission system operators' defence and restoration plans; carrying out regional outage planning coordination; and carrying out post-operation and post-disturbances analysis. (EU Regulation 2019/943, 2019a, 2019b, 2019c)

Regulatory Authority A regulatory authority means a regulatory authority designated by each Member State pursuant to Article 57(1) of Directive (EU) 2019/944. Member States shall guarantee the independence of the regulatory authority and shall ensure that it exercises its powers impartially and transparently. (EU Regulation 2019/944, 2019a, 2019b, 2019c)

Remedial Action A remedial action means any measure applied by a TSO or several TSOs, manually or automatically, in order to maintain operational security. (EU Regulation 2015/1222, 2015)

Scheduled Exchange Scheduled exchange means an electricity transfer scheduled between geographic areas, for each market time unit and for a given direction. (EU Regulation 2015/1222, 2015)

Scheduled Exchange Calculator Scheduled exchange calculator means the entity or entities with the task of calculating scheduled exchanges. (EU Regulation 2015/1222, 2015)

Shadow Auctions Shadow auctions are day-ahead explicit auctions performed in case of failure of market coupling (implicit allocations). Only capacity rights are auctioned, electricity is to be purchased directly through power exchanges. (Joint Allocation Office, 2022)

Shared Order Book A shared order book is a module in the continuous intraday coupling system collecting all matchable orders from the participating nominated electricity market operators in the single intraday coupling and performing continuous matching of those orders. (EU Regulation 2015/1222, 2015)

Shipping Agent A shipping agent means the entity or entities with the task of transferring net positions between different central counter parties. (EU Regulation 2015/1222, 2015)

Single Day-Ahead Coupling Single day-ahead coupling means the auctioning process where collected orders are matched and cross-zonal capacity is allocated simultaneously for different bidding zones in the day-ahead market. (EU Regulation 2015/1222, 2015)

Single Intraday Coupling Single intraday coupling means the continuous process where collected orders are matched and cross-zonal capacity is allocated simultaneously for different bidding zones in the intraday market. (EU Regulation 2015/1222, 2015)

Third Energy Package The Third Energy Package is a set of two European directives and three regulations from 2009, aimed at improving the functioning of the internal energy market and resolving certain structural problems. Only the gas part of the third energy package rules (from 2009) is still in force. The current electricity market rules were adopted as part of the Clean energy for all Europeans package in 2019. (European Commission, 2024)

Transparency Transparency is essential for the implementation of the Internal Electricity Market (IEM) and for the creation of efficient, liquid and competitive wholesale markets. It is also critical for creating a level playing field between market participants and avoiding the scope for market power (if it exists) to be abused. Through Regulation (EU) No 543/2013, the establishment of a central information transparency platform was mandated. (EU Regulation 2013/543, 2013)

Transmission System Operators Transmission system operator means a natural or legal person who is responsible for operating, ensuring the maintenance of and, if necessary, developing the transmission system in a given area and, where applicable, its interconnections with other systems, and for ensuring the long-term ability of the system to meet reasonable demands for the transmission of electricity. (EU Regulation 2019/944, 2019a, 2019b, 2019c)

70% Rule (or Margin Available for Cross-zonal Trade (MACZT)) The Clean Energy Package includes several legislative acts, one of which is the revised Electricity Regulation where the 70% rule is introduced, requiring TSOs to provide at least 70% ('minimum 70% target') of the available capacity for cross-zonal trade. (EU Regulation 2019/943, 2019a, 2019b, 2019c). ACER issued a report evaluating the progress of cross-zonal capacities in 2022 and their role in achieving the European Union's energy objectives and market integration. (ACER, 2023)

References

ACER. (2019). *Annex 1 of the ACER decision on the methodology for pricing intraday cross-zonal capacity.* s.l.:s.n.

ACER. (2023). *Cross-zonal capacities and the 70% margin available for cross-zonal electricity trade (MACZT).* s.l.: ACER.

ACER. (2024). *The key role of NRAs.* [Online] Available at: https://www.acer.europa.eu/remit/cooperation/national-regulatory-authorities. Accessed 29 June 2024.

Council Decision 2006/500/EC. (2006). *2006/500/EC: Council decision of 29 May 2006 on the conclusion by the European community of the energy community treaty.* EU: EUR-Lex.

ENTSO-E. (2000). *Definitions of transfer capacities in liberalised markets.* [Online] Available at: https://eepublicdownloads.entsoe.eu/clean-documents/pre2015/ntc/entsoe_transferCapacityDefinitions.pdf. Accessed 11 May 2024.

ENTSO-E. (2016). *Explanatory note of the common long term capacity.* [Online] Available at: https://eepublicdownloads.entsoe.eu/clean-documents/nc-tasks/SEE%20-%2010.1%20-%201%20TSOs_Explanatory_document.pdf. Accessed 09 May 2024.

ENTSO-E, Pre-2015. (2015). *Definitions of transfer capacities in liberalised electricity markets.* [Online] Available at: https://eepublicdownloads.entsoe.eu/clean-documents/pre2015/ntc/entsoe_NTCusersInformation.pdf. Accessed 09 May 2024.

EU Directive 2003/87/EC. (2003). *Directive 2003/87/EC of the European parliament and of the council of 13 October 2003 establishing a scheme for greenhouse gas emission allowance trading within the community and amending council directive 96/61/EC (Text with EEA relevance).* EU: EUR-Lex.

EU Directive 2023/959. (2023). *Directive (EU) 2023/959 of the parliament & council of May 2023 amending directive 2003/87/EC establishing a system for GHG emission allowance trading within the EU & decision (EU) 2015/1814 on the establishment & operation of a market stability reserve.* EU: EUR-Lex.

EU Regulation 2009/713. (2009). *Regulation (EC) No 713/2009 of the European parliament and of the council of 13 July 2009 establishing an agency for the cooperation of energy regulators (Text with EEA relevance).* EU: EUR-Lex.

EU Regulation 2009/714. (2009). *Regulation (EC) No 714/2009 of the European parliament and of the council of 13 July 2009 on conditions for access to the network for cross-border exchanges in electricity and repealing regulation (EC) No 1228/2003 (Text with EEA relevance).* EU: EUR-Lex.

EU Regulation 2012/648. (2012). *Regulation (EU) No 648/2012 of the European parliament and of the council of 4 July 2012 on OTC derivatives, central counterparties and trade repositories. Text with EEA relevance.* EU: EUR-Lex.

A. Estermann et al. (eds.), *European Electricity Market Coupling*, Future of Energy,
https://doi.org/10.1007/978-3-031-86315-8

EU Regulation 2013/543. (2013). *REGULATION (EU) No 543/2013 of June 2013 on submission & publication of data in electricity markets & amending Annex I to regulation (EC) No 714/2009 of the European parliament and of the counci.* EU: EUR-Lex.

EU Regulation 2015/1222. (2015). *Commission regulation (EU) 2015/1222 of 24 July 2015 establishing a guideline on capacity allocation and congestion management (Text with EEA relevance).* EU: EUR-Lex.

EU Regulation 2017/2195. (2017). *Commission regulation (EU) 2017/2195 of 23 November 2017 establishing a guideline on electricity balancing (Text with EEA relevance).* EU: EUR-Lex.

EU Regulation 2019/942. (2019). *Regulation (EU) 2019/942 of the European parliament and of the council of 5 June 2019 establishing a European Union agency for the cooperation of energy regulators (recast) (Text with EEA relevance.).* EU: EUR-Lex.

EU Regulation 2019/943. (2019). *Consolidated text: Regulation (EU) 2019/943 of the European parliament and of the council of 5 June 2019 on the internal market for electricity (recast) (Text with EEA relevance).* EU: EUR-Lex.

EU Regulation 2019/944. (2019). *Directive (EU) 2019/944 of the European parliament and of the council of 5 June 2019 on common rules for the internal market for electricity and amending directive 2012/27/EU (recast) (Text with EEA relevance.).* EU: EUR-Lex.

EU Regulation 2023/956. (2023). *Regulation (EU) 2023/956 of the European parliament and of the council of 10 May 2023 establishing a carbon border adjustment mechanism (Text with EEA relevance).* EU: EUR-Lex.

Eurofound. (2011). *European industrial relations dictionary.* [Online] Available at: https://www.eurofound.europa.eu/en/european-industrial-relations-dictionary/comitology. Accessed 09 May 2024.

European Commission. (2019). *Clean energy for all Europeans.* [Online] Available at: https://energy.ec.europa.eu/topics/energy-strategy/clean-energy-all-europeans-package_en. Accessed 09 May 2024.

European Commission. (2024). *Third energy package.* [Online] Available at: https://energy.ec.europa.eu/topics/markets-and-consumers/market-legislation/third-energy-package_en. Accessed 11 May 2024.

Glowacki Law Firm. (2015). *Emissions - EUETS.* [Online] Available at: https://emissions-euets.com/internal-electricity-market-glossary/779-coordinated-net-transmission-capacity-approach. Accessed 09 May 2024.

IATE. (2024). *European union terminology.* [Online] Available at: https://iate.europa.eu/search/result/1680783622616/1 [Accessed 09 May 2024].

Joint Allocation Office. (2022). *Shadow auction guide.* [Online] Available at: https://www.jao.eu/sites/default/files/2022-11/ShadowAuctionsPackage_0_0.pdf. Accessed 11 May 2024.

Next Kraftwerke. (2019). *What does intraday trading mean?.* s.l.:s.n.

Nord Pool, NASDAQ OMX Commodities. (2013). *Glossary of electricity market related terms.* [Online] Available at: https://www.energiforetagen.se/globalassets/energiforetagen/det-erbjuder-vi/kurser-och-konferenser/elhandel/lathund-ce.pdf. Accessed 11 May 2024.

The manufacturer's authorised representative in the EU is Springer Nature Customer Service Centre GmbH, Europaplatz 3, 69115 Heidelberg, Germany. If you have any concerns regarding our products, please contact ProductSafety@springernature.com

Printed and bound by CPI Group (UK) Ltd, Croydon, CR0 4YY

30/09/2026

02225594-0002